Chikako Tanaka

Patrick L. McGeer

Yasuo Ihara

(Editors)

Neuroscientific Basis of Dementia

Springer Basel AG

Editors

Prof. Dr. Chikako Tanaka
Hyogo Institute for Aging Brain and
Cognitive Disorders
520, Saisho-ko
Himeji 670-0981
Japan

Prof. Dr. Patrick L. McGeer
Kinsmen Laboratory of Neurological Research
University of British Columbia
2255 Wesbrook Mall
Vancouver, B.C., V6T 1Z3
Canada

Prof. Dr. Yasuo Ihara
Department of Neuropathology
Faculty of Medicine
University of Tokyo
7-3-1 Hongo, Bunkyo-ku
Tokyo 113-0033
Japan

A CIP catalogue record for this book is available from the Library of Congress, Washington, D.C., USA

Deutsche Bibliothek Cataloging-in-Publication Data
Neuroscientific basis of dementia / ed. by Chikako Tanaka ... - Basel ; Boston ; Berlin : Birkhäuser, 2001
 ISBN 978-3-0348-9482-1   ISBN 978-3-0348-8225-5 (eBook)
 DOI 10.1007/978-3-0348-8225-5
The publisher and editor can give no guarantee for the information on drug dosage and administration contained in this publication. The respective user must check its accuracy by consulting other sources of reference in each individual case.

Printed on acid-free paper produced from chlorine-free pulp. TFC ∞
Cover design: Micha Lotrovsky
Cover illustration: Colocalization of a protein kinase (PKN, red) and a microtubule-associated protein (tau, green) in the CA1 pyramidal neurons of hippocampus in patients with Alzheimer disease. Tau was probed with a phospho-tau-specific monoclonal antibody (AT8). The picture is reproduced by kind permission of Toshio Kawamata (p. 121ff).

9 8 7 6 5 4 3 2 1

Table of contents

Pathogenesis of dementia – presenilin and amyloid

Diagnosis and therapeutics of dementia

List of contributors

Haruhiko Akiyama, Tokyo Institute of Psychiatry, 2-1-8 Kamikitazawa, Setagaya-ku, Tokyo 156-8585, Japan; e-mail: akiyama@prit.go.jp

Anja Leona Biere, Amgen, Inc., One Amgen Center Drive, Thousand Oaks, CA 91320-1799, USA

Jacek Biernat, Max-Planck-Unit for Structural Molecular Biology, Notkestrasse 85, D-22603 Hamburg, Germany

Thomas D. Bird, Geriatric Research Education and Clinical Center, Veterans Affairs Puget Sound Health Care System, Seattle Division, Seattle, WA 98108, USA

Lawrence F. Cahill , Department of Neurobiology and Behavior and Center for the Neurobiology of Learning and Memory, University of California, Irvine, CA 92697-3800, USA; e-mail: lfcahill@uci.edu

Keun-A Chang, Department of Pharmacology, College of Medicine and Neuroscience Research Institute, Medical Research Center, Seoul National University and Biomedical Brain Research Center, NIH, Seoul 110-799, Korea

Se Hoon Choi, Department of Pharmacology, College of Medicine and Neuroscience Research Institute, Medical Research Center, Seoul National University and Biomedical Brain Research Center, NIH, Seoul 110-799, Korea

De-Hua Chui, Department of Demyelinating Disease and Aging, National Institute of Neuroscience, National Center of Neurology and Psychiatry, 4-1-1 Ogawahigashi, Kodaira, Tokyo 187-8551, Japan

Martin Citron, Amgen, Inc., One Amgen Center Drive, Thousand Oaks, CA 91320-1799, USA; e-mail: mcitron@amgen.com

Ian D'Souza, Geriatric Research Education and Clinical Center, Veterans Affairs Puget Sound Health Care System, Seattle Division, Seattle, WA 98108, USA

John E. Duda, Center for Neurodegenerative Disease Research, Department of Pathology and Laboratory Medicine, University of Pennsylvania School of Medicine, 3600 Spruce St., 3rd Floor, Malony, PA 19104, USA; e-mail: johnduda@mail.med.upenn.edu

Nicholas C. Fox, Dementia Research Group, Institute of Neurology, The National Hospital for Neurology and Neurosurgery, Queen Square, London WC1N 3BG, UK

Kohji Fukunaga, Department of Pharmacology, Kumamoto University School of Medicine, 2-2-1 Honjo, Kumamoto 860-0811, Japan

Michel Goedert, Laboratory of Molecular Biology, Hills Road, Cambridge CB2 2QH, UK; e-mail: mb@mrc-lmb.cam.ac.uk

Richard J. Harvey, Dementia Research Group, Institute of Neurology, Hiroshi Hasegawa, Hyogo Institute for Aging Brain and Cognitive Disorders, 520 Saisho-ko, Himeji 670-0981, Japan

Masato Hasegawa, Department of Neuropathology and Neuroscience, Graduate School of Pharmaceutical Sciences, University of Tokyo, 7-3-1 Hongo, Bunkyo-ku, Tokyo 113-0033, Japan

Takeshi Hashimoto, Hyogo Institute for Aging Brain and Cognitive Disorders, 520 Saisho-ko, Himeji 670-0981, Japan

Toshiyuki Honda, Laboratory for Alzheimer's Disease, RIKEN Brain Science Institute, 2-1 Hirosawa, Wako-shi, Saitama 351-0198, Japan

Ming Hong, Center for Neurodegenerative Disease Research, Department of Pathology and Laboratory Medicine, University of Pennsylvania School of Medicine, 3600 Spruce St., 3rd Floor, Malony, PA 19104, USA

Yasuo lhara, Department of Neuropathology, Faculty of Medicine, University of Tokyo, 7-3-1 Hongo, Bunkyo-ku, Tokyo 113-0033, Japan; e-mail: yihara@m.u-tokyo.ac.jp

Kazuhiko Ikeda, Department of Ultrastructure, Tokyo Metropolitan Institute of Psychiatry, 2-1-8 Kamikitazawa, Setagaya-ku, Tokyo 156-8585, Japan; e-mail: kazikeda@artemis.prit.go.jp

Kenji Ikeda, Tokyo Institute of Psychiatry, 2-1-8, Kamikitazawa, Setagaya-ku, Tokyo 156-8585, Japan

Kazunori Imaizumi, Department of Anatomy and Neuroscience, Osaka University, Graduate School of Medicine, 2-2 Yamadaoka, Suita, Osaka 565-0871, Japan, Japan

Takeshi Ishihara, Center for Neurodegenerative Disease Research, Department of Pathology and Laboratory Medicine, University of Pennsylvania School of Medicine, 3600 Spruce St., 3rd Floor, Malony, PA 19104, USA

Yoshinori Itoh, Department of Internal Medicine, Yokufukai Geriatric Hospital, Tokyo 168-0071, Japan

Nobuhisa Iwata, Laboratory for Proteolytic Neuroscience, RIKEN Brain Science Institute, 2-1 Hirosawa, Wako-shi, Saitama 351-0198, Japan; e-mail: iwatan@brain.riken.go.jp

Takeshi Iwatsubo, Department of Neuropathology and Neuroscience, Graduate School of Pharmaceutical Sciences, University of Tokyo, 7-3-1 Hongo Bunkyo-ku, Tokyo 113-0033, Japan; e-mail: iwatsubo@mol.f.u-tokyo.ac.jp

John C. Janssen, Dementia Research Group, Institute of Neurology, The National Hospital for Neurology and Neurosurgery, Queen Square, London WC1N 3BG, UK

Sung-Jin Jeong, Department of Pharmacology, College of Medicine and Neuroscience Research Institute, Medical Research Center, Seoul National University and Biomedical Brain Research Center, NIH, Seoul 110-799, Korea

Yuka Jinno, Department of Clinical Neuroscience and Psychiatry, Osaka University, Graduate School of Medicine, 2-2 Yamadaoka, Suita, Osaka 565-0871, Japan

Taiichi Katayama, Department of Anatomy and Neuroscience, Osaka University, Graduate School of Medicine, 2-2 Yamadaoka, Suita, Osaka 565-0871, Japan

Hideshi Kawakami, Third Department of Internal Medicine, Hiroshima University School of Medicine, Kasumi 1-2-3, Minami-ku, Hiroshima 734-8551, Japan

Toshio Kawamata, Hyogo Institute for Aging Brain and Cognitive Disorders, 520 Saisho-ko, Himeji 670-0981, Japan; e-mail: kawamata@hiabcd.go.jp

Hye-Sun Kim, Department of Pharmacology, College of Medicine and Neuroscience Research Institute, Medical Research Center, Seoul National University and Biomedical Brain Research Center, NIH, Seoul 110-799, Korea

Sung-Soo Kim, Department of Pharmacology, College of Medicine, Kang Won National University, Korea; e-mail: kssly@netian.com

Sung-Su Kim, Department of Anatomy, College of Medicine, Chung Ang University, Korea; e-mail: sungsu@cau.ac.kr

Jun-Ho Lee, Department of Pharmacology, College of Medicine and Neuroscience Research Institute, Medical Research Center, Seoul National University and Biomedical Brain Research Center, NIH, Seoul 110-799, Korea

Yasumasa Kokubo, Department of Neurology, Mie University School of Medicine, 2-174 Edobashi, Tsu 514-8507, Japan

Osamu Komure, Department of Neurology, Utano National Hospital, 8 Ondoyama-cho, Narutaki Ukyo-ku, Kyoto 616-8255, Japan

Hiromi Kondo, Tokyo Institute of Psychiatry, 2-1-8, Kamikitazawa, Setagaya-ku, Tokyo 156-8585, Japan

Satoshi Kotorii, Department of Demyelinating Disease and Aging, National Institute of Neuroscience, National Center of Neurology and Psychiatry, 4-1-1 Ogawahigashi, Kodaira, Tokyo 187-8551, Japan

Takashi Kudo, Department of Clinical Neuroscience and Psychiatry, Osaka University, Graduate School of Medicine, 2-2 Yamadaoka, Suita, Osaka 565-0871, Japan; e-mail: kudo@psy.med.osaka-u.ac.jp

Sadako Kuno, Department of Neurology, Utano National Hospital, 8 Ondoyama-cho, Narutaki Ukyo-ku, Kyoto 616-8255, Japan

Shigeki Kuzuhara, Department of Neurology, Mie University School of Medicine, 2-174 Edobashi, Tsu 514-8507, Japan; e-mail: kuzuhara@clin.medic.mie-u.ac.jp

Jean-Pyo Lee, Department of Pharmacology, College of Medicine and Neuroscience Research Institute, Medical Research Center, Seoul National University and Biomedical Brain Research Center, NIH, Seoul 110-799, Korea

Jun-Ho Lee, Department of Pharmacology, College of Medicine and Neuroscience Research Institute, Medical Research Center, Seoul National University and Biomedical Brain Research Center, NIH, Seoul 110-799, Korea

Virginia M.-Y. Lee, Center for Neurodegenerative Disease Research, Department of Pathology and Laboratory Medicine, University of Pennsylvania School of Medicine, 3600 Spruce St., 3rd Floor, Malony, PA 19104, USA; e-mail: vmylee@mail.med.upenn.edu

Jie Liu, Department of Pharmacology, Kumamoto University School of Medicine, 2-2-1 Honjo, Kumamoto 860-0811, Japan

Jean-Claude Louis, Amgen, Inc., One Amgen Center Drive, Thousand Oaks, CA 91320-1799, USA

Kiyoshi Maeda, Hyogo Institute for Aging Brain and Cognitive Disorders, 520 Saisho-ko, Himeji 670-0981, Japan

Eckhard Mandelkow, Max-Planck-Unit for Structural Molecular Biology, Notkestrasse 85, D-22603 Hamburg, Germany; e-mail: mand@mpasmb.desy.de

Eva-Maria Mandelkow, Max-Planck-Unit for Structural Molecular Biology, Notkestrasse 85, D-22603 Hamburg, Germany

Hirofumi Maruyama, Third Department of Internal Medicine, Hiroshima University School of Medicine Kasumi 1-2-3, Minami-ku, Hiroshima 734-8551, Japan

Naomi Matoh, Laboratory of Molecular Clinical Chemistry, Institute for Chemical Research, Kyoto University, Gokasho Uji, Kyoto 611-0011, Japan; e-mail: naomi@scl.kyoto-u.ac.jp

Masaaki Matsushita, Department of Neuropathology, Tokyo Metropolitan Institute of Psychiatry, Tokyo 156-0057, Japan

Edith G. McGeer, Kinsmen Laboratory of Neurological Research, Department of Psychiatry, University of British Columbia, 2255 Wesbrook Mall, Vancouver, B.C. V6T 1Z3, Canada

Patrick L. McGeer, Kinsmen Laboratory of Neurological Research, Department of Psychiatry, University of British Columbia, 2255 Wesbrook Mall, Vancouver, B.C., V6T 1Z3, Canada; email: mcgeerpl@interchange.ubc.ca

Yasuyo Mimori, Third Department of Internal Medicine, Hiroshima University School of Medicine Kasumi 1-2-3, Minami-ku, Hiroshima 734-8551, Japan

Eishichi Miyamoto, Department of Pharmacology, Kumamoto University School of Medicine, 2-2-1 Honjo, Kumamoto 860-0811, Japan; e-mail: emiyamot@gpo.kumamoto-u.ac.jp

Koho Miyoshi, Hyogo Institute for Aging Brain and Cognitive Disorders, 520 Saisho-ko, Himeji 670-0981, Japan

Hirdehiro Mizusawa, Department of Neurology, Tokyo Medical and Dental University, Tokyo 113-8519, Japan

Etsuro Mori, Hyogo Institute for Aging Brain and Cognitive Disorders, 520 Saisho-ko, Himeji 670-0981, Japan; e-mail: mori@hiabcd.go.jp

Hiroshi Mori, Department of Neuroscience, Osaka City University Medical School, 1-4-3 Asahimachi, Abeno-ku, Osaka 545-8585, Japan; e-mail: mori@med.osaka-cu.ac.jp

Maho Morishima-Kawashima, Department of Neuropathology, Faculty of Medicine, University of Tokyo, 7-3-1 Hongo, Bunkyo-ku, Tokyo 113-0033, Japan; e-mail: maho@m.u-tokyo.ac.jp

Hideyuki Mukai, Department of Biology, Faculty of Science, Kobe University, Kobe 654, Japan

Dominique Muller, Department of Pharmacology, Centre Medical Universitaire, Université de Genève, Faculté de Medecine, CH 1211 Genève 4, Switzerland

Ohoshi Murayama, Laboratory for Alzheimer's Disease, RIKEN Brain Science Institute, 2-1 Hirosawa, Wako-shi, Saitama 351-0198, Japan

Toshitaka Nabeshima, Department of Neuropsychopharmacology and Hospital Pharmacy, Nagoya University Graduate School of Medicine, Showa-ku, Nagoya 466-8560, Japan; e-mail: tnabeshi@ med.nagoya-u.ac.jp

Masamichi Nakai, Hyogo Institute for Aging Brain and Cognitive Disorders, 520 Saisho-ko, Himeji 670-0981, Japan; e-mail: nakai@hiabcd.go.jp

Shigenobu Nakamura, Third Department of Internal Medicine, Hiroshima University School of Medicine   Kasumi 1-2-3, Minamiku, Hiroshima 734-8551 Japan; e-mail: nakamura@mcai.med.hiroshima-u.ac.jp

Yu Nakamura, Department of Clinical Neuroscience and Psychiatry, Osaka University Graduate School of Medicine, 2-2 Yamadaoka, Suita, Osaka 565-0871, Japan

Yuka Nakano, Department of Social and Environmental Medicine, Osaka University, Graduate School of Medicine, 2-2 Yamadaoka, Suita, Osaka 565-0871, Japan

Yoshio Nanba, Department of Geriatric Medicine, Graduate School of Medicine, Tokyo University, 7-3-1 Hongo, Bunkyo-ku, Tokyo 113-8655 Japan; e-mail: ynanba-tky@umin.ac.jp

Linda Narhi, Amgen, Inc., One Amgen Center Drive, Thousand Oaks, CA 91320-1799, USA

Yugo Narita, Department of Neurology, Mie University School of Medicine, 2-174 Edobashi, Tsu 514-8507, Japan

Takashi Nishikawa, Department of Clinical Neuroscience and Psychiatry, Osaka University Graduate School of Medicine, 2-2 Yamadaoka, Suita, Osaka 565-0871, Japan

Tsuyoshi Nishimura, Department of Human and Cultural Sciences, Koshien University, Japan

Yasutomi Nishizuka, Biosignal Research Center, Kobe University, 1-1 Rokkoudai-cho, Nada-ku, Kobe 657-8501, Japan; e-mail: nisizuka@kobe-u.ac.jp

Taketoshi Ono, Department of Physiology, Faculty of Medicine, Toyama Medical and Pharmaceutical University, 2630 Sugitani, Toyama 930-0194, Japan; e-mail: onotake@ms.toyama-mpu.ac.jp

Yoshitaka Ono, Department of Biology, Faculty of Science, Kobe University, Kobe 654, Japan

Eiichi Otomo, Department of Internal Medicine, Yokufukai Geriatric Hospital, Tokyo 168-0071, Japan

Mieko Otsuka, Department of Neurology, Omiya Medical Center, Jichi Medical School, 1-847 Amanuma-cho, Omiya City, Saitama 330-8503, Japan; e-mail: motsuka@omiya.jichi.ac.jp

Yasuyoshi Ouchi, Department of Geriatric Medicine, Graduate School of Medicine, Tokyo University, 7-3-1 Hongo, Bunkyo-ku, Tokyo 113-8655, Japan; e-mail: bxr05600@nifty.ne.jp

Cheol Hyoung Park, Department of Pharmacology, College of Medicine and Neuroscience Research Institute, Medical Research Center, Seoul National University and Biomedical Brain Research Center, NIH, Seoul 110-799, Korea

Parvoneh Poorkaj, Geriatric Research Education and Clinical Center, Veterans Affairs Puget Sound Health Care System, Seattle Division, Seattle, WA 98108, USA

Jong-Cheol Rah, Department of Pharmacology, College of Medicine and Neuroscience Research Institute, Medical Research Center, Seoul National University, and Biomedical Brain Research Center, NIH, Seoul 100-799, Korea

Martin N. Rossor, Dementia Research Group, Institute of Neurology, The National Hospital for Neurology and Neurosurgery, Queen Square, London WC1N 3BG, UK; e-mail: m.rossor@dementia.ion.ucl.ac.uk

Naruhiko Sahara, Department of Neuroscience, Osaka City University Medical School, 1-4-3 Asahimachi, Abeno-ku, Osaka 545-8585, Japan; e-mail: nsahara@med.osaka-cu.ac.jp

Takaomi C. Saido, Laboratory for Proteolytic Neuroscience, RIKEN Brain Science Institute, 2-1 Hirosawa, Wako-shi, Saitama 351-0198, Japan; e-mail: email: saido@brain.riken.go.jp

Naoaki Saito, Biosignal Research Center, Kobe University, Rokkodai, Nada-ku, Kobe 657-0013, Japan

Niu San-Yu, Hyogo Institute for Aging Brain and Cognitive Disorders, 520 Saisho-ko, Himeji 670-0981, Japan

Ryogen Sasaki, Department of Neurology, Mie University School of Medicine, 2-174 Edobashi, Tsu 514-8507, Japan; e-mail: stssasak@east.ncc.go.jp

Satosi Sasaki, Department of Epidemiology, National Cancer Research Center, 6-5-1 Kashiwanoha, Kashiwa City, Chiba 227-8577, Japan

Naoya Sato, Department of Anatomy and Neuroscience, Osaka University, Graduate School of Medicine, 2-2 Yamadaoka, Suita, Osaka 565-0871, Japan

Shinji Sato, Laboratory for Alzheimer's Disease, RIKEN Brain Science Institute, 2-1 Hirosawa, Wako-shi, Saitama 351-0198, Japan

Carlos A. Saura, Department of Neurobiology, Pharmacology and Physiology, University of Chicago, Chicago, IL 60637, USA

Gerard D. Schellenberg, Geriatric Research Education and Clinical Center, Veterans Affairs Puget Sound Health Care System, 1660 S. Colombian Way Seattle, Seattle WA 98108-1597, USA; e-mail: zachdad@u.washington.edu

Yuko Segawa, Department of Clinical Neuroscience and Psychiatry, Osaka University Graduate School of Medicine, 2-2 Yamadaoka, Suita, Osaka 565-0871, Japan

Dennis J. Selkoe, Department of Neurology and Program in Neuroscience, Harvard Medical School and Center for Neurologic Diseases, Brigham and Women's Hospital, 77, Ave. Louis Pasteur, HIM 730, Boston, MA 02115-5716, USA

Ji-Heui Seo, Department of Pharmacology, College of Medicine and Neuroscience Research Institute, Medical Research Center, Seoul National University and Biomedical Brain Research Center, NIH, Seoul 110-799, Korea

Kazuhiro Shinosaki, Department of Clinical Neuroscience and Psychiatry, Osaka University Graduate School of Medicine, 2-2 Yamadaoka, Suita, Osaka 565-0871, Japan

Keiro Shirotani, Department of Demyelinating Disease and Aging, National Institute of Neuroscience, National Center of Neurology and Psychiatry, 4-1-1 Ogawahigashi, Kodaira, Tokyo 187-8551, Japan

Nobuyuki Sodeyama, Department of Neurology, Tokyo Medical and Dental University, Tokyo 113-8519, Japan

Karsten Stamer, Max-Planck-Unit for Structural Molecular Biology, Notkestrasse 85, D-22603 Hamburg, Germany

John Stevens, Department of Radiology, The National Hospital for Neurology and Neurosurgery, Queen Square, London WC1N 3BG, UK

Peter H. St George-Hyslop, Department of Medicine, Division of Neurology, The Toronto Hospital, University of Toronto, 6 Queen's Park Crescent West, Toronto, Ontario M5S 3H2, Canada; e-mail: p.hyslop@utoronto.ca

Naomi Suematsu, Department of Pathology, Yokufukai Geriatric Hospital, Tokyo 168-0071, Japan

Yoo-Hun Suh, Department of Pharmacology, College of Medicine and Neuroscience Research Institute, Medical Research Center, Seoul National University and Biomedical Brain Research Center, NIH, Seoul 110-799, Korea; e-mail : yhsuh@plaza.snu.ac.kr.

Xiaoyan Sun, Laboratory for Alzheimer's Disease, RIKEN Brain Science Institute, 2-1 Hirosawa, Wako-shi, Saitama 351-0198, Japan

Takeshi Tabira, Department of Demyelinating Disease and Aging, National Institute of Neuroscience, National Center of Neurology and Psychiatry, 4-1-1 Ogawahigashi, Kodaira, Tokyo 187-8551, Japan

Keikichi Takahashi Department of Demyelinating Disease and Aging, National Institute of Neuroscience, National Center of Neurology and Psychiatry, 4-1-1 Ogawahigashi, Kodaira, Tokyo 187-8551, Japan; e-mail: takahasi@ncnp.go.jp

Junichi Takamatsu, Division of Clinical Research, Kikuchi National Hospital, 208 Fukuhara Goushi-cho, Kikuti-gun, Kumamoto 861-1116, Japan

Akihiko Takashima, RIKEN Brain Science Institute, 2-1 Hirosawa, Wako-shi, Saitama 351-0198, Japan; e-mail: kenneth@brain.riken.go.jp

Junji Takeda, Department of Social and Environmental Medicine, Osaka University, Graduate School of Medicine, 2-2 Yamadaoka, Suita, Osaka 565-0871, Japan

Masatoshi Takeda, Department of Clinical Neuroscience, Osaka University, Graduate School of Medicine, 2-2 Yamadaoka Suita, Osaka 565-0871, Japan; e-mail: mtakeda@psy.med.osaka-u.ac.jp

Masanori Takehashi, Laboratory of Molecular Clinical Chemistry, Institute for Chemical - Research, Kyoto University, Gokasho Uji, Kyoto 611-0011, Japan; e-mail: takehashi@scl.kyoto-u.ac.jp

Ryoi Tamura, Department of Physiology, Faculty of Medicine, Toyama Medical and Pharmaceutical University, 2630 Sugitani, Toyama 930-0194, Japan; e-mail: rtamura@ms.toyama-mpu.ac.jp

Chikako Tanaka, Hyogo Institute for Aging Brain and Cognitive Disorders, 520 Saisho-ko, Himeji 670-0981, Japan; e-mail: chikako@hiabcd.go.jp

Seigo Tanaka, Laboratory of Molecular Clinical Chemistry, Institute for Chemical Research, Kyoto University, Gokasho Uji, Kyoto 611-0011, Japan; e-mail: seigo@scl.kyoto-u.ac.jp

Toshihisa Tanaka, Department of Clinical Neuroscience and Psychiatry, Osaka University, Graduate School of Medicine, 2-2 Yamadaoka Suita, Osaka 565-0871, Japan

Taizo Taniguchi, Hyogo Institute for Aging Brain and Cognitive Disorders, 520 Saisho-ko, Himeji 670-0981, Japan

Satoshi Tanimukai, Hyogo Institute for Aging Brain and Cognitive Disorders, 520 Saisho-ko, Himeji 670-0981, Japan

Eiko Tanno, Tokyo Institute of Psychiatry, 2-1-8, Kamikitazawa, Setagaya-ku, Tokyo 156-8585, Japan

Akira Terashima, Hyogo Institute for Aging Brain and Cognitive Disorders, 520 Saisho-ko, Himeji 670-0981, Japan

Gopal Thinakaran, Department of Neurobiology, Pharmacology and Physiology, University of Chicago, Chicago, IL 60637, USA; e-mail: gopal@uchicago.edu

Masaya Tohyama, Department of Anatomy and Neuroscience, Osaka University, Graduate School of Medicine, 2-2 Yamadaoka, Suita, Osaka 565-0871, Japan

Hiromasa Toji, Third Department of Internal Medicine, Hiroshima University School of Medicine Kasumi 1-2-3, Minami-ku, Hiroshima 734-8551, Japan

Taisuke Tomita, Department of Neuropathology and Neuroscience, Graduate School of Pharmaceutical Sciences, University of Tokyo, 7-3-1 Hongo Bunkyo-ku, Tokyo 113-0033, Japan

Takami Tomiyama, Department of Neuroscience, Osaka City University Medical School, 1-4-3 Asahimachi Abeno-ku, Osaka 545-8585, Japan; e-mail: tomi@med.osaka-cu.ac.jp

Bernhard Trinczek, Max-Planck-Unit for Structural Molecular Biology, Notkestrasse 85, D-22603 Hamburg, Germany

John Q. Trojanowski Center for Neurodegenerative Disease Research, Department of Pathology and Laboratory Medicine, University of Pennsylvania School of Medicine, 3600 Spruce St., 3rd Floor, Malony, PA 19104, USA; e-mail: trojanow@mail.med.upenn.edu

Kunihiro Ueda, Laboratory of Molecular Clinical Chemistry, Institute for Chemical Research, Kyoto University, Gokasho Uji, Kyoto 611-0011, Japan; e-mail: ueda@scl.kyoto-u.ac.jp

Akira Ueki, Department of Neurology, Omiya Medical Center, Jichi Medical School, 1-847 Amanuma-cho, Omiya City, Saitama 330-8503 Japan; e-mail: uekia@omiya.jichi.ac.jp

Elizabeth K. Warrington, Dementia Research Group, Institute of Neurology, The National Hospital for Neurology and Neurosurgery, Queen Square, London WC1N 3BG, UK

Jette Wypych, Amgen Inc., One Amgen Center Drive, Thousand Oaks, CA 91320-1799, USA

Weiming Xia, Department of Neurology and Program in Neuroscience, Harvard Medical School and Center for Neurologic Diseases, Brigham and Women's Hospital, 77, Ave. Louis Pasteur, HIM 730, Boston, MA 02115-5716, USA; e-mail: xia@cnd.bwh.harvard.edu

Keiko Yagi, Biosignal Research Center, Kobe University, Rokkodai, Nada-ku, Kobe 657-0013, Japan, and Department of Clinical Pharmacy, Kobe Pharmaceutical University, Kobe 658-8558, Japan

Kiyofumi Yamada, Department of Neuropsychopharmacology and Hospital Pharmacy, Nagoya University Graduate School of Medicine, Showa-ku, Nagoya 466-8560, Japan; e-mail: k-yamada@med.nagoya-u.ac.jp

Masahito Yamada, Department of Neurology, Kanazawa University School of Medicine, 13-1 Takara-mach, Kanazawa 920-8640, Japan; e-mail: m-yamada@med.kanazawa-u.ac.jp

Michiko Yamada, Radiation Effects Research Foundation, Kasumi 1-2-3, Minami-ku, Hiroshima 734-8551, Japan

Katsuhiko Yanagisawa, Department of Dementia Research, National Institute for Longevity Sciences, 36-3 Gengo, Morioka, Obu 474-8522, Japan; e-mail: katuhiko@nils.go.jp

Koji Yasojima, Kinsmen Laboratory of Neurological Research, Department of Psychiatry, University of British Columbia, 2255 Wesbrook Mall, Vancouver, B.C. V6T 1Z3, Canada

Minoru Yasuda, Hyogo Institute for Aging Brain and Cognitive Disorders, 520 Saisho-ko, Himeji 670-0981, Japan; e-mail: yasuda@hiabcd.go.jp

Bin Zhang, The Center for Neurodegenerative Disease Research, Department of Pathology and Laboratory Medicine, University of Pennsylvania School of Medicine, 3600 Spruce St., 3rd Floor, Malony, PA 19104, USA

Preface

The lengthening survival time of populations around the world is pushing dementia into the forefront of medicine. The chapters in this timely volume are the official record of papers presented at the International Symposium on Dementia held in Kobe, Japan, September 11–13, 1999. Doctor Yasuo Ihara of the Program Committee, and Doctors Chikako Tanaka and Toshio Kawamata of the Organizing Committee masterfully put together a Symposium that gives a comprehensive picture of dementia, from molecular biology to therapeutics. The papers, initially for the benefit of the attendees of the Symposium, are now available to a much wider audience through this publication. As readers will discover, the chapters are written by leading authorities in their fields, each of whom has presented a clear and up-to-date summary of their particular area. The sum of this volume is much greater than the individual parts, because the reader can develop a broad understanding of dementia from the ways in which the individual papers integrate into an overview of the field.

The volume opens with chapters by Nishizuka and McGeer et al., which deal with the phenomena of membrane lipid signaling and neuroinflammation. Although these are important to dementia, they have implications in broader fields of medicine. The next section, dealing with memory and its impairment, has chapters by Cahill on long-term memory, Mori on amygdala damage, Tamura and Ono on the hippocampus and Miyamoto et al. on kinases in hippocampal long-term potentiation. The volume then turns to pathogenesis. There are eight chapters involving the protein tau which aggregates to cause neuronal degeneration in Alzheimer disease and other degenerative neurological disorders where neurofibrillary tangles and other tau deposits occur. These chapters include one on transgenic mice by Ishihara et al., two on tau mutations by Schellenberg et al. and Yasuda et al., one on the ALS-PDC complex of the Kii peninsula by Kuzuhara et al., one on an apparently new NFT entity by Yamada et al. and three on cellular pathology by Mandelkow et al., Sahara et al. and Kawamata et al. The next section contains three chapters on dementia related to synuclein by Duda et al., Tanaka et al. and Citron et al. The role of the presenilins and amyloid precursor protein in Alzheimer disease is next covered in a series of thirteen chapters. These include seven on the presenilins by St. George-Hyslop, Thinakaran et al., Iwatsubo and Tomita, Xia and Selkoe, Kudo et al., Takashima et al. and Takahashi et al. They are followed by six chapters on various aspects of beta-amyloid peptide by Suh et al., Akiyama et al., Nakai et al., Morishima-Kawashima and Ihara, Yanagisawa, and Saido and Iwata. Finally, there are five chapters on the diagnosis and therapeutics of dementia by Rossor et al., Takeda et al., Ueki et al., Nakamura et al., and Nabeshima and Yamada.

The reader will find a rich intellectual array in this volume. Key aspects of a vast and scattered literature have been neatly drawn together to provide an integrated view of one of the major medical problems of our time.

Patrick L. McGeer

Special lectures

Neuroscientific Basis of Dementia
C. Tanaka, P.L. McGeer, Y. Ihara (eds)
© 2001 Birkhäuser Verlag Basel/Switzerland

A tale of protein kinase C and membrane lipid signaling

Yasutomi Nishizuka

Biosignal Research Center, Kobe University, Kobe, Japan

Introduction

Many receptors transduce extracellular signals through inositol phospholipid hydrolysis, but this signaling pathway was fully accepted only about one decade ago. For many years the cell membrane was thought to be a biologically inactive semi-permeable barrier that splits exterior and interior cellular compartments. A major constituent of the cell membrane is phospholipid, which consists of a water-insoluble lipid portion and a water-soluble polar head group. The phospholipid with inositol as the polar head group, "inositol phospholipid", is a relatively minor component. It comprises approximately less than 8% of total phospholipids in cell membranes. A small portion of this inositol phospholipid contains an additional phosphate at the position 4 (PI-4-P) or 3 (PI-3-P), two phosphates at the positions of 4 and 5 (PI-4,5-P2), or 3 and 4 (PI-3,4-P2), and sometimes 3 phosphates at the positions 3, 4 and 5 (PI-3,4,5-P3). These polyphospho-inositides are key players in cell signaling.

Inositol was first identified in the brain tissue by Wooley as early as 1941. In the next year, Folch and Wooley described the chemical structure of inositol phospholipid. In the late 1940s, Folch found inositol phospholipid that contains multiple phosphates in the inositol portion. These phosphate groups turn over rapidly when the brain function is stimulated. Folch's preparation was a mixture of polyphospho-inositides.

In 1953, Mabel Hokin and Lou Hokin at Sheffield University found that acetylcholine induces rapid hydrolysis of inositol phospholipids in some excretory tissues such as the pancreas. Subsequent studies by many investigators clarified the biochemical pathway of the degradation and synthesis of inositol phospholipids. The potential role of this lipid hydrolysis in the receptor function was suggested by Drell in 1969, but no obvious evidence for this assumption was available at that time. In 1975, Bob Michel in Nottingham wrote an important review article in *Biochim Biophys Acta*, and suggested that this receptor-induced lipid hydrolysis may open the Ca^{2+}-gate. But again, no evidence was available to support this hypothesis.

Nearly 20 years ago, our research group in Kobe found a unique enzyme in the brain tissue. This enzyme is normally inactive, but is activated by diacylglycerol, one of the products of inositol phospholipid hydrolysis. We call this enzyme "Protein Kinase C (PKC)", because Ca^{2+} is necessary for its full activation. We suspected then that the enzyme could be activated by extra-cellular signals, such as hormones, neurotransmitters and growth factors, which induce inositol phospholipid hydrolysis. To prove this possibility, we prepared a cell-mem-

brane permeable diacylglycerol. Around that time, we also found that a powerful tumor-pro-
moting phorbol ester, TPA, mimics diacylglycerol action, and directly activates PKC within
the cell. It became possible to demonstrate, by the simultaneous addition of either permeable
diacylglycerol or phorbol ester and Ca^{2+} ionophore, that PKC activation and Ca^{2+} increase
together could fully reproduce physiological cell responses such as lysosomal enzyme release
and T-cell activation. A few years later, Michael Berridge in Cambridge announced that inositol triphosphate (IP_3), the other half of the molecule, mobilizes Ca^{2+} from its internal store
(see [1] for review).

The family of protein kinase C and related enzymes

Today, PKC is known to be a large family of proteins with closely related structures, but
slightly distinct properties and functions. All PKC isoforms consist of a N-terminal regulatory region and a C-terminal catalytic region, which have serine/threonine kinase activity
(Fig. 1). The catalytic regions are very similar, and there is no clear-cut difference *in vitro* in
enzymatic specificity of the catalytic regions. However, these isoforms are divided into three
subgroups, according to the properties of the regulatory regions. cPKC isoforms, the initially identified classical PKC enzymes, have a C-1 domain. This C-1 domain consists of tandem repeats of Zn-finger-like structures, which bind diacylglycerol. Thus, the enzymes
translocate to the membrane, where diacylglycerol is produced. The C-2 domain binds Ca^{2+}
and makes the enzyme sensitive to "Ca^{2+} increase". nPKC isoforms show similar properties

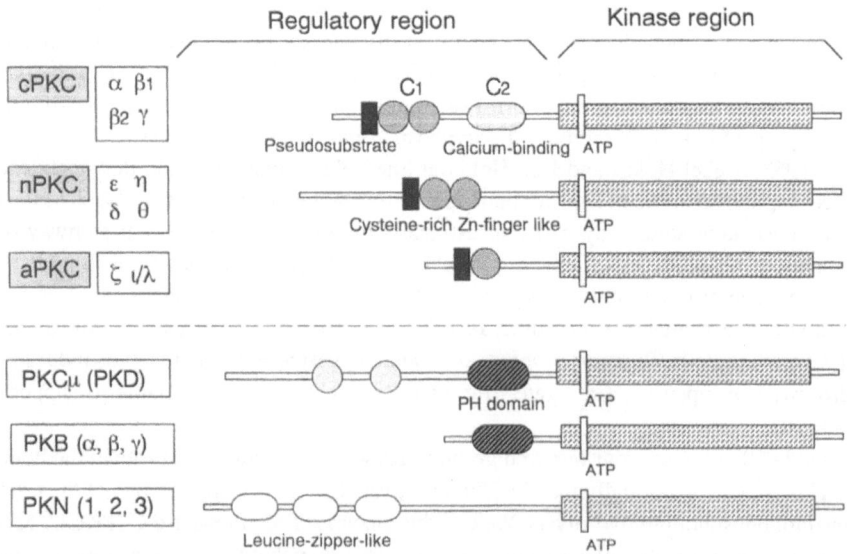

Figure 1. The mammalian PKC family and related protein kinases. (See footnote at page 5 for further explanan-
tions.)

but are not sensitive to Ca^{2+}, because the C-2 domain is absent. aPKC isoforms do not respond to diacylglycerol or to Ca^{2+} (see [2] for review).

On the other hand, the three related enzymes, PKD, PKB and PKN, identified during the studies of PKC enzymes, show properties more or less similar to those of the PKC family. PKCμ, another name for PKD, has a PH domain and is activated by phorbol ester. PKB was isolated by several groups of investigators: Hemmings in Switzerland, Tsichlis in Philadelphia and Coffer in Utrecht. The PKB γ-isoform was isolated by Kikkawa in Kobe. The PKB isoforms show similarities to both PKC and PKA, and thus are termed PKB. The PKB isoforms have a PH domain and are activated by phosphatidylinositol-3-kinase (PI-3-kinase) products, $PI-3,4-P_2$, as described later. PKN was isolated by Ono, again in our group, and by Parker in London. This enzyme is activated by free fatty acids, and has a unique leucine zipper-like structure.

In short, the regulatory regions of these enzymes have very unique structures, and bind several lipid mediators to exhibit their catalytic activities.

Lipid messengers for protein kinase activation

It became clear subsequently that, in addition to phospholipase C, other phospholipases are also activated upon cell surface receptor stimulation, and generate many lipid messengers or mediators for cell signaling (see [2] for review). Phospholipase C activation produces diacylglycerol and IP_3, and the mechanism of this enzyme activation has been well clarified by Sue-Goo Lee and others. Phospholipase A_2, particularly cytosolic enzymes, and phospholipase D are activated by various signals such as growth factors, and produce lipid mediators, including free fatty acids and lyso-phospholipids. The activation mechanism of these phospholipases is another focus of cell signaling research.

More recently, the PI-3-kinase signaling pathway has attracted great attention. This group of enzymes produces inositol phospholipids containing additional phosphate at position 3. This PI-3-kinase pathway produces phosphatidylinositol-3,4-bisphosphate $(PI-3,4-P_2)$ and phosphatidylinositol-3,4,5-trisphosphate $(PI-3,4,5-P_3)$. These lipids activate several kinases including PDK-1, PKB and aPKC. This pathway mediates the signals from insulin and some growth and trophic factors. Inositol phospholipid, PI, is also shown to be converted to PI-3-phosphate by the action of a unique class III PI-3 kinase. PI-3-P binds to proteins having a FYVE (Fab1, YOTB/ZK632.12, Vac1, EEA1) finger domain, which play roles in membrane fusion and exocytosis (see [3] for review).

Abbreviations used in the table and figures are: PI, phosphatidylinositol; PIP_2, PI-4,5-bisphosphate; PIP_3, PI-3,4,5-trisphosphate; IP_3, inositol-1,4,5-trisphosphate; PC, phosphatidylcholine; PS, phosphatidylserine; PE, phosphatidylethanolamine; SM, sphingomyelin; DG or DAG, diacylglycerol; PA, phosphatidic acid; lyso-PA, lyso-phosphatidic acid; AA, arachidonic acid; FFA, free fatty acid; PLA_2, phospholipase A_2; $cPLA_2$, cytosolic phospholipase A_2; $iPLA_2$, Ca^{2+}-insensitive phospholipase A_2; PLC, phospholipase C; PLD, phospholipase D; SMase, sphingomyelinase; PKA, protein kinase A; PKC, protein kinase C; PTPase, phosphotyrosine protein phosphatase; S/T-ppase, phosphoserine/phosphothreonine phosphoprotein phosphatase; and PDK-1, polyphosphoinositide-dependent protein kinase-1.

In short, in response to extracellular signals, major membrane phospholipids are metabolized rapidly to produce various lipid messengers and mediators for the control of intracellular events. Table 1 summarizes the results so far obtained by this and several other laboratories. PKC isoforms respond differently to various lipid mediators. For instance, the classical PKC isoforms are activated by diacylglycerol, and sometimes by fatty acids and lyso-PC. aPKC isoforms are activated by PI-3,4,5-P_3, or indirectly through phosphorylation by PI-3,4,5-P_3-dependent kinase, PDK-1. Several other lipids produced in the membrane activate many PKC isoforms and related signaling molecules, including PKB and PKN.

In general, the N-terminal half of the molecule regulates the C-terminal catalytic kinase activity. In addition, through interaction with lipid components, the N-terminal half plays essential roles in translocation and targeting of the enzyme to each specific intracellular compartment. Further, the PKC molecule *per se* also plays roles in controlling functional proteins through protein-protein interaction rather than protein phosphorylation. For instance, it is known that the N-terminal half of PKC directly activates phospholipase D without its phosphorylation (see [4] for review).

Table 1. Proposed lipid mediators for PKC, PKB and PKN activation. (See footnote at page 5 for further explanation)

Phospholipid	Enzyme	Mediator	Proposed target of mediator action
PI (PIP$_2$)	PLCs	DG, IP$_3$	cPKC nPKC Ca^{2+} mobilization
	PI-3-kinases	PIP$_3$, PI-(3,4)-P$_2$	PDK-1 aPKC PKB, PLD$_1$, p70S6K, etc.
PC	cPLA, iPLAs	AA, FFAs	Eicosanoids cPKC nPKC PKN, PLD$_2$ etc.
		Lyso-PA	cPKC PAF formation
		PA, DG	cPKC nPKC Raf-1 etc.
		Lyso-PA	cPKC mitogenic
PS, PE	iPLAs	FFAs	cPKC nPKC PKN, PLD$_2$ etc.
SM	SMase	Ceramide Sphingosine-P	PKCζ ?

Translocation and targeting of protein kinases

Several distinct isoforms of the PKC family and related enzymes activated as described above, are targeted to each distinct cellular compartment, such as plasma membrane, Golgi membrane, nuclear membrane or cytoskeleton, where each isoform performs its specific

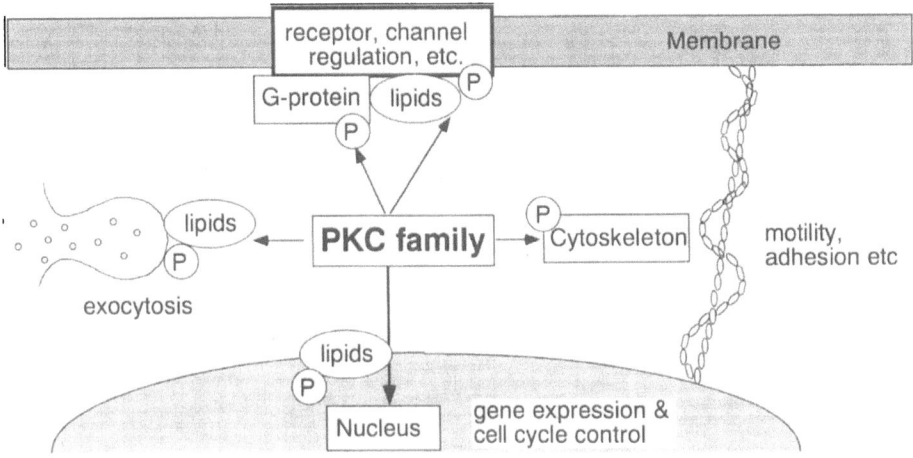

Figure 2. Targeting and functions of PKC.

function (Fig. 2). To visualize such dynamic movement and targeting of the enzyme mole-cule within the cell, Saito in our research group prepared a fusion protein of the PKC mole-cule and a green fluorescence protein (GFP), and expressed it in Chinese hamster ovary (CHO) cells [5].

In resting CHO cells, both γ- and ε-isoforms are localized in the cytoplasm. When stimu-lated by phorbol ester or by a physiological signal, purinergic signal ATP for instance, both isoforms are rapidly translocated to the plasma membrane. With phorbol ester stimulation, the translocation is relatively slow and irreversible. In contrast, with ATP stimulation, the translo-cation is almost instantaneous, and the enzyme soon returns back to the cytoplasm. When stimulated by membrane-permeable diacylglycerol, the pattern of translocation was exactly identical with that caused by receptor stimulation (Fig. 3). The isoform diffused back to the cytoplasm within a minute. Tracer experiments show that this permeable diacylglycerol added to the cell is metabolized very rapidly, and disappears within a minute. On the other hand, arachidonic acid stimulation induces translocation of ε-PKC to the Golgi membrane. This translocation to the Golgi is apparently irreversible, and not induced by prostaglandins, suggesting that free fatty acid itself induces this unique translocation of the ε-PKC isoform.

Another action of free fatty acids is to synergize with diacylglycerol to activate PKC enzymes to induce cellular responses (see [2] for review). Our earlier studies have shown that free fatty acids, especially unsaturated fatty acids, frequently synergize with diacylglyc-erol to cause exocytosis and release reactions. For instance, platelets respond to simultane-ous addition of diacylglycerol and Ca^{2+} ionophore, and release serotonin. This release reac-tion is greatly enhanced by further addition of unsaturated fatty acids. Fatty acid alone shows no activity. In cell-free enzymatic reactions, diacylglycerol and free fatty acids in fact syn-ergize to activate PKC isoforms. Consistent with such observations, translocation of PKC to

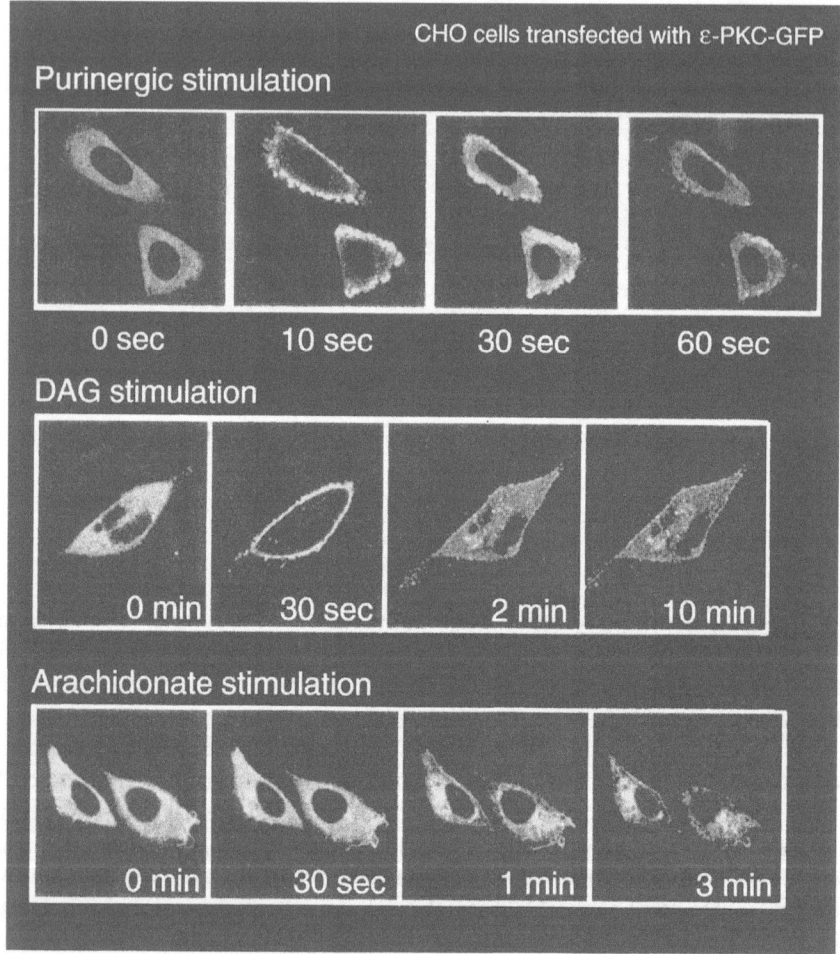

Figure 3. Translocation of PKC to membrane. (See footnote at page 5 for further explanation)

membranes is enhanced by free fatty acids in the presence of diacylglycerol, as visualized by the GFP-fluorescence.

Translocation of some PKC isoforms to membranes frequently shows oscillation. The molecular mechanism of this oscillation is not clear. This translocation of the enzyme does not appear to synchronize with Ca^{2+} oscillation. It seems plausible that lipid messengers, such as diacylglycerol, are oscillating after receptor stimulation. In short, lipid-protein interaction for PKC activation appears to be more complicated than expected before.

Destination and mechanism of translocation

The mechanism of PKC targeting to each specific membrane site is one of the major focuses of current interest (see [6] for review). Some years ago Mochly-Rosen proposed that there may be a specific PKC receptor protein at membranes. When the cell is stimulated, the PKC molecule is activated and translocates to this specific receptor protein. Such a receptor protein is reported for the β- and ε-isoforms.

An additional major question in this field of research is how each PKC isoform recognizes its substrate protein. One possibility emerging in recent years is that intracellular localization of protein kinase, its substrate, phosphatase and other signaling molecules are associated together, consisting of a unique functional unit. For example, a scaffold protein or anchoring protein associated with membranes serves as a talking place of several signaling pathways (Fig. 4). In a post-synaptic signaling protein complex, for instance, an anchoring protein binds PKC, protein kinase A (PKA), their substrates and phosphatases. Some L-type Ca^{2+} channels are bound to a unique A-kinase-anchoring protein (AKAP). Scaffold proteins are sometimes associated with a series of protein kinases in cascade. Such a protein kinase complex is proposed to exist in yeast, and recently in mammalian cells.

Ono in our research group in Kobe identified a huge protein of 450 kDa molecular size, uniformly expressing in human tissues. This protein is localized to the centrosome throughout the cell cycle, and to the Golgi membrane at interphase. The protein has many unique domains which selectively bind several signaling molecules, including PKA, PKN, probably PKC, and protein phosphatases. These signaling molecules are presumably associated together to interact, to counteract, or to co-operate for control of the centrosome function. In fact, PKN translocates to the nucleus, when cells are stimulated by stresses such as heat

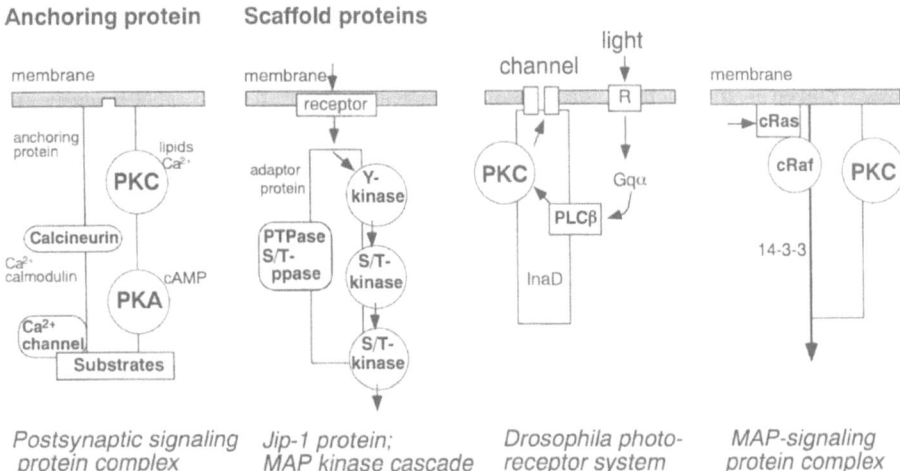

Figure 4. Hypothesis of protein kinases and signaling protein complexes. (See footnote at page 5 for further explanation)

shock or U.V. irradiation. Work done by Ono has shown that major substrates of PKN appear to be skeleton proteins in the cytoplasm and nucleus, including vimentin, caldesmon, neurofilament and Tau protein.

Cell survival and death

On the other hand, the function of PKB is becoming clearer. Insulin, insuline-like growth factor 1 (IGF1) and some growth and trophic factors activate class I and II PI-3-kinases and produce PI-3,4-P_2. This PI-3,4-P_2 binds to the PH domain of the PKB molecule to exhibit its catalytic activity. Several cell survival signals activate PKB through the PI-3-kinase pathway. PKB, once activated in this way, phosphorylates a group of unique transcription factors, called "Forkhead" transcription factors. This factor was first identified by genetic studies with *C. elegans*, as DAF-16, and later in human cells as AFX and FKHR proteins.

In growing cells the Forkhead proteins are phosphorylated by PKB, and cannot get into the nucleus through nuclear pores. Once PKB is inactivated by removal of survival signals or by inhibiting PI-3-kinase with Wortmannin or LY294002, then the Forkhead proteins are dephosphorylated, get into the nucleus, and transcribe some apoptotic genes, resulting in DNA fragmentation and cell death.

To visualize the translocation of Forkhead protein from cytoplasm to nucleus, Kikkawa prepared a fusion protein of AFX and GFP, and expressed it in CHO cells [7]. In growing cells, the AFX-GFP protein is localized in the cytoplasm, because it is phosphorylated. If, however, the cells were treated with PI-3-kinase inhibitor, Wortmannin or LY294002, the GFP fusion protein was dephosphorylated and translocated into the nucleus. The AFX Forkhead transcription factor is composed of about 500 amino acids, and has three sites that can be phosphorylated by PKB: one threonine and two serine residues. All Forkhead proteins so far identified have the three consensus sequences. If these three sites were eliminated by point mutation to alanine, then the mutant AFX protein could not be phosphorylated, and always stayed in the nucleus. Flow cytometric analysis indicates that cells that have the mutated non-phosphorylatable Forkhead protein undergo DNA fragmentation, resulting in cell death after prolonged incubation. These and other results indicate that the PKB enzyme is linked to the PI-3-kinase signaling pathway, negatively regulates Forkhead transcription factors and prevents cell death (Fig. 5). It may be noted that the Forkhead protein shows the gene transcription activity irrespective of the phosphorylation of these three sites, and that the phosphorylation reaction itself serves as a signal for transport of the protein through nuclear pores. It has been shown recently that PKB also phosphorylates I-κB-kinase, and then activates NF-κB. This reaction serves as a cell survival signal.

On the other hand, evidence obtained for some haematopoietic cells suggests that the δ-PKC isoform acts in an opposing way. Tumor necrosis factor (TNF), anti-FAS antibody and stresses, such as oxidative stresses, cause activation of δ-PKC, partly due to its proteolytic activation. This δ-PKC activation inhibits DNA-dependent protein kinase, and prevents repair of DNA damage. Thus, it is attractive to suggest that PKB and δ-PKC sometimes function in opposing ways for cell survival and cell death.

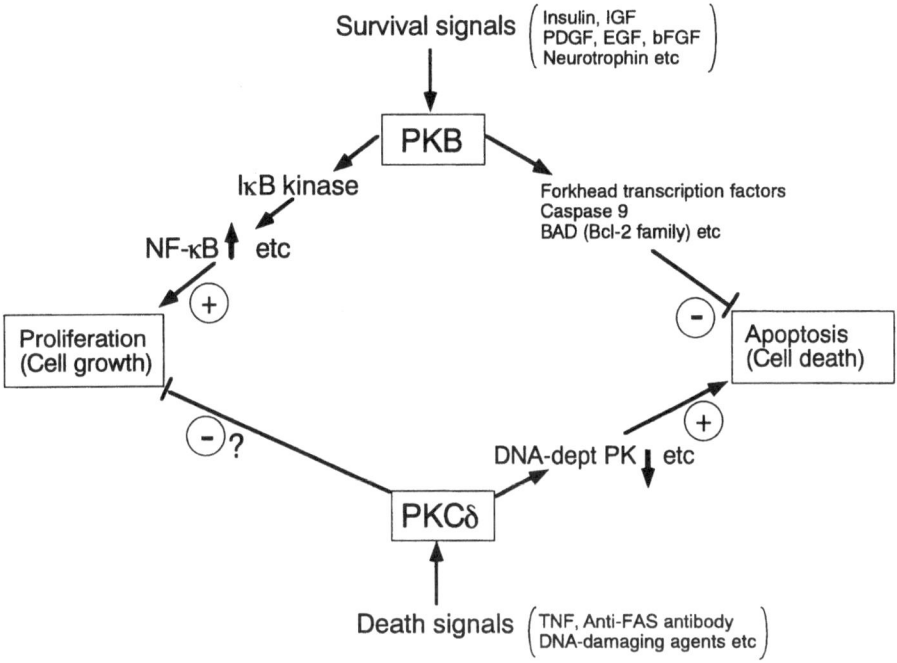

Figure 5. Opposing roles of PKB and δ-PKC (hypothesis). (See footnote at page 5 for further explanation)

Upon stimulation of the receptor, δ-PKC is normally activated by lipid mediators, and translocates to the membrane. In addition to this mechanism, oxidative stresses, such as H_2O_2, are able to activate this enzyme. This process appears to be independent of lipid mediators, but tyrosine residues of the enzyme are phosphorylated concomitantly. At present, the causal relationship between tyrosine phosphorylation and δ-PKC activation is not fully understood.

Control of nuclear PKC and the cell cycle

Another important function of the PKC family is to regulate transcription factors and other proteins which are involved in cell cycle control. Undoubtedly, coordinated actions of multiple signaling pathways and kinases are needed for the regulation of nuclear functions. Vimentin is one of the nuclear proteins and its phosphorylation appears to play a critical role in cell cycle regulation. Inagaki prepared a series of antibodies, each recognizing selectively the site of the phosphate that is covalently linked to specific serine residues in vimentin. The serine residue is phosphorylated selectively by each specific protein kinase [8]. Therefore, it is possible to estimate the specific time point of the cell cycle when each protein kinase plays its role. By this procedure, it has been shown that cdc-2-kinase is active at prometa- and meta-phase, PKC is active at meta- and ana-phase, and that Rho-kinase is active at ana- and telo-phase during the cell cycle. On the other hand, calmodulin-dependent protein kinase appears to be active throughout the cell cycle. Another series of experiments suggests that PKC becomes active at the late G_1-phase or the beginning of the S-phase.

There are many possibilities for the mechanism of activation of PKC in the nucleus. Numerous reports have proposed that PKC may be translocated into the nuclear compartment from the cytoplasm, although no obvious nuclear targeting signal motif is found in the PKC molecules. The mechanism and specificity of this translocation are not clearly understood at present. It is alternatively proposed that PKC itself is always present in the nucleus, and that this nuclear PKC is activated by nucleus-specific inositol phospholipid hydrolysis. This possibility has been studied extensively, and a nucleus-specific phospholipase C is shown to increase dramatically in the G1/S phase transition. It has been proposed that nuclear PKC may be activated by a unique lipid mediator, phosphatidyl-glycerol.

It has been shown that diacylglycerol-kinase ζ, which exists in the nucleus, is phosphorylated by nuclear PKC itself. This phosphorylated form of the diacylglycerol-kinase ζ does not change its enzymatic activity, but is removed from the nucleus. As a result, the nuclear diacylglycerol concentration is increased, and more PKC is activated in the nucleus. In summary, further studies are necessary to clarify the mechanism of nuclear PKC activation, which appears to be important for gene expression and cell cycle control.

Conclusion

In the last nearly two decades, knowledge of the PKC family and its related enzymes has expanded rapidly. Today, several research fields in biology and medicine are integrated and directly involved in the research on protein phosphorylation. The activation of the PKC family is clearly coupled to the dynamic metabolism of cell membrane lipids. Targeting of the enzyme to specific intracellular compartments, as well as its association with other signaling molecules and substrate proteins, are important research projects today. Various cytoskeleton proteins appear to be intimately related to the signal pathways involving the PKC family and other protein kinases. Publications related to the PKC family that appear in major journals of biology and medicine are still steadily and rapidly increasing. Numerous reports describe therapeutic applications, such as drug designs for diabetic complications, cancer chemotherapy, pain therapy, and treatment of Parkinson's disease. It is hoped that more clinical and therapeutic intervention will develop in the years to come.

Acknowledgements
Most of the results described above were obtained by the collaborators in our research group at Kobe University. In particular, the dynamic translocation of the PKC isoforms by several lipid mediators was extensively studied by Saito, Sakai, Shirai and Tsujishita. The structure and functions of PKC and PKB enzymes and stress-dependent activation of PKC were investigated by Kikkawa and Konishi. Identification and nuclear translocation of a unique enzyme, PKN, were clarified by Ono, Takahashi and Mukai. I wish to express my deep gratitude to Prof. C. Tanaka, editor of this proceeding, for giving me this excellent opportunity.

References

1 Nishizuka Y (1984) The role of protein kinase C in cell surface signal transduction and tumour promotion. *Nature* 308: 693–698
2 Nishizuka Y (1995) Protein·kinase C and lipid signaling for sustained cellular responses. *FASEB J* 9: 484–496
3 Toker A, Cantley LC (1997) Signalling through the lipid products of phosphoinositide-3-OH kinase. *Nature* 387: 673–676
4 Morris AJ, Hammond SM, Colley C, Sung T-C, Jenco JM, Sciorra VA, Rudge SA, Frohman MA (1997) Regulation and functions of phospholipase D. *Biochem Soc Trans* 25: 1151–1157
5 Sakai N, Sasaki K, Ikegaki N, Shirai Y, Ono Y, Saito N (1997) Direct visualization of translocation of gamma-subspecies of protein kinase C in living cells using fusion proteins with green fluorescent protein. *J Cell Biol* 139: 1465–1476
6 Mochly-Rosen D, Gordon AS, (1998) Anchoring proteins for protein kinase C: a means for isozyme. *FASEB J* 12: 35–42
7 Takaishi H, Konishi H, Matsuzaki H, Ono Y, Shirai Y, Saito N, Kitamura T, Ogawa W, Kasuga M, Kikkawa U, Nishizuka Y, (1999) Regulation of nuclear translocation of Forkhead transcription factor AFX by protein kinase B. *Proc Natl Acad Sci USA* 96: 11836–11841
8 Kamei Y, Inagaki N, Nishizawa M, Tsutsumi O, Taketani Y, Inagaki M (1998) Visualization of mitotic radial lineage cells in the developing rat brain by cdc2 kinase-phosphorylated vimentin. *Glia* 23: 191–199

Neuroscientific Basis of Dementia
C. Tanaka, P.L. McGeer, Y. Ihara (eds)
© 2001 Birkhäuser Verlag Basel/Switzerland

Complement, neuroinflammation and neuronal degeneration in Alzheimer disease

Patrick L. McGeer, Edith G. McGeer and Koji Yasojima

Kinsmen Laboratory of Neurological Research, Department of Psychiatry, University of British Columbia, Vancouver, Canada

Introduction

It is now established that the lesions of Alzheimer disease (AD) are characterized by the presence of a broad spectrum of inflammatory molecules. They include complement proteins and their regulators, inflammatory cytokines, acute phase reactants and numerous proteases and protease inhibitors [1, 2]. Several of these inflammatory products are neurotoxic and may contribute in a major way to the progressive neuronal loss of AD. Probably the most dangerous to host tissue is the complement system. It can be activated *in vitro* by several molecules found in AD lesions, including β-amyloid protein and C-reactive protein. Activated complement fragments richly decorate AD lesions. The membrane-attack complex of the complement is observed attached to damaged neurites. Many activated microglia are also seen clustering around AD lesions. Like all phagocytes, activated microglia produce large amounts of free radicals, glutamate and other potentially neurotoxic compounds. This is a local rather than a systemic immune reaction, with the brain cells making the inflammatory components [2]. The hypothesis that chronic inflammation may be contributing to neuronal death in AD is supported by epidemiological studies indicating that patients taking anti-inflammatory drugs, particularly of the nonsteroidal type, have a substantially reduced risk of AD [3]. In a pilot, 6-month, double-blind clinical trial, indomethacin appeared to arrest the disease [4].

Although the evidence is not as strong, there are also data indicating the possible damaging role of a chronic inflammatory state in conditions such as myocardial infarct [5] and stroke [6]. Studies on infarcted heart tissue, for example, have shown the presence of all elements of the complement cascade, including the membrane-attack complex [7]. Again the evidence suggests that the inflammatory components are generated locally.

A remaining question is how intense the inflammatory reaction is in affected tissues in AD, myocardial infarct or stroke, as compared with that in a well-recognized immune reaction such as occurs in osteoarthritic joints. Since the inflammatory components appear to be locally produced, one approach to this question is to measure relative levels of their mRNAs in affected and unaffected tissue. Here we will review some evidence, obtained in our laboratory, suggesting that the relative intensity of inflammation is comparable for AD brain, osteoarthritic joints, atherosclerotic plaques and myocardial infarcts.

Complement proteins

We have already published on the greatly elevated levels of the mRNAs for the complement proteins in affected regions of AD brain [8] and of infarcted human heart [9]. Preliminary evidence indicates similarly significant elevations of the levels of the mRNAs for complement proteins in atherosclerotic plaques (Fig. 1A). In these individuals there was no elevation of the mRNAs for complement proteins in the liver (Fig. 1A), emphasizing the local nature of the immune reaction.

Endogenous inhibitors of the complement system also exist as a protective mechanism against attack on host tissue. Two such inhibitors are C1 inhibitor and CD59. The former regulates the activity of C1r and C1s, arresting the complement cascade at the C1 dissociation level. The latter is expressed on the external surface membrane of host cells. It binds to C5b678, preventing insertion of the complex into the membrane and subsequent addition of multiple molecules of C9 to produce a lytic hole. It seemed worthwhile to determine whether these are also upregulated during inflammation. Measurements have not yet been done on infarcted heart or atherosclerotic plaques, but our results on AD brain [10] suggest the upregulation of the mRNAs for these inhibitors is far less than that for the mRNAs of the complement proteins (Fig. 1B).

C-reactive protein

C-reactive protein (CRP) is an acute phase reactant which is locally produced and increases are found in association with AD lesions. Deposited CRP can activate complement *in vitro* by binding to C1q. It is of interest that CRP does this by binding to the collagen-like tail region of the C1q A chain, specifically in the 14–26 region and 76–92 regions [11]. This is unlike antibodies which bind to the globular head region. The importance of this distinction is that drugs may be found which inhibit non-immunoglobulin complement activation without affecting the immune response to antibodies.

CRP also appears related to myocardial infarct since it has been shown that apparently healthy individuals with high serum levels of CRP have a several-fold increase in risk of heart attacks [12]. Moreover, following a myocardial infarct or stroke, the increase in serum CRP levels reflects the size of the lesion and predicts survival or death [12].

Measurements of the levels of the mRNA for CRP in normal and inflamed tissue revealed significant elevations in affected regions of AD brain, in atherosclerotic plaques, in osteoarthritic joints and in infarcted heart tissue. Again, as with the mRNAs for complement proteins, levels in the livers of persons dying with these conditions were normal (Fig. 2).

Microglia/macrophage markers

Activated macrophages of all types, including brain microglia, display upregulated levels of a wide variety of surface markers. Among these are the MHC class II glycoprotein HLA-DR

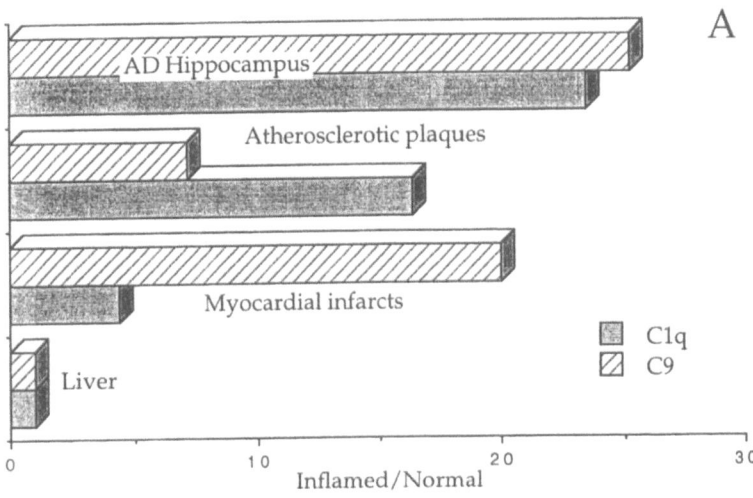

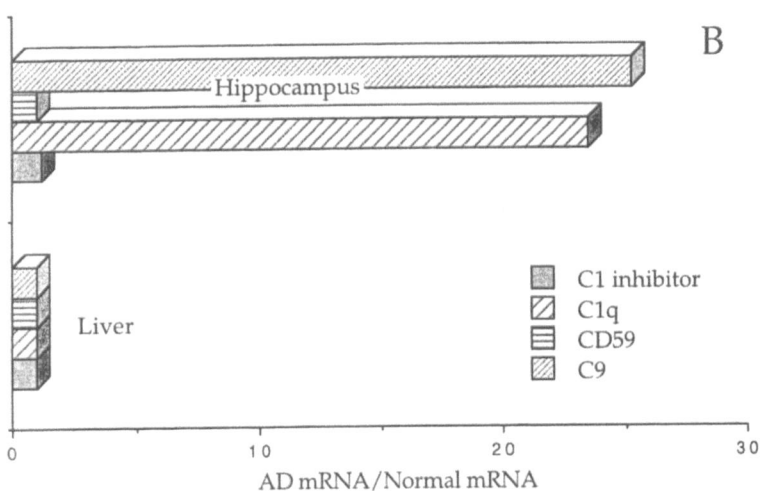

Figure 1. (A) Elevation of mRNAs for complement proteins C1q and C9, the first and last components of the complement cascade, in AD hippocampus, atherosclerotic plaques and infarcted heart tissue (ratio of level in inflamed to normal tissue shown), and lack of elevation in the liver of persons dying with these conditions. (B) Relatively slight elevation of the mRNAs for the complement inhibitors C1 inhibitor and CD59 in AD hippocampus compared to the large elevations in the mRNAs for the complement proteins.

and the complement receptor CD11b. Immunohistochemical detection of large numbers of HLA-DR-positive microglia around the lesions in AD brain was the first evidence reported of a chronic inflammatory reaction in such brain [13, 14]. Immunohistochemical work has

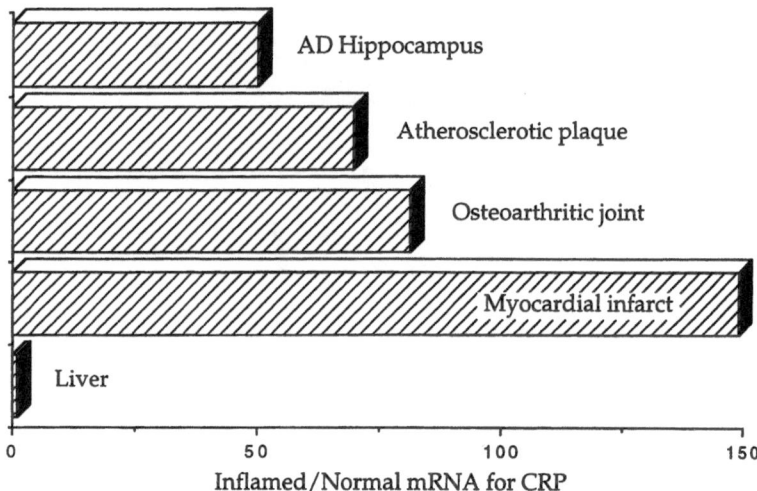

Figure 2. Elevation of the mRNA for C-reactive protein in AD hippocampus, atherosclerotic plaques, osteoarthritic joints and infarcted heart tissue. In each case, the graph shows the ratio of inflamed to normal tissue of the same type. The liver data come from persons dying with these conditions as compared to those without evidence of any inflammatory disease.

also demonstrated that such microglia are profusely decorated with complement receptors [15]. It is therefore not surprising that we have found a significant elevation of the mRNAs for these microglial markers in affected regions in AD brain (Fig. 3). The elevation seems somewhat greater than that occurring in atherosclerotic plaques which, at least for CD11b, is apparently comparable to that seen in osteoarthritic joints (Fig. 3). We have not yet determined the levels of the mRNA for HLA-DR in osteoarthritic and normal joints, nor have we looked at either of these mRNAs in normal and infarcted heart tissue. We would expect, however, that the elevation of the mRNA for HLA-DR in osteoarthritic joints would be comparable to that of the mRNA for CD11b. We would also expect that neither of these mRNAs would show a pronounced elevation in infarcted heart tissue because large numbers of activated macrophages are not seen histologically in such conditions.

Conclusions

Much evidence suggests that chronic inflammation is a major driving force in the most important diseases of our time: AD, heart attacks and strokes. The inflammation in each case is not believed to be a primary cause but rather a secondary phenomenon which takes over the pathology (Fig. 4). Prompt intervention with anti-inflammatory treatment might halt or at least inhibit the progressive tissue destruction seen in these conditions. At the very least, this evidence should inspire a sustained search for new and better types of anti-inflammato-

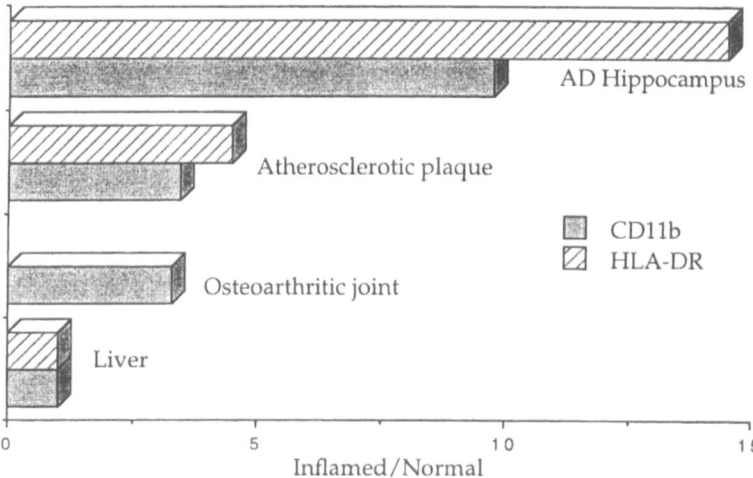

Figure 3. Elevation of the mRNAs for the macrophage/microglia markers HLA-DR and CD11b in AD hippocampus and atherosclerotic plaques and of the latter in osteoarthritic joints. Again, no significant change is seen in the liver of persons dying with such conditions.

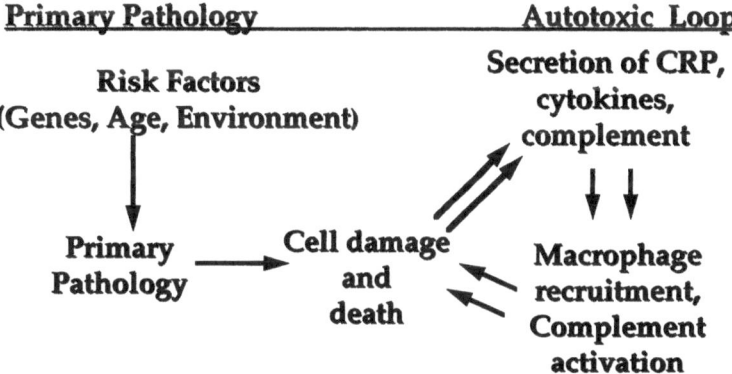

Figure 4. Hypothesized way in which inflammation and autotoxicity take over the pathology in conditions starting with a variety of primary pathologies that lead to initial cell death.

ry drugs. The complement system appears to play a key role and may be a prime target for new drug development [16].

Acknowledgements
Our work on Alzheimer disease has been supported by grants from the Alzheimer Societies of B.C. and Canada, and the Jack Brown and Family A.D. Research Fund, as well as donations from individual British Columbians.

References

1 McGeer PL, McGeer EG (1995) The inflammatory response system of brain: implications for therapy of Alzheimer and other neurodegenerative diseases. *Brain Res* 21: 195–218
2 McGeer EG, McGeer PL (1999) Brain inflammation in Alzheimer disease and the therapeutic implications. *Curr Pharmacol Design* 5: 821–826
3 McGeer PL, Schulzer M, McGeer EG (1996) Arthritis and anti-inflammatory agents as negative risk factors for Alzheimer disease: A review of seventeen epidemiological studies. *Neurology* 47: 425–432
4 Rogers J, Kirby LC, Hempelman SR, Berry DL, McGeer PL, Kaszniak AW, Zalinski J, Cofield M, Mansukhani L, Willson P et al (1993) Clinical trial of indomethacin in Alzheimer's disease. *Neurology* 43: 1609–1611
5 Lucchesi BR, Kilgore KS (1997) Complement inhibitors in myocardial ischemia/reperfusion injury. *Immunopharmacology* 38: 27–42
6 Stoll G, Jander S, Schroeter M (1998) Inflammation and glial responses in ischemic brain lesions. *Prog Neurobiol* 56: 149–171
7 Webster S, Lue L-F, Brachova L, Tenner AJ, McGeer PL, Terai K, Walker DG, Bradt B, Cooper NR, Rogers J (1997) Molecular and cellular characterization of the membrane-attack complex, C5b-9, in Alzheimer's disease. *Neurobiol Aging* 18: 415–421
8 Yasojima K, Schwab C, McGeer EG, McGeer PL (1999) Upregulated production and activation of the complement system in Alzheimer disease brain. *Amer J Pathol* 154: 927–936
9 Yasojima K, Schwab C, McGeer EG, McGeer PL (1998) Human heart generates complement proteins that are upregulated and activated after myocardial infarction. *Circ Res* 83: 860–869
10 Yasojima K, McGeer EG, McGeer PL (1999) Complement regulators C1 inhibitor and CD59 do not significantly inhibit complement activation in Alzheimer disease. *Brain Res* 833: 297–301
11 Jiang H, Robey FA, Gewurz H (1992) Localization of sites through which C-reactive protein binds and activates complement to residues 14–26 and 76–92 of the human C1q A chain. *J Exp Med* 175: 1373–1379
12 Lagrand WK, Visser CA, Hermens WT, Niessen HW, Verheugt FW, Wolbink GJ, Hack CE (1999) C-reactive protein as a cardiovascular risk factor: more than an epiphenomenon? *Circulation* 100: 96–102
13 McGeer PL, Itagaki S, Tago H, McGeer EG (1987) Reactive microglia in patients with senile dementia of the Alzheimer type are positive for the histocompatibility glycoprotein HLA-DR. *Neurosci Lett* 79: 195–200
14 Rogers J, Luber-Narod J, Styren SD, Civin WH (1988) Expression of immune system-associated antigen by cells of the human central nervous system. Relationship to the pathology of Alzheimer disease. *Neurobiol Aging* 9: 339–349
15 Akiyama H, McGeer PL (1990) Brain microglia constitutively express β-2 integrins. *J Neuroimmunol* 30: 81–93
16 McGeer EG, McGeer PL (1998) The future use of complement inhibitors for the treatment of neurological diseases. *Drugs* 55: 739–746

Memory and its impairment

Neuroscientific Basis of Dementia
C. Tanaka, P.L. McGeer, Y. Ihara (eds)
© 2001 Birkhäuser Verlag Basel/Switzerland

Neurobiological mechanisms by which emotional arousal influences long-term memory formation

Larry Cahill

Department of Neurobiology and Behavior and, Center for the Neurobiology of Learning and Memory, University of California, Irvine, CA, USA

"There is a tradition in the study of memory that, like a great many others, can be traced to Ebbinghaus (1885). The tradition is that the materials used in memory experiments are exceedingly boring."

Robert Hendersen [1]

Introduction

Hermann Ebbinghaus is properly credited with an outstanding scientific achievement: with the publication of a single monograph involving a single subject (himself), he created the formal science of memory, overturning the *zeitgeist* of his time, which held that a formal science of memory was impossible. His use of simple and tractable to-be-remembered stimuli (nonsense syllables) was critical to his success. But his use of these stimuli also began, as noted above by Hendersen, a widespread tradition of studying human memory without reference to the emotional significance of stimuli and events, with which memory storage mechanisms presumably evolved to operate. We are in recent years witnessing an apparent change in this state of affairs. There appears to be a growing awareness among both psychologists and neurobiologists that understanding memory mechanisms cannot be fully achieved, and in fact may be hindered, by ignoring the emotional reasons influencing whether memories are made or lost.

Over one hundred years after Ebbinghaus, we are still trying to understand the mechanisms by which the brain stores memory in relation to emotional arousal [2]. One view which has received considerable support holds that stress-hormone activation produced during and after emotionally arousing events, together with the amygdala complex, are key elements by which emotional arousal helps determine long-term memory storage. Findings from extensive research involving both human and animal subjects now converge on this view. Here I summarize some of the key findings supporting this view, including new evidence from human subject studies.

Modulation of memory by endogenous stress hormones

The hypothesis that endogenous stress hormones and the amygdala interact to influence memory storage is anchored in a very large, consistent, and growing body of animal research, reviewed in detail elsewhere [3, 4]. Only some key points are addressed here. A substantial amount of research suggests that memory formation can be modulated (enhanced or impaired) by administration of physiological doses of endogenous stress hormones (most commonly epinephrine) soon after a training event. These effects are time- and dose-dependent, that is, they affect memory only when given soon after training, and the dose-response function is typically an "inverted-U". Moderate doses enhance memory, while doses above or below that median range do not, and may even impair memory.

Recently, we have found that post-learning adrenergic stimulation with epinephrine can also enhance memory consolidation in healthy humans (Cahill and Alkire, in preparation). Subjects in this study viewed a series of pictures, and received immediately afterward an infusion of either placebo, or one of two physiological doses of epinephrine, for 3 min. One week later, their memory for the slides viewed was assessed in a free recall test. Epinephrine produced no effect on recall of the slides when considered as a whole. However, epinephrine significantly enhanced, in a dose-dependent fashion, recall of the initial slides in the series. A follow-up investigation revealed that these same initial slides produced a significantly higher electrodermal skin response in the subjects than did any of the remaining slides. Thus, it appears that the post-learning infusion of epinephrine acted in a retrograde manner to enhance memory storage of those pictures (the initial pictures) that had been encoded with a moderate degree of associated arousal. By demonstrating for the first time that a physiological dose of a natural stress hormone can enhance memory consolidation, these findings provide compelling new support for the view that endogenous stress hormones modulate memory storage after emotionally arousing events which induce their release [2, 4]. The findings also raise significant new issues for future research with human subjects, such as the nature of the interaction between initial encoding and post-learning adrenergic activation in memory consolidation.

Systemic adrenergic blockade and memory

Given the large number of studies demonstrating memory enhancement with post-learning adrenergic stimulation, it is somewhat surprising that there have been no clear demonstrations that systemic adrenergic blockade after learning impairs memory consolidation. Such a demonstration is required by the view that adrenergic activation after an emotionally stressful event modulates memory storage for the event. Therefore, we recently examined the role of post-learning adrenergic blockade on memory formation in rats [5]. Rats were trained in a stressful learning task assumed to produce substantial adrenergic activation ("Morris" water maze training). In this task, rats are trained to swim to a hidden platform always located in the same spatial position in a large circular tank. Immediately after training, either saline or the $\beta 1 \beta 2$ adrenergic blocker propranolol (5 mg/kg, i.p. (intra-peritoneally)) was administered.

Memory for the location of the platform was tested one day later. Propranolol produced a very mild, although statistically significant impairment of retention in this task when the group was considered as a whole. However, analysis of the effect of the drug on "good learners" (rats whose performance was better than the median on the final acquisition trial) *versus* "poor learners" (rats whose performance was below the median on the final acquisition trial) revealed a clearly differential effect. Propranolol clearly impaired memory in the "good learners," but had no significant effect on memory of the "poor learners." These findings are presented in Figure 1. By providing the first unequivocal demonstration that post-learning adrenergic blockade can impair memory storage, these findings fill a notable gap in the support for the hypothesis that post-learning adrenergic activation modulates memory consolidation.

In humans, several studies have found that adrenergic blockade (before learning, in each study to date) also has selective effects on memory consistent with a primary role for the adrenergic system in modulating memory for emotionally arousing events [6–9]. In one study [6] subjects received either a placebo or propranolol one hour before viewing either an

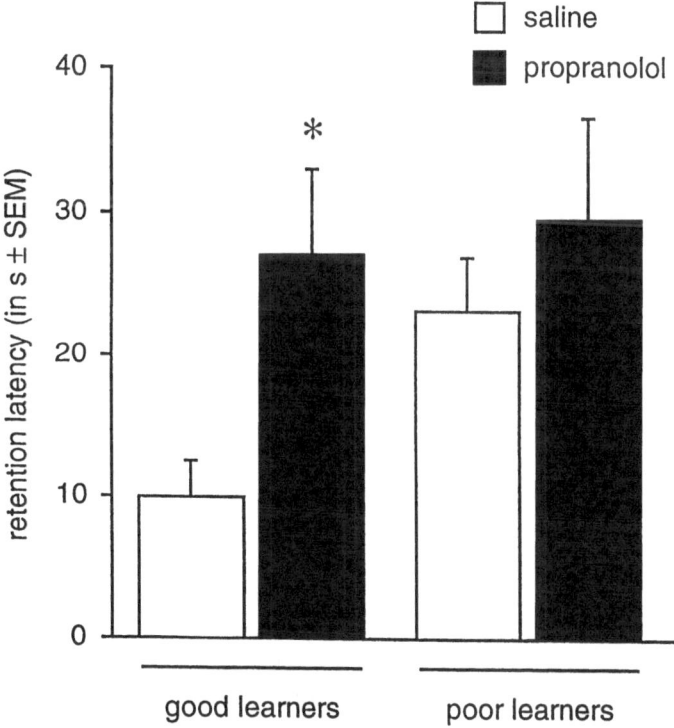

Figure 1. Retention latencies for saline- and propranolol-treated rats in the "Morris" swim maze. Lower scores indicate better retention. Administration of propranolol after training had no effect on 24-hour retention in "poor learners," but significantly impaired retention in "good learners." Reprinted with permission from Cahill, Pham, and Setlow, in press.

emotionally arousing or a relatively emotionally neutral narrated slide show. One week later, memory for the show viewed was assessed with both free recall and recognition memory (multiple choice) tests. Propranolol had no significant effect compared to placebo controls on memory for the emotionally neutral story. However, propranolol treatment significantly impaired memory for the emotionally arousing story. The lack of effect of the drug on memory for the neutral story means that the impairing effect on memory for the emotional story cannot easily be attributed to non-specific sedative or attentional effects of the drug. Also, as the subject's self-assessed emotional reaction to the stories was unaffected by the drug, the impairing effects on memory also cannot be attributed to a drug-induced lack of emotional responsiveness to the stories. Thus, the findings clearly suggest, consistent with substantial animal subject research, that adrenergic activation is primarily involved with the modulation of memory by emotional arousal.

The impairing effect of propranolol on memory for emotional material in humans has been confirmed in a report by van Stegeren and colleagues [8], who used the same stimuli used by Cahill et al. [6]. Propranolol freely crosses the blood-brain barrier, and therefore can act at both peripheral and central receptors. Substantial evidence from animal subject studies indicates that peripheral adrenergic activation also influences memory storage. To test whether the impairing effect of propranolol on memory was due to peripheral or central actions, van Stegeren et al. [8] gave a separate group of subjects nadolol, which also acts at β receptors like propranolol, but whose actions are restricted primarily to the peripheral nervous system. Unlike propranolol, nadolol had no effect on memory for the emotionally arousing story. Thus, it appears that the impairing effect of propranolol in this situation is due primarily, if not exclusively, to its action at central adrenergic receptors. However, it remains fully possible, and even likely, given the animal subject research, that an impairing effect on long-term memory of blocking peripheral adrenergic receptors will be detected in other, perhaps more emotionally arousing, training conditions. This important possibility remains to be experimentally explored.

The role of the amygdala

The brain region most clearly implicated to date as a site of action of stress hormones in modulating memory for emotionally arousing events is the amygdala. Manipulations of the amygdala (in particular the basolateral nucleus), such as lesions or intra-amygdala drug infusions, influence the effects of systemically administered hormones. For example, basolateral amygdala lesions block both memory-enhancing and memory-impairing effects of virtually every drug and hormone tested to date [3, 4]. The conclusion appears unavoidable that modulation of memory by peripheral drugs and hormones somehow critically involves basolateral amygdala function.

If a primary function of the amygdala is to interact with endogenous stress hormones to influence memory storage, then impairing effects of amygdala lesions on memory should be most evident for relatively arousing (i.e., sympathetic nervous system-activating) learning situations that cause stress hormone release. This implication was tested in a series of experiments involving rats with large lesions of the amygdala complex induced by injection of the excitatory amino acid N-methyl-D-aspartic acid (NMDA) [10]. Lesioned and control rats were trained in simple appetitively and aversively motivated tasks. In these experiments, amygdala lesions had no effect on learning when relatively mild reinforcements (water, sucrose, quinine) were used, but significantly impaired retention when a more arousing reinforcement (footshock) was used. On the basis of these findings, Cahill and McGaugh [10] concluded that "*the degree of arousal produced by the unconditioned stimulus, and not the aversive nature* per se, *determines the level of amygdala involvement*" in a learning situation. According to this view, the amygdala should be important for long-term memory formation whenever the learning conditions are sufficiently emotionally arousing (i.e., sympathetic nervous system-activating) to engage the amygdala, independently of whether the particular emotions involved are positive or negative. This conclusion now receives further support from human brain-imaging experiments.

Studies involving human subjects now confirm the view derived from animal research [3] that the amygdala is primarily involved with the enhanced, or modulated, declarative memory associated with emotional arousal. Several studies involving subjects with amygdala lesions indicate that the amygdala is involved specifically with enhanced memory associated with emotionally arousing material [11–13]. In contrast, amnesic subjects with intact amygdalae show relatively intact enhancement of memory for emotional material despite their overall impaired memory performance, further implicating the amygdala in the effect of emotion on memory [14].

Importantly, amygdala damage does not appear to block emotional reactions to the material; rather, it affects the enhanced long-term memory normally associated with emotional arousal. One subject even spontaneously described to the experimenters her strong negative emotional reaction to a particular aversive stimulus, yet failed to demonstrate enhanced recall of that stimulus as seen in control subjects [13]. Accordingly, we have suggested that the amygdala in humans may be more important for the translation of an emotional reaction into heightened long-term memory, than for the production of emotional reactions *per se* [15].

Findings from recent human brain-imaging experiments provide strong confirmation of a selective amygdala role in declarative memory for emotional events. For example, using Positron Emission Tomography (or PET) Cahill et al. [15] found that activity of the right amygdala while viewing a series of relatively emotionally arousing films correlated very highly ($r = 0.93$) with long-term recall of the films, yet activity of the same region in the same subjects, while viewing a relatively emotionally neutral series of films did not correlate with recall. These findings are shown in Figure 2. This basic finding regarding the amygdala has been strongly confirmed in a second PET study [16].

A necessary corollary of the view that the amygdala functions especially to influence memory for emotionally arousing events is that its activity should be generally unrelated to memory for non-emotional events or material. Although it is always difficult to prove a negative, evidence from several PET studies supports this important implication. In one recent study, amygdala activity failed to correlate with long-term memory for a series of non-emotional words, despite the fact that activity of hippocampal regions correlated very highly with memory in the same conditions [17].

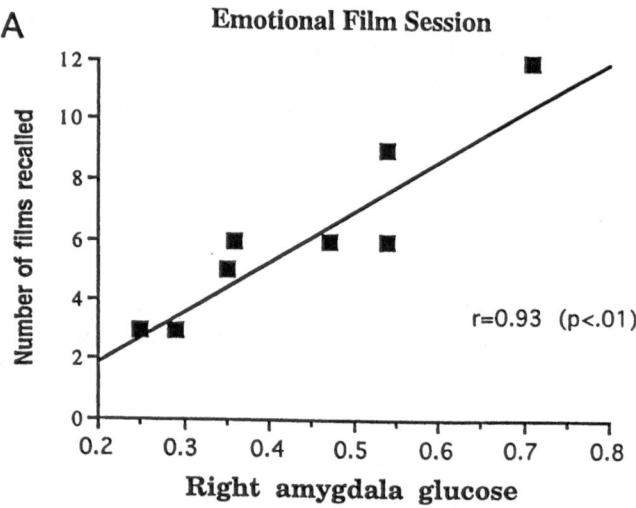

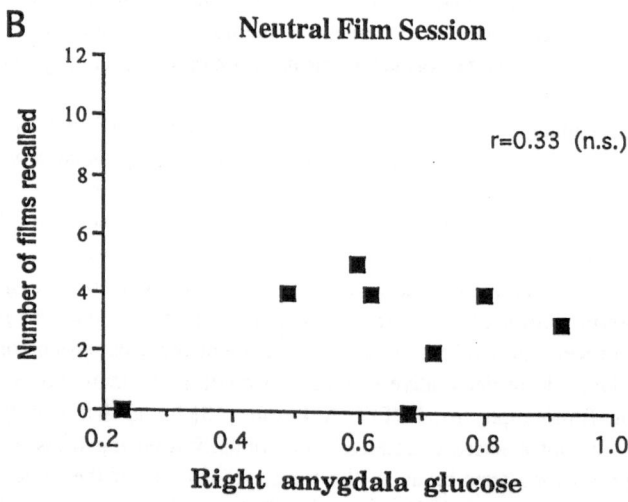

Figure 2. (A) Amygdala activity while watching a series of emotionally arousing films correlated very highly with long-term (3-week) recall of the films. (B) Amygdala activity in the same subjects while viewing a series of relatively emotionally neutral films did not correlate significantly with recall. From Cahill et al., 1996, with permission.

Finally, a study of memory in Alzheimer's Disease (AD) patients provides striking new evidence for the "memory modulation" view of amygdala function. Mori and colleagues [18] studied the memory of AD patients who had experienced a highly emotional event, the Kobe earthquake. Using magnetic resonance imaging (MRI), they measured the volume of amygdala and hippocampus in the patients. They found that amygdala volume correlated significantly with retention of the emotionally salient personal events experienced during the earthquake, and not with retention of widely disseminated semantic knowledge of the earthquake (such as its size on the Richter scale). These findings suggest that the modulatory role for the amygdala in emotionally influenced memory is very a robust phenomenon.

Thus, there exists a striking convergence of animal subject studies with both neuropsychological and brain-imaging studies in humans, pointing to the conclusion that the amygdala modulates declarative memory storage for emotionally arousing events [4, 10]. Relevance of this theory for understanding and treating dementia has already been demonstrated.

Mechanisms by which peripheral epinephrine may influence amygdala function

Epinephrine does not readily cross the blood-brain barrier. Thus, it is presumed to affect memory storage by effects initiated at peripheral receptors. There are two leading candidate mechanisms to account for the presumably peripheral effects of epinephrine on central memory storage processes. The first is that epinephrine may modulate memory by its well-established influence on glucose levels [19]. Many studies involving both animal and human subjects indicate that peripherally administered glucose can modulate memory consolidation. Additionally, a recent report indicates the emotionally arousing story used by Cahill et al. [6] also produces a slight but significant increase in blood glucose levels, consistent with a modulatory action of glucose in that paradigm [20].

A second well-supported candidate explanation of how epinephrine can act peripherally to influence central memory storage processes is that it does so via activating vagal afferents. For example, inactivation of either the vagus nerve, or the nucleus of the solitary tract (NTS) in which vagal input enters the central nervous system, blocks the modulatory action of epinephrine on memory in rats [21]. Furthermore, direct stimulation of the vagus enhances memory in rats [22], and the effect is due to ascending (towards the brain), rather than descending (towards the periphery) influences of the stimulation. Enhanced memory following vagal stimulation is also now reported in humans [23].

A third possible, but to my knowledge unexplored, mechanism by which epinephrine may influence central memory storage processes is by a *direct* action on one or more of the circumventricular organs, such as the *area postrema*, at which gaps in the blood-brain barrier allow direct action of blood-borne substances on brain function. The *area postrema* has been implicated in some arousing learning situations [24], and projects to the NTS [25]. Additional work is needed to address potential direct action of epinephrine on brainstem regions in modulating memory.

Lastly, it is distinctly possible that a combination of the above-mentioned candidate mechanisms may underlie epinephrine's effects on memory. A recent study indicates, for example, that activity of the NTS is influenced by blood glucose levels [26], clearly suggesting an interaction between the "glucose" and "vagus nerve" explanations of the mnemonic effects of epinephrine. Future work should be directed at exploring such potential interactions.

Amygdala modulation of memory in other brain regions

There is strong evidence that the amygdala influences memory storage processes occurring in other brain regions, as can be inferred from the evidence mentioned above of its time-limited involvement in memory. For example, there is now strong evidence that stimulation of the amygdala can modulate both hippocampal-related and caudate nucleus-related memory processes [27, 28]. The basolateral amygdala projects prominently to both hippocampal structures and the caudate. Furthermore, pharmacological stimulation of the amygdala with drugs known to modulate memory functionally affects both these regions [4, 29]. Manipulations of the basolateral amygdala (stimulation or lesions) modulate long-term potentiation in the hippocampus [30]. Finally, even the memory-enhancing effects of direct hippocampal stimulation can be blocked by amygdala lesions [31].

The basolateral amygdala may also modulate memory storage indirectly, via its influence on diffusely-projecting modulatory nuclei such as the nucleus basalis. Interestingly, stimulation of the basolateral amygdala activates the cortical electro encephalogram (EEG) [32], and appears to do so via the *nucleus basalis* [33]. On the basis of such evidence, it has been suggested that basolateral amygdala modulation of long-term memory storage may occur, at least in part, via modulation of the magnitude and/or duration of cortical activation produced by the *nucleus basalis* after an emotionally arousing event [34].

An emerging picture of memory modulation

From evidence such as that briefly described so far, a picture of the neurobiological mechanisms of memory modulation is emerging, presented schematically in Figure 3. Emotionally arousing events activate peripheral sympathetic stress hormone release. Activation of vagal afferents, at a minimum, carries modulatory influences back to the amygdala via the NTS. The amygdala in turn influences memory storage processes occurring in other brain regions. By this view, storage of long-term memory for emotionally significant events results from neural processes in both the body and the brain. Thus, an appealing feature of this theory is that, unlike the currently dominant views of memory storage that focus exclusively on central nervous system events to the complete exclusion of everything else, it places the mnemonic functions of the brain back in the context of the functions of the body with which they evolved and are inextricably functionally linked.

Are indelible "emotional memories" formed and stored in the amygdala?

A hypothesis widely advanced in recent years holds that the amygdala (in particular its lateral/basolateral nuclei, L/BL) is a site in which a presumed separate form of memory – "emotional memory" – is formed and "indelibly" stored [35, 36]. The formation of a CS-US (conditioned stimulus-unconditioned stimulus) association during Pavlovian conditioning involving a fear-inducing US is widely referred to by the misnomer "fear conditioning."

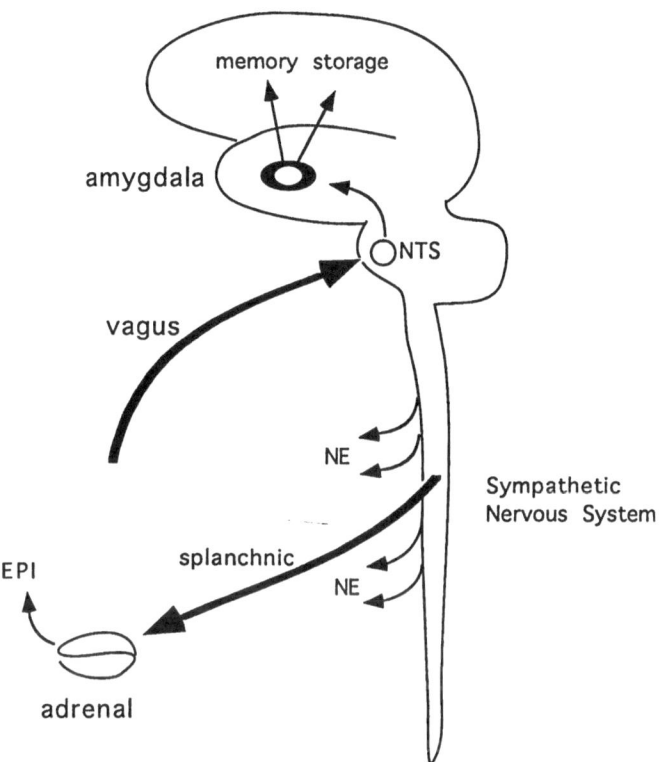

Figure 3. Schematic depiction of some key elements in our current understanding of neural mechanisms of memory modulation. An interaction between peripheral and central nervous system elements appears critical. Reprinted with permission from Cahill, 1999.

There are at present many critical and unresolved issues involving the experimental support for the "fear conditioning" view [37].

A critical feature of the "fear conditioning" hypothesis is that, as stated recently by Ledoux and colleagues, "the amygdala is the site of CS-US convergence or plasticity for...fear conditioning" [38]. By this hypothesis, animals with damage to the amygdala (or more accurately, the L/BL) should be completely unable to acquire Pavlovian fear conditioning. However, most studies to date examining amygdala involvement in "fear conditioning" have used only a few training trials massed in a single session. It therefore remains possible that animals with L/BL lesions will acquire Pavlovian "fear conditioning" if given appropriate training.

We recently re-examined the hypothesized necessary role of the L/BL in "fear conditioning" [39]. Rats with complete, excitotoxin lesions of the L/BL and control rats received two trials (footshocks) of Pavlovian contextual "fear conditioning" on day one of training.

Retention of the training was assessed 24 h later by placing the rats back in the training environment and measuring freezing behavior. L/BL-lesioned rats initially exhibited the expected nearly total loss of retention. However, administration of even a single footshock during the retention test revealed a very clear "savings" of retention from the training day. Control experiments showed that the freezing observed after the footshock during the retention test was indeed evidence of retention, and was specific to the box in which the training occurred. These results indicate that rats with complete L/BL lesions can acquire Pavlovian "fear conditioning" in only two trials. Note that these findings do not contradict those of earlier studies – no prior study gave footshock during the long-term retention test. Thus it appears that, consistent with findings from many previous studies (see [37]) the L/BL, while no doubt *involved* in "fear conditioning" in some manner, is not required for its acquisition.

Closing comment

Recent years have witnessed a shift in memory research away from the tradition traceable to Ebbinghaus wherein memory is studied without reference to emotion. Through an examination of memory formation in relation to emotional salience, an understanding of neurobiological mechanisms of memory modulation appears to be emerging (Fig. 3). This understanding is already influencing studies of dementia [18], and will, it is hoped, continue to do so.

Acknowledgements
Some of the research discussed in this article was supported by NIMH Grant MH57508-02 to L. Cahill

References

1 Hendersen RW (1985) Fearful Memories: The motivational significance of forgetting. *In*: FR Brush, JB Overmier (eds): *Affect, Conditioning, and Cognition: Essays on the Determinants of Behavior*. Lawrence Erlbaum Associates, Hillsdale, NJ
2 Gold PE, McGaugh JL (1975) A single-trace, two process view of memory storage processes. *In*: J Deutsch, D Deutsch (eds): *Short-term memory*. Academic Press, New York, 355–378
3 McGaugh JL, Cahill L, Roozendaal B (1996) Involvement of the amygdala in memory storage: Interaction with other brain systems. *Proc Natl Acad Sci USA* 93: 13508–13514
4 Cahill L, McGaugh JL (1998) Mechanisms of emotional arousal and lasting declarative memory. *Trends Neurosci* 21: 294–299
5 Cahill, L, Pham, C, and Setlow, B. Impaired memory consolidation in rats produced with β-adrenergic blockade. *Neurobiol Learn Memory*; *in press*
6 Cahill L, Prins B, Weber M, McGaugh JL (1994) β-adrenergic activation and memory for emotional events. *Nature* 371: 702–704
7 Nielson KA, Jensen R (1994) *Beta*-adrenergic receptor antagonist antihypertensive medications impair arousal-induced modulation of working memory in elderly humans. *Behav Neur Biol* 62: 190–200
8 van Stegeren A, Everaerd W, Cahill L, McGaugh JL, Goeren L (1998) Memory for emotional events: Differential effects of centrally *versus* peripherally acting beta-blocking agents. *Psychopharmacology* 138: 305–310
9 O'Carroll RE, Drysdale E, Cahill L, Shajahan P, Ebmeier KP (1999) Stimulation of the noradrenergic system enhances and blockade reduces memory for emotional material in man. *Psychol Med* 29: 1083–1088
10 Cahill L, McGaugh JL (1990) Amygdaloid complex lesions differentially affect retention of tasks using appetitive and aversive reinforcement. *Behav Neurosci* 104: 532–543

11 Babinsky R, Calabrese P, Durwen HF, Markowitsch HJ, Brechtelsbauer D, Heuser L, Gehlen W (1993) The possible contribution of the amygdala to memory. *Behav Neurol* 6: 167–170

12 Cahill L, Babinsky R, Markowitsch H, McGaugh JL (1995) The amygdala and emotional memory. *Nature* 377: 295–296

13 Adolphs R, Cahill L, Schul R, Babinsky R (1997) Impaired declarative memory for emotional stimuli following bilateral amygdala damage in humans. *Learn Memory* 4: 291–300

14 Hamann S, Cahill L, McGaugh JL, Squire L (1997) Intact enhancement of declarative memory by emotional arousal in amnesia. *Learn Memory* 4: 301–309

15 Cahill L, Haier R, Fallon J, Alkire M, Tang C, Keator D, Wu J, McGaugh JL (1996) Amygdala activity at encoding correlated with long-term, free recall of emotional information. *Proc Natl Acad Sci USA* 93: 8016–8021

16 Hamann S, Elt T, Grafton S, Kilts C (1999) Amygdala activity related to enhanced memory for pleasant and unpleasant aversive stimuli. *Nat Neurosci* 2: 289–293

17 Alkire M, Haier R, Fallon J, Cahill L (1998) Hippocampal, but not amygdala, activity at encoding correlates with long-term, free recall of non-emotional information. *Proc Natl Acad Sci USA* 95: 14506–14510

18 Mori E, Ikeda M, Hirono N, Kitagaki H, Imamura T, Shimomura T (1999) Amygdalar volume and emotional memory in Alzheimer's disease. *Amer J of Psychiat* 156: 216–222

19 Gold PE (1995) Role of glucose in regulating the brain and cognition. *Amer J Clin Nutr* 61(suppl): 987S–95S

20 Parent MB, Varnhagen C, Gold PE (1999) A memory-enhancing emotionally arousing narrative increases blood glucose levels in human subjects. *Psychobiology* 27: 386–396

21 Packard M, Williams C, Cahill L, McGaugh JL (1995) The anatomy of a memory modulatory system: From periphery to brain. *In*: NE Speer, LP Speer, ML Woodruff (eds): *Neurobehavioral Plasticity: Learning, Development, and Response to Brain Insults.* Lawrence Erlbaum Associates, Hillsdale, NJ

22 Clark KB, Smith DC, Hassert DL, Browning RA, Naritoku DK, Jensen RA (1998) Posttraining electrical stimulation of vagal afferents with concomitant vagal efferent inactivation enhances memory storage processes in the rat. *Neurobiol Learn Memory* 70: 364–373

23 Clark K, Naritoku D, Smith D, Browning R, Jensen R (1999) Enhanced recognition memory following vagus nerve stimulation in human subjects. *Nat Neurosci* 2: 94–98

24 Arnedo M, Gallo M, Aguero A, Puerto A (1990) Effects of medullary afferent vagal axotomy and area postrema lesions on short-term and long-term NaCl-induced taste aversion learning. *Physiol Behav* 47: 1067–1074

25 Parent A (1996) *Carpenter's Human Neuroanatomy.* Williams and Wilkins, Baltimore

26 Dallaporta M, Himmi T, Perrin J, Orsini JC (1999) Solitary tract nucleus sensitivity to moderate changes in glucose level. *Neuroreport* 10: 2657–2660

27 Packard M, Cahill L, McGaugh JL (1994) Amygdala modulation of hippocampal-dependent and caudate nucleus-dependent memory processes. *Proc Natl Acad Sci USA* 91: 8477–8481

28 Packard M, Teather L (1998) Amygdala modulation of multiple memory systems: Hippocampus and caudate-putamen. *Neurobiol Learn Memory* 69: 163–203

29 Cahill L (1999) A neurobiological perspective on emotionally influenced, long-term memory. *Semin Clin Psychol* 4: 266–273

30 Ikegaya Y, Saito H, Abe K (1996) The basomedial and basolateral amygdaloid nuclei contribute to the induction of long-term potentiation in the dentate gyrus *in vivo. Eur J Neurosci* 8: 1833–1839

31 Roozendaal B, McGaugh JL (1997) Basolateral amygdala lesions block the memory-enhancing effect of glucocorticoid administration in the dorsal hippocampus of rats. *Eur J Neurosci* 9: 76–83

32 Kaada BR (1972) Stimulation and regional ablation of the amygdaloid complex with reference to functional representations. *In*: B Eleftheriou (ed.) *The Neurobiology of the Amygdala.* Plenum Press, New York, 205–282

33 Dringenberg H, Vanderwolf C (1996) Cholinergic activation of the electrocorticogram: an amygdaloid activating system. *Exp Brain Res* 108: 285–296

34 Cahill L. Emotional modulation of long-term memory storage in humans: Adrenergic activation and the amygdala. *In*: J. Aggleton (ed.): *The Amygdala: A Functional Analysis.* Oxford University Press, Oxford, UK; *in press*

35 LeDoux JE (1995) Emotion: clues from the brain. *Annu Rev Psychol* 46: 209–235

36 Maren S, Fanselow MS (1996) The amygdala and fear conditioning: has the nut been cracked? *Neuron* 16: 237–240

37 Cahill L, Weinberger NM, Roozendaal B, McGaugh JL (1999) Is the amygdala a locus of "conditioned

fear"? Some questions and caveats. *Neuron* 23: 227–228

38 Schafe G, Nadel N, Sullivan G, Harris A, LeDoux J (1999) Memory consolidation for contextual and auditory fear conditioning is dependent on protein synthesis, PKA, and MAP kinase. *Learn Memory* 6: 97–110

39 Cahill L, Vazdarjanova A, Setlow B. The basolateral amygdala is involved with, but not necessary for, rapid acquisition of Pavlovian "fear conditioning". *Eur J Neurosci; in press*

Neuroscientific Basis of Dementia
C. Tanaka, P.L. McGeer, Y. Ihara (eds)
© 2001 Birkhäuser Verlag Basel/Switzerland

Amygdalar damage and memory impairment in Alzheimer's disease

Etsuro Mori

Hyogo Institute for Aging Brain and Cognitive Disorders, Himeji, Japan

Introduction

Neuropathological changes in Alzheimer's disease (AD) include neurofibrillary tangles, senile plaques and neuronal and synaptic loss, which predominantly affect regions of the limbic/paralimbic medial temporal structures and associated areas of the neocortex [1, 2]. There is a significant correlation between dementia severity in AD and these *post mortem* neuropathological findings [3]. These neuropathological changes in the medial temporal lobe may account for memory impairment in AD [4], which is not only the earliest clinical symptom but a central and prominent feature throughout the course of AD. *In vivo* neuroimaging techniques have demonstrated a relationship between memory impairment and the degree of medial temporal damage as expressed by atrophy on magnetic resonance imagings (MRI) and hypoperfusion or hypometabolism demonstrated by positron emission tomography (PET) [5].

The hippocampus is a central component of the medial temporal lobe memory system, and its structural integrity is necessary for declarative memory [6]. Destruction restricted to the bilateral hippocampi is known to produce memory deficits in animals and in humans [7]. There is neuroimaging evidence for loss of hippocampal tissue in human diseases associated with memory impairment [8]. The amygdala is not a part of the medial temporal lobe memory system for the processing of declarative memory [6] and recent studies have indicated that the additive effect of amygdalar damage upon hippocampal damage is not a consequence of the amygdalar damage itself but due to involvement of the adjacent cortices (periamygdalar, entorhinal and perirhinal cortices) covering the surface of the amygdala and parahippocampal gyrus [6]. Nevertheless, there is evidence indicating that amygdalar lesions produce memory deficits in human and animals [9, 10]. Studies on animal models have indicated that the neural system underlying behavior associated with emotional situations involves the amygdala and structures with which it is connected [11]. A couple of studies in patients with Urbach-Wiethe disease, in which the amygdala is selectively damaged, have also indicated a central involvement of the amygdala in memory for emotional events in humans [12, 13]. However, studies in humans have been limited, and the role of the amygdala in memory function is still controversial. The role of the amygdaloid complex, hippocampus, subiculum and parahippocampal gyrus in memory impairment in AD patients remains unsettled. Based on our recent studies, we delineate the involvement of the amygdala in memory impairment caused by AD. Studies on autopsied cases of advanced AD have also demonstrated that amygdalar atrophy is related to the Alzheimer pathology [14]. They demonstrated that the amygdalar atrophy in AD is due to loss of neuronal somata and processes, the accumulation of neuritic plaques and neurofibrillary tangles as well as extensive gliosis in discrete amygdalar subnuclei and in particular the loss of large nerve cells in the magnocellular basolateral amygdalar nuclei group. The morphological deformation relates to intrinsic damage to the amygdalar subnuclei and their reciprocal circuitry with other brain regions. A strong correlation between the decrease in the volume of hippocampal formation subdivisions (dentate gyrus, CA4-1 and subiculum) and the decrease in the total number of neurons in AD have also been demonstrated in pathological studies [15]. These pathological findings indicates a cause of neuronal loss for amygdalar and hippocampal volumetric loss, and provide a rationale for the studies of MRI volumetry-function correlation in AD.

Amygdalar volume and visual memory

Using high-resolution MRI and a semi-automated image analysis technique, we measured volumes of the medial temporal structures (amygdaloid complex, hippocampal formation, subiculum and parahippocampal gyrus), and examined correlations between atrophy of each structure and memory dysfunction in patients with AD [16]. Observed volumes were normalized for individual intracranial volume with the covariance method. To assess memory function, we used the ADAS (Alzheimer disease assessment scale) word recall subtest and WMS-R (Wechsler memory scale-revised). The performance on all the test items was significantly lower in the patients than in the healthy controls. Step-wise regression analyses disclosed that the volume of the right amygdala specifically predicted visual memory, and that the volume of the left subiculum specifically predicted verbal memory function. The subjects were 46 Japanese patients with mild to moderate AD who fulfilled the NINCDS/ADRDA (National Institute of Neurological and Communicative Disorders and Stroke/Alzheimer's Disease and Related Disorders Association) criteria for probable AD. The mean (SD) age was 70.3 (7.1) years (range = 44 to 79) for 27 women and 19 men; the mean educational attainment was 8.7 (2.3) years (range = 6–16), and the mean MMSE (mini-mental state examination) score was 19.6 (3.5) (range = 12–26). All patients were right-handed.

Patients with AD showed poor performance on verbal and non-verbal memory tests, and the MRI volumetry demonstrated a significant volume reduction of the medial temporal lobe structures. Volumes of the amygdaloid complex and of the subiculum correlated with memory performance. Step-wise regression analyses revealed that the volume of the right amygdaloid complex specifically predicted visual memory function (WMS-R figure memory, partial $r = 0.34$, $p = 0.03$; WMS-R visual reproduction, partial $r = 0.47$, $p = 0.001$) and to some extent verbal memory function (WMS-R verbal paired associates, partial $r = 0.34$, $p = 0.026$), and that the volume of the left subiculum specifically predicted verbal memory function (ADAS word recall, partial $r = 0.32$, $p = 0.038$; WMS-R logical memory, partial $r = 0.35$, $p = 0.02$). Atrophy of the hippocampus did not predict severity of memory impairment. An association between amygdalar volume and performance on visual memory tests has been demonstrated in a study of age-associated memory impairment. An active role for the amygdala in human memory is supported by the studies of patients with bilateral damage to the amygdala due to Urbach-Wiethe disease; in particular, a significant defect in visual memory has been reported. Our results suggested that the amygdala and subiculum together further increased the severity of memory impairment attributable to hippocampal damage in AD. However, this does not necessarily disagree with the view of the central role of the hippocampus in memory processes. Although bilateral lesions restricted to the hippocampi produce memory impairment in animals and in humans, the deficits are limited. Our previous volumetric study in amnesic patients after *herpes simplex* encephalitis indicated that hippocampal damage resulting in less than 50% volume loss did not result in severe, lasting amnesia. As the Alzheimer pathology invariably affects the hippocampal formation, damage to the hippocampus might be an indispensable condition for occurrence of memory deficits in Alzheimer patients.

Memory of an emotional event: the Kobe earthquake

Everyday experience suggests that highly emotional events are often the most memorable and both naturalistic and experimental studies have demonstrated that emotional arousal enhances declarative memory in humans. Studies on animal models and in patients with Urbach-Wiethe disease, in which the amygdala is selectively damaged, have indicated that the neural system underlying emotional memory involves the amygdala. We encountered a devastating earthquake on January 17, 1995, in Kobe. The Kobe Earthquake, measuring 7.2 on the Richter scale, caused more than 6,000 deaths and enormous structural damage. For most residents, this earthquake forcibly and simultaneously exposed them to fear of death. We noticed that those with AD, who could not recall any events minutes or hours ago, vividly described their own terrible experience of the earthquake. The tragedy, in turn, provided us with an opportunity that none other than nature can make to study the relationship between emotion and memory [17].

Our institute, located at Himeji, approximately 50 km west from the origin of the earthquake, serves an area covering the whole Hyogo prefecture, including the stricken district; it fortunately escaped direct damage from the earthquake. We studied memory of the earthquake in 51 outpatients with probable AD who experienced the earthquake at their home in the greater Kobe area. Memory retention for the earthquake 2 months after the disaster was assessed by a semi-structured interview, which was compared with their memory of a MRI examination given after the earthquake. Functional severity was very mild in 7 patients, mild in 24 patients, moderate in 16 patients and severe in 4 patients as determined by the CDR (clinical dementia rating). The mean MMSE was 17. A semi-structured interview was developed to assess patients' memory of the earthquake and their memory of a MRI examination. The first 3 items were arranged to examine whether patients remembered the occurrence of the earthquake. The next 3 items measured the depth of personal memory of the earthquake. The last 3 items assessed semantic knowledge of the earthquake. One point was given for each correct answer. The comparable questions were given to evaluate memory of the MRI experience. MRI was carried out without sedatives and anaesthetics.

The mean interval between the earthquake and the memory survey was 59 days, and the mean interval between the MRI study and the survey was 18 days. 86% of the patients remembered the earthquake, while 31% of the patients remembered the MRI experience. This difference was statistically significant ($\chi^2 = 26.0$, p < 0.001, McNemar test). All the patients who remembered the MRI also remembered the earthquake. The logistic regression analysis revealed that any one of the variables of sex, age, education, disease duration and either cognitive or functional impairment did not significantly predict the recall of the earthquake. Recall performance of details concerning personal events around the earthquake was very good. However, factual content or semantic knowledge of the earthquake was mostly lost. These results suggest that fearful emotion reinforces memory retention of the episode in the absence of its. However, the reinforcement of memory by emotion is limited to high self-reference of materials. General knowledge is certainly of low self-referential and emotional value and is insignificant in an individual's life.

Memory of the earthquake and amygdala

Subsequently, to elucidate the relationship between medial temporal damage and impaired memory for emotional events in patients with AD, correlation with memories of the earthquake and amygdalar and hippocampal atrophy quantified by MRI were studied in 36 patients who experienced the 1995 Kobe earthquake [18]. Memories of events surrounding the earthquake were examined as an index of emotional memory by using a semi-structured interview, and amygdalar and hippocampal volumes were quantified by MRI. In this study, the amygdala was defined to include the amygdala proper, *gyrus semilunaris* and *gyrus ambiens*, and the hippocampus was defined to include the hippocampus proper, dentate gyrus and subiculum. The effects of the atrophy of the structures on the recall performance were determined by using a multiple regression analysis.

Patients' performance on the emotional and standardized memory test was defective. The mean normalized volumes of the whole brain, amygdala and hippocampus were significantly smaller in the patients than in the normal subjects. The Pearson correlation coefficients between total emotional memory score and normalized amygdalar volume (r = 0.52, p = 0.001) and between the score and hippocampal volume (r = 0.49, p = 0.002) are both significant. However, in multiple regression models in which both amygdalar and hippocampal volumes were entered into independent variables, the partial correlation coefficient between amygdalar volume and total emotional memory score was significant (r = 0.43, p = 0.01). When scores of event recall and personal memory were separately analysed, the partial correlation coefficient between amygdalar volume and each memory score was significant (r = 0.40, p = 0.02 and r = 0.45, p = 0.01, respectively). The correlations remained significant, even when factors of age, sex, education, whole brain volume and dementia severity were covaried. However, hippocampal volume was not a significant predictor of each memory score. On the other hand, non-personal factual knowledge of the earthquake, which was very poorly recalled by the patients, was correlated with neither amygdalar volume nor hippocampal volume. Our results indicate that impairment of emotional event memory in patients with AD are related to intensity of amygdalar damage, and provide evidence of the amygdala's involvement in emotional memory in humans. These findings are consistent with those in experimental studies on patients with Urbach-Wiethe disease who had selective bilateral lesions of the amygdala, in which a dysfunction in memorizing emotional materials but not in memorizing neutral materials has been demonstrated [13]. By using positron emission tomography of glucose metabolism, Cahill et al. [19] demonstrated that the glucose metabolic rate in the right amygdala at encoding correlated with long-term, free recall of emotionally arousing information. Memory for personally experienced, time- and place-specific events belongs to episodic memory, while memory for facts that lack particular time and space contexts is involved more in semantic memory. Less involvement of emotion in general knowledge would account for the less active participation of the amygdala in general knowledge. These results suggest that memory of real-life emotional episodes, like other aspects of episodic memory function, is defective in AD patients, while emotional arousal enhances memory for all aspects of an event except details irrelevant to the individual's life. Emotional memory deficits in AD are related to the intensity of amygdalar damage and less to the intensity of hippocampal damage. It is likely that the amygdala is involved mainly in the declarative process of personal memory. This provides further evidence that the amygdala plays a central role in emotional memory in humans.

Experimental emotional memory

It is impossible to control the quality and intensity of emotional involvement in memory for real-life episodes. Of course, we never expect tragedy like the earthquake. To examine emotional memory in experimental settings, we can virtually delegate emotional context to test materials.

We utilized a modified version of an emotional memory test developed by Cahill et al. In this paradigm, subjects viewed 11 color photographs, accompanied by either an emotionally arousing story, or a closely matched but more emotionally neutral story. The two stories are identical except for the second phase of each story; one is emotionally flavored and the other not. The two stories are closely matched in content, complexity, style and comprehensibility. Immediately after the story presentation, the subjects were asked to rate the strength of their emotional responses to the story on a scale of 1–4. Five minutes later, the subjects were given a recall test. While the photographs were presented one by one again in the same order, subjects were asked to answer the questions about the story line. The subjects were 34 patients with mild probable AD and 10 normal elderly subjects matched on age, sex and educational attainment. The 34 AD patients and 10 normal subjects were divided into two comparable groups. The test was repeated at an interval of 2 weeks, for one group the arousal story first and the other the neutral story first.

There was no significant difference in the emotion rating between patients and controls. The percent of correct responses was analysed by using three-way ANOVA for repeated measures with one between factor (group) and two within factors (story and phase). The effects of group, story, phase and story x phase interaction were significant. The group x story interaction, a group x phase interaction and a group x story x phase interaction were not significant. This result indicated that declarative memory for emotional material was enhanced in AD similarly to healthy subjects. The magnitude of enhancement of recollection for emotional material in patients with AD was similar to that in normal subjects, probably because of a ceiling effect in normal subjects. We are now working on studying a determinant factor for the emotional memory function in AD. The relationship between visual memory function and emotional memory is also of interest.

Conclusion

In conclusion, both hippocampus and amygdala are atrophic in AD patients. Emotional involvement enhances memory in AD patients. Amygdalar atrophy is a determinant of emotional memory in AD patients. The amygdala is also involved in memory of visual materials, and the relationship between visual memory function and emotional memory should be pursued. Finally, as for the clinical implications of those findings, we will eventually know how emotion enhances memory and how the amygdala is involved in the process. Then, there would be a possibility of drug intervention to increase memory function in AD. Emotional involvement is probably of use in memory rehabilitation. At least, it is evident that efforts such as introducing meaningful stimulation and avoiding trauma are very important in the care of patients.

References

1 Hyman BT, Van Hoesen GW, Damasio AR (1990) Memory-related neural systems in Alzheimer's disease: an anatomic study. *Neurology* 40: 1721–1730
2 Scott SA, DeKosky ST, Scheff SW (1991) Volumetric atrophy of the amygdala in Alzheimer's disease: quantitative serial reconstruction. *Neurology* 41: 351–356
3 Terry RD, Masliah E, Salmon DP, Butters N, DeTeresa R, Hill R, Hansen LA, Katzman R (1991) Physical basis of cognitive alterations in Alzheimer's disease: synapse loss is the major correlate of cognitive impairment. *Ann Neurol* 30: 572–580
4 Hyman BT, Van Horsen GW, Damasio AR, Barnes CL (1984) Alzheimer's disease: cell-specific pathology isolates the hippocampal formation. *Science* 225: 1168–1170
5 Ishii K, Kitagaki H, Kono M, Mori E (1996) Decreased medial temporal oxygen metabolism in Alzheimer's disease shown by positron emission tomography. *J Nucl Med* 37: 1159–1165
6 Squire LR, Zola-Morgan S (1991) The medial temporal lobe memory system. *Science* 20: 1380–1386
7 Zola-Morgan S, Squire LR, Amaral DG (1986) Human amnesia and the medial temporal region: Enduring memory impairment following a bilateral lesion limited to field CA1 of the hippocampus. *J Neurosci* 6: 2950–2967
8 Yoneda Y, Mori E, Yamadori A, Yamashita H (1994) MRI volumetry of medial temporal lobe structures in amnesia following herpes simplex encephalitis. *Eur Neurol* 34: 243–252
9 Aggleton JP (1992) The effects of amygdala lesions in humans: a comparison with findings from monkeys. *In*: JP Aggleton (ed.): *The Amygdala: Neurobiological Aspects of Emotion, Memory, and Mental Dysfunction*. Wiley-Liss, New York, 485–503
10 Mishkin M (1978) Memory in monkeys is severely impaired by combined but not by separate removal of amygdala and hippocampus. *Nature* 273: 297–298
11 Davis M (1992) The role of the amygdala in conditioned fear. *In*: JP Aggleton (ed.): *The Amygdala: Neurobiological Aspects of Emotion, Memory, and Mental Dysfunction*. Wiley-Liss, New York, 255–305

12 Cahill L, McGaugh JL (1998) Mechanisms of emotional arousal and lasting declarative memory. *Trends Neurosci* 21: 294–299
13 Markowitsh HJ, Calabrese P, Würker M, Durwen HF, Kessler J, Babinsky R, Brechtelsbauer D, Heuser L, Gehlen W (1994) The amygdala's contribution to memory: a study on two patients with Urbach-Wiethe disease. *Neuroreport* 27: 1349–1352
14 Scott SA, DeKosky ST, Sparks DL, Knox CA, Scheff SW (1992) Amygdala cell loss and atrophy in Alzheimer's disease. *Ann Neurol* 32: 555–563
15 Bobinski M, Wegiel J, Wisniewski HM, Tarnawski M, Bobinski M, Reisberg B, De Leon MJ, Miller DC (1996) Neurofibrillary pathology – correlation with hippocampal formation atrophy in Alzheimer disease. *Neurobiol Aging* 17: 909–919
16 Mori E, Yoneda Y, Yamashita H, Hirono N, Ikeda M, Yamadori A (1997) Medial temporal structures relate to memory impairment in Alzheimer's disease: an MRI volumetric study. *J Neurol Neurosurg Psychiat* 63: 214–221
17 Ikeda M, Mori E, Hirono N, Imamura T, Shimomura T, Ikejiri Y, Yamashita H (1998) Amnestic people with Alzheimer's disease who remembered the Kobe Earthquake. *Brit J Psychiat* 172: 425–428
18 Mori E, Ikeda M, Hirono N, Kitagaki H, Imamura T, Shimomura T (1999) Amygdalar volume and emotional memory in Alzheimer's disease. *Amer J Psychiat* 156: 216–222
19 Cahill L, Haier RJ, Fallon J, Alkire M, Tang C, Keator D, Wu J, McGaugh JL (1996) Amygdala activity at encoding correlated with long-term, free recall of emotional information. *Proc Natl Acad Sci USA* 93: 8016–8021

Neuroscientific Basis of Dementia
C. Tanaka, P.L. McGeer, Y. Ihara (eds)
© 2001 Birkhäuser Verlag Basel/Switzerland

Neural substrate for spatial memory in the monkey hippocampus

Ryoi Tamura and Taketoshi Ono

Department of Physiology, Toyama Medical and Pharmaceutical University, Toyama, Japan

Involvement of hippocampal formation in spatial memory

Involvement of the hippocampal formation (HF) in memory has been extensively studied since Scoville and Milner's report on patient H.M. who developed amnesia following surgical resection of the bilateral HF [1]. However, HF damage does not necessarily impair all types of memory. Squire divided memory into 2 broad categories: declarative memory and non-declarative memory [2]. Declarative memory refers to conscious recollections of facts and events and depends on the integrity of the medial temporal lobe, including, among other structures, the HF. HF involvement in non-declarative memory is thought to be minimal.

Spatial context is an elemental component of memory of an event, referred to as episodic memory. Evidence suggests that damage to the HF impairs spatial memory in rodents, monkeys and humans. Recent functional brain-imaging studies have demonstrated that the HF is activated during performance of spatial tasks even in humans (e.g., [3]). At the neuronal level, correlates of HF neuronal activity with extrapersonal place have been well documented in studies on rodents since O'Keefe reported rat place cells [4]; the activity of place cells increases when the rat is in a specific location within a given environment. However, in the 1980 s, little was known concerning the spatial correlates of HF neurons in primates. Therefore, in the past 15 years, we have attempted to identify spatial correlates of monkey HF neurons using a motorized cab on which the monkey could change its location by pressing bars or manipulating a joystick [5, 6].

Location-related activity of monkey HF neurons

In 1993, we reported the monkey HF neurons that increase activity when the monkey is in a specific location in the experimental room during performance of a spatial task in which the monkey itself drives the cab [5]. Figure 1A shows a schematic drawing of our experimental setup. Aside from the rear wall (made of steel, and containing 2 speakers and 2 cab lamps), the walls and ceiling of the cab were all one-way mirrors. Therefore, as it was dark outside the cab (room lights off), the monkey could not see the scene outside the cab. The lower part of the front wall contained 5 bars. The monkey sat in a chair with its head fixed painlessly in a stereotaxic apparatus, facing +Y direction. Several landmarks (the entrance door, table, refrigerator, etc.) were placed in the room for the monkey. In the behavioral task (spatial moving task), the monkey was required to drive the cab within a 2.5 × 2.5 m driving space by pressing one of 4 bars: the middle bar corresponded to forward movement, the near-left and near-right bars corresponded to leftward movement and rightward movement, respectively, and the far-right bar corresponded to backward movement (the far-left bar was usual-

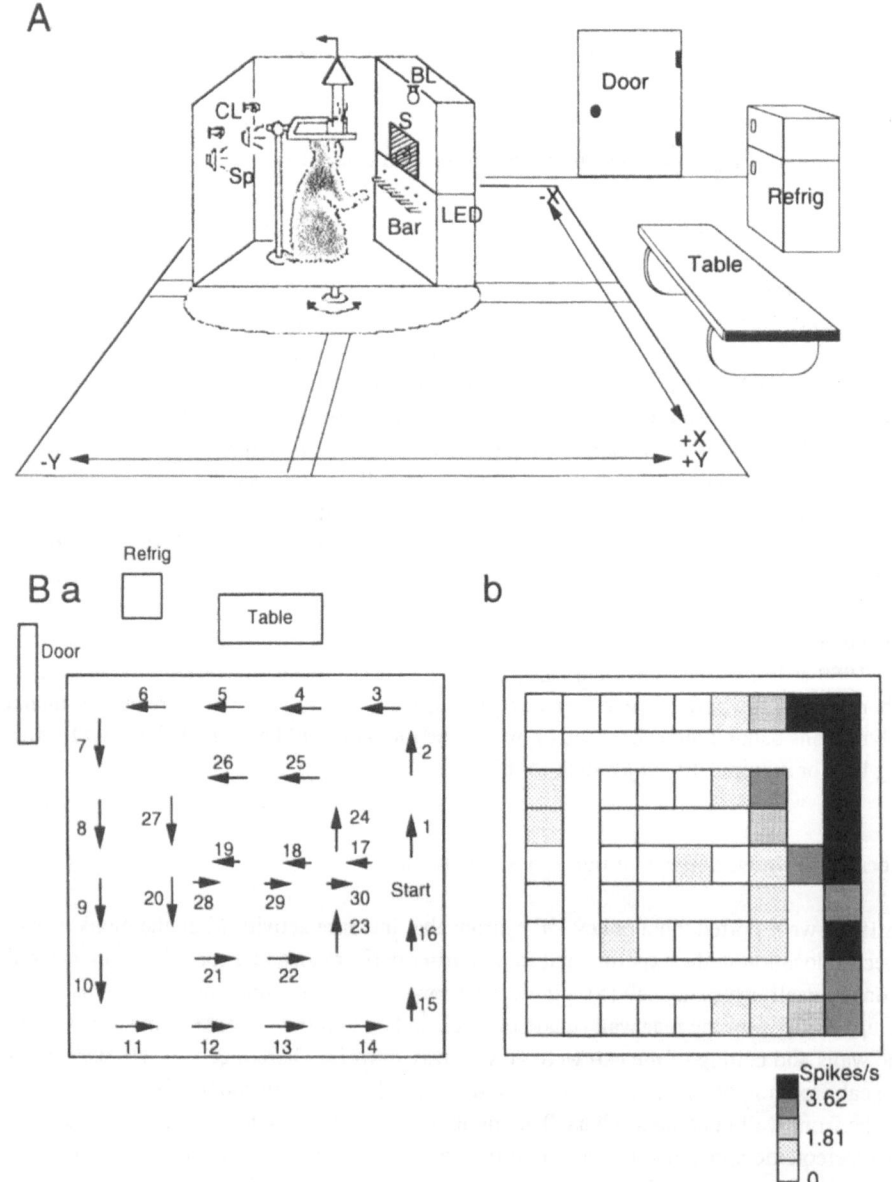

ly not used). A cue tone at one of 4 different frequencies warned of the initiation of a trial. The frequency of the cue tone indicated the bar to be pressed and thus also indicated movement direction (see figure legend for details). As long as the monkey continued to press the appropriate bar, the cab moved one quarter of the driving space, and at the end of this drive, the monkey was rewarded with a drop of juice or a piece of food.

Figure 1B shows an example of HF neurons displaying location-related activity. We first tested this neuron under the lighted-room (illuminated) condition during the performance of the task. As shown in Figure 1Ba, the monkey drove the cab in a counter-clockwise fashion to various locations within the driving space. The neuron significantly increased firing in and around Locations 1–3, whereas firing did not increase markedly at other locations (Fig. 1Bb). We then tested the same neuron under different conditions to determine the effects of visual input from outside of the cab and of movement fashions (the monkey itself drove the cab or was translocated passively by the experimenter, driving in a clockwise or counter-clockwise fashion, etc.) on the location-related activity of this neuron (Fig. 2). Since this neuron had a firing field (i.e., the area in which the location-related neuron showed a significant increase in the activity) at Locations 1–3, only the outer pathway in the driving space was used for this test. The neuron displayed similar location-related activity when tested under the same conditions as in the original test (Fig. 2A). However, location-related activity was absent under both of the following conditions: when the monkey was not able to see the scene outside the cab (Fig. 2B) or when the cab was translocated passively (i.e., without performing the task) by the experimenter and the monkey was able to see outside the cab (Fig. 2C). On the other hand, when the monkey drove the cab in the clockwise fashion under the lighted-room condition, the neuron again displayed location-related activity (Fig. 2D). Therefore, the responsiveness of this neuron was thought to depend on at least 2 factors: both visual inputs (scene of the room) from outside the cab and the monkey itself actively driving the cab.

During the experiment, we recorded a total of 238 neurons from the monkey HF. Of these neurons, 79 (33.2%) showed location-related responses.

Figure 1. Experimental setup (A) and an example of location-related HF neurons (B). (A) Spatial moving apparatus (cab) and arrangement in the experimental room. Monkeys were seated in the cab with their heads fixed in a stereotaxic apparatus and performed a task (spatial moving task). In this task, a 0.5-s cue tone at one of 5 different frequencies started each trial: a 5200 Hz tone was the signal for the monkey to press the far right bar to move the cab backward; a 1700 Hz tone was the signal for the monkey to press the right bar to move the cab to the right; a 1100 Hz tone was the signal for the monkey to press the middle bar to move the cab forward; a 620 Hz tone was the signal for the monkey to press the left bar to move the cab to the left; and a 210 Hz tone (usually not used) was the signal for the monkey to press the far left bar to move the cab backward. The tone was followed, after an 0.7-s delay, by an 0.5-s LED light over the corresponding bar. The front bay lamp (BL) turned on after an 0.8-s delay to show food or an object associated with juice. If the monkey pressed the appropriate bar, the cab moved toward the goal location as long as the monkey held the bar down. At the goal, the front shutter opened and the monkey got a piece of food or a drop of juice. CL, cab lamp; Sp, speaker; S, shutter; Refrig, refrigerator. (B) Movement trail of the cab (a) and firing rate map for one location-related HF neuron (b) during the performance of the spatial moving task. Number in Ba indicates the trial number. Gray scale tables to the right of the firing map (Bb) indicate calibration for firing rate.

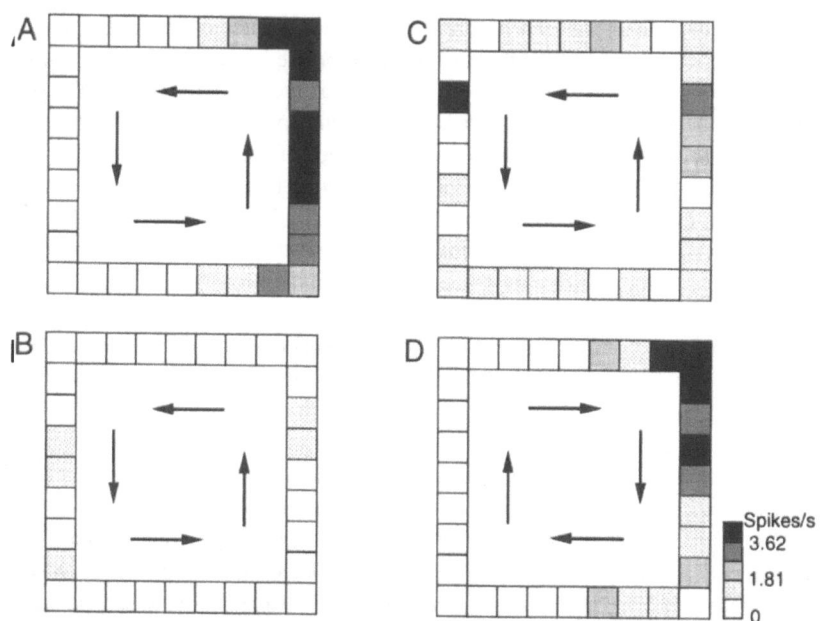

Figure 2. Activity of a location-related HF neuron under different conditions. The location-related activity of the HF neuron shown in Figure 1 was tested under 4 different conditions. Only the outer pathway was used in this test. A: The monkey itself drove the cab in a counter-clockwise fashion under the lighted-room condition. B: The monkey itself drove the cab in a counter-clockwise fashion under the lights-off condition. C: The monkey was passively translocated in a counter-clockwise fashion under the lighted-room condition. D: The monkey itself drove the cab in a clockwise fashion under the lighted-room condition. Other descriptions were as for Figure 1.

Context-dependent activity-change of location-related HF neurons

The importance of the HF for contextual processing in the brain has also been suggested. We recently studied the relationship between location-differential firing of HF neurons and context [6]. In this study, we manipulated context by changing behavioral tasks. The apparatus was almost identical to that used in the above experiment. The main differences were an LCD monitor placed in front of the monkey, and a joystick to control cab movement instead of the bars. Figure 3A shows the task paradigm for this experiment. We used 4 tasks, which were called RN/TC, RN/P-TC, VN/P-TC and VN/P tasks (for details, see legend of Figure 3A; readers can also refer to our paper [6]). The basic behavior required in these tasks was to either move the cab to one of 4 target areas in the driving space (the RN/TC and RN/P-TC tasks) or move the pointer to one of 4 target areas on the LCD monitor screen (the RN/P-TC, VN/P-TC and VN/P tasks).

Figure 3B shows an example of task-dependency of a location-related HF neuron. This neuron was active when the monkey moved the cab toward the right anterior part of the driv-

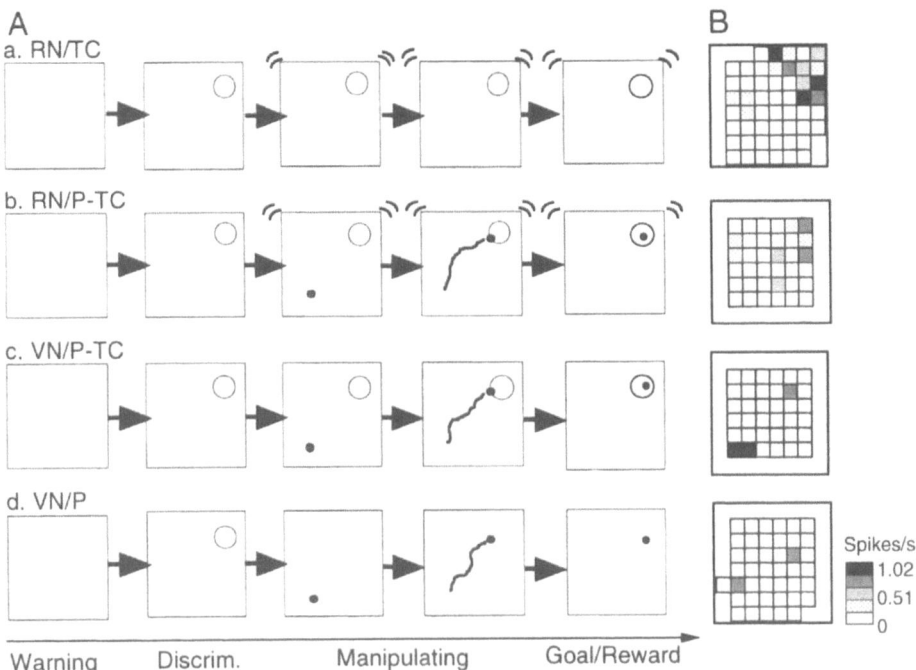

Figure 3. Experimental paradigm (A) and an example of a contextual effect on location-related activity of an HF neuron (B). (A) Sequence of real navigation without a pointer (a, RN/TC), real navigation with a pointer (b, RN/P-TC), virtual navigation with a pointer (c, VN/P-TC), and virtual navigation with a pointer and disappearance of TC (d, VN/P) tasks. RN, real navigation; VN, virtual navigation; P, cursor pointer; TC, target circle. In the RN/TC task, after the cab had been placed at one of the 4 corners (the left posterior area is the starting point in this figure), the task was initiated by simultaneous presentation of a warning tone and a 20 x 20 cm square frame on the monitor screen (warning phase). This square corresponded to the 2.5 x 2.5 m driving space. Two seconds after the onset of the warning tone, TC appeared in one of the 3 corners on the screen other than the corner corresponding to the starting point (TC appears at the upper right corner in this figure). TC indicated the goal location where the monkey was required to drive. After 2 s of a TC-discrimination phase, the monkey could drive the cab to the goal location by manipulating the joystick (manipulating phase). Once the monkey entered the goal location, a drop of juice was given as a reward. In the RN/P-TC task, a cursor pointer appeared during the manipulating phase, which indicated the cab location on the monitor. Therefore, the monkey could perform the task not only by using the same strategy as that in the RN/TC task but also by controlling the pointer movement on the LCD monitor screen. The VN/P-TC task was the same as the RN/P-TC task except for the absence of cab movement. In the VN/P task, TC disappeared during the manipulating phase. Therefore, the monkey had to remember the location of TC during this phase. (B) Firing rate map of one HF neuron that showed firing fields in the RN/TC and VN/P-TC tasks. The neuron was tested in the RN/TC (a), the RN/P-TC (b), VN/TC (c) and VN/P-TC (d) tasks. Other descriptions were as for the previous figures.

ing space during the performance of the RN/TC task. The activity of this neuron also increased significantly when the cursor pointer was in the lower left corner on the LCD screen during the performance of the VN/P-TC task. We should emphasize that the firing

field during the performance of the PN/TC task did not overlap with that on the corresponding LCD screen during the performance of the VN/P-TC task. This neuron did not respond
at all in the RN/P-TC and VN/P tasks. Given the difference in activity of this neuron in different tasks and that the locations of the monkey and cursor pointer were the same in the
driving space and on the monitor screen, the activity of this neuron was interpreted to be
task-dependent. In this experiment, a total of 389 neurons were recorded in the HF, and 166
of these neurons showed location-related activity. Of the 166 location-related neurons, 68
responded in only one of the 4 tasks, while 98 responded in 2 tasks or more. However, in the
majority of these 98 neurons, the firing field did not overlap at all, as shown in Figure 3B.

Conclusion

The results of our two studies [5, 6] provide neurophysiological evidence of primate HF
involvement in spatial information processing. In some unknown context, each HF neuron
may be assigned to represent a given location in an environment, with the result that a set of
HF neurons, assigned to represent one location, becomes active once the animal comes into
that location. Such neuron sets, each of which represents a specific location in a given environment, may conjunctively form an environmental map (cognitive map) in the HF, as proposed by O'Keefe and Nadel [7]. However, within a different context, each HF neuron may
be re-assigned to represent a different location. As a result, the location that was originally
coded by the previous set of neurons in the previous context may be represented by a completely different set of neurons in the new context. This suggests that several environmental
maps are ready to form in the HF, and a suitable map image for a given context is loaded as
an ensemble group of HF neurons.

Acknowledgement
This work was partly supported by the Japanese Ministry of Education, Science and Culture, Grants-in-Aid for
Scientific Research 08279105, 11308033, and 11680805.

References

1 Scoville WB, Milner B (1952) Loss of recent memory after bilateral hippocampal lesions. *J Neurol
   Neurosurg Psychiat* 20: 11–21
2 Squire LR (1987) *Memory and brain.* Oxford University Press, New York
3 Maguire EA, Burgess N, Donnett JG, Frackowiak RS, Frith CD, O'Keefe J (1998) Knowing where and
   getting there: a human navigation network. *Science* 280: 921–924
4 O'Keefe J, Dostrovsky J (1971) The hippocampus as a spatial map. Preliminary evidence from unit activity in the freely-moving rat. *Brain Res* 34: 171–175
5 Ono T, Nakamura K, Nishijo H, Eifuku S (1993) Monkey hippocampal neurons related to spatial and non-
   spatial functions. *J Neurophysiol* 70: 1516–1529
6 Matsumura N, Nishijo H, Tamura R, Eifuku S, Endo S, Ono T (1999) Spatial- and task-dependent neuronal responses during real and virtual translocation in the monkey hippocampal formation. *J Neurosci* 19:
   2381–2393
7 O'Keefe J, Nadel L (1978) *The hippocampus as a cognitive map.* Clarendon, Oxford

Neuroscientific Basis of Dementia
C. Tanaka, P.L. McGeer, Y. Ihara (eds)
© 2001 Birkhäuser Verlag Basel/Switzerland

Involvement of CaM kinase II and mitogen-activated protein kinase in hippocampal long-term potentiation

Eishichi Miyamoto[1], Jie Liu[1], Kohji Fukunaga[1] and Dominique Muller[2]

[1] *Department of Pharmacology, Kumamoto University School of Medicine, Kumamoto, Japan*
[2] *Department of Pharmacology, Université de Genève, Faculté de Médicine, Genève, Switzerland*

Introduction

In 1973, Bliss and his coworkers [1] reported that tetanic stimulation of the perforant pathway of presynaptic fibres resulted in high responses of granule cells of postsynaptic neurons to electric stimulation. The experiments were performed *in vivo*, using rabbits. This phenomenon was called long-term potentiation (LTP). LTP was considered to be important to elucidate the molecular mechanism of learning and memory [2], because learning and memory may be based on long-lasting potentiation of synaptic efficacy. Since learning and memory were supposed to be stored in specific compounds such as RNA, peptides or proteins in the 1960s, Bliss et al. proposed a new concept. Since then, the molecular mechanisms of LTP have been studied by many investigators, using hippocampal slices *in vitro*. The development of techniques made the study much easier.

At the end of the 1980s, several inhibitor experiments were performed. Addition of Ca^{2+} chelator, calmodulin antagonists and protein kinase (PK) inhibitors to the medium of rat hippocampal slices inhibited tetanic stimulation-induced LTP in the CA1 area of the hippocampus. Microinjection of a peptide of PKC inhibitory domain, calmidazolium (a calmodulin antagonist), and a peptide of Ca^{2+}/calmodulin-dependent PK II (CaM kinase II) inhibitory domain into pyramidal cells of the CA1 area of the hippocampus inhibited the induction of LTP by tetanic stimulation [3, 4]. Administration of the inhibitors after LTP induction had no effect on LTP. These results suggest that the above compounds are mainly related to induction of LTP, but not to maintenance of LTP.

In the 1990s, various knockout mice were produced by the gene-targeting method to examine which component is important for induction of LTP. Mice deficient in N-methyl-D-aspartic acid (NMDA) glutamate receptors [5], PKCγ subunit [6], CaM kinase II α subunit [7] and Fyn showed both impairment of LTP induction by using hippocampal slices and lack of spatial learning by using whole mice. These results suggest that the above components are involved in LTP induction and that LTP induction is closely correlated to spatial learning in the hippocampus.

In the present study, we focused on involvement of CaM kinase II and mitogen-activated PK (MAP kinase) in induction of LTP in the CA1 area of the hippocampus.

Increases in the activity of CaM kinase II during LTP induction

LTP was induced by application of theta-burst patterned stimulation to hippocampal slices and organotypic cultures [8]. In each slice, two stimulation electrodes were placed in the stratum of the CA1 area. Stimulation was applied simultaneously to the two inputs. After the induction of LTP, the CA1 area of the hippocampus was dissected and homogenized in the buffer containing a detergent. CaM kinase II was assayed for activity in the supernatant. The homogenizing buffer contained protease inhibitors, a PKC inhibitory peptide and protein phosphatase inhibitors. Low-frequency stimulation was given to control slices.

As shown in Figure 1, the induction of LTP resulted in increases in both the Ca^{2+}/CaM-independent and Ca^{2+}/CaM activities of CaM kinase II as substrates with syntide 2 and synapsin I. The effect could be detected as early as 3–5 min after stimulation and was present 60 min later. Prior to stimulation, the slices were treated with AP5, a specific inhibitor of the NMDA glutamate receptor. Under these conditions, LTP was not induced and the elevation of the CaM kinase II activity was not observed.

Hippocampal slices were prelabeled with [^{32}P]orthophosphate for 30 min [9]. Then stimulation was applied to the slices to induce LTP. Incorporation of [^{32}P]phosphate into the α and β subunits of CaM kinase II was determined in LTP-induced and control slices (Fig. 1). Autophosphorylation of both the α and β subunits increased during the induction of LTP. The treatment of slices with AP5 prevented the induction of LTP and abolished the increase in autophosphorylation of CaM kinase II.

Transient activation of 42-kDa MAP kinase during LTP induction

Our previous report showed that stimulation of glutamate receptors with NMDA and glutamate produced the increase in MAP kinase activity using primary cultures of hippocampal neurons [10, 11]. Furthermore, Sweatt's group (USA) reported that MAP kinase is activated during LTP induction and that addition of PD098059, a MAP kinase and ERK kinases (MEK) inhibitor, inhibited LTP induction as well as MAP kinase activation during LTP induction [12]. Therefore, we examined whether MAP kinase is activated during LTP induction [13]. Immediately after tetanic stimulation, increases in the activity of 42-kDa MAP kinase but not 44-kDa MAP kinase were observed in LTP-induced slices. The maximal peak of MAP kinase activity was observed 3 min after stimulation and decreased to the basal level within 30 min. The results were consistent with those previously reported. The control sample with low-frequency stimulation showed no increase in MAP kinase activity.

Inhibitory effects of PD098059 on LTP induction

It has been well documented that the phosphorylation of MAP kinase is regulated by MEK, MAP kinase kinase, which occurs upstream of MAP kinase. Thus, to test the physiological role of 42-kDa MAP kinase in LTP, we used the specific inhibitor of MEK, PD098059, to

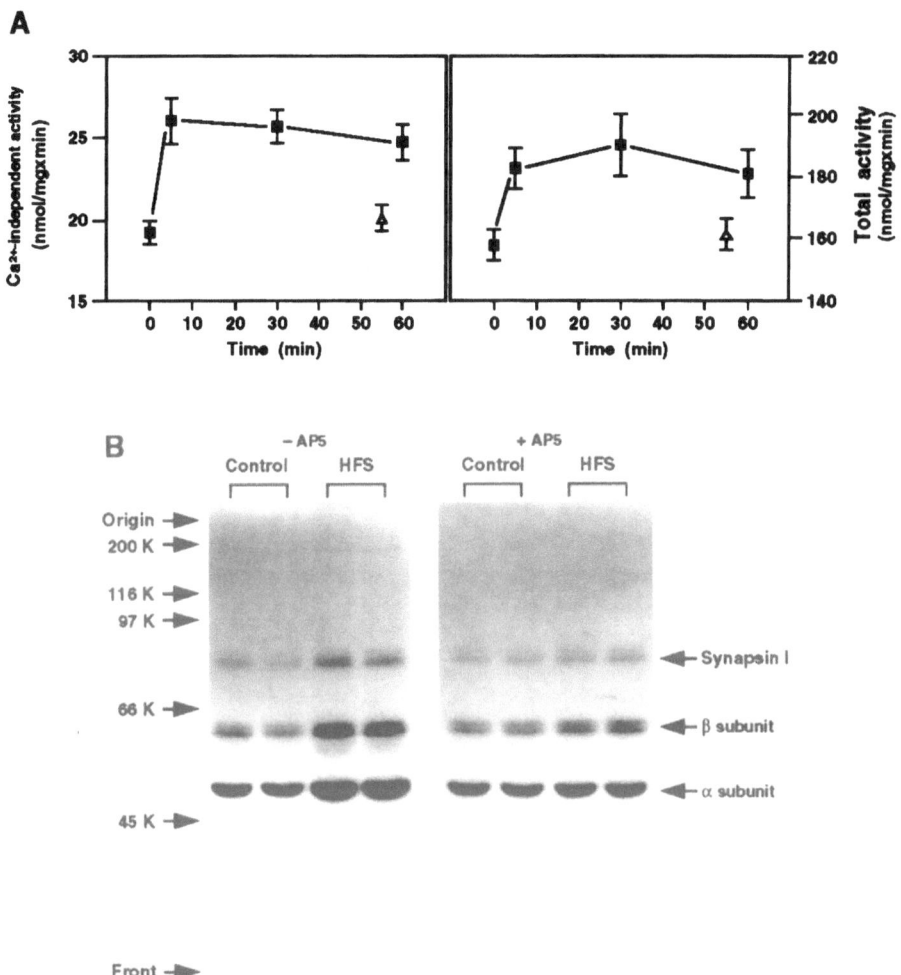

Figure 1. Increases in Ca^{2+}/CaM-independent and dependent activities and autophosphorylation of CaM kinase II during LTP induction [8, 9]. A) Time course of Ca^{2+}/CaM-independent (left panel) and dependent (right panel) activities after LTP induction. CaM kinase II activities after LTP induction (filled squares) and low-frequency stimulation (open triangles).

indirectly abolish the activation of MAP kinase. In the LTP experiments, we first used a relatively higher concentration of 50 µM PD098059 because more than 90% blocking effect on MAP kinase has been observed with use of this concentration. Before high frequency stimulation, PD 098059 was applied to slices for 90 min of the preincubation period. During the drug application, we found no change in basal synaptic transmission of the slices. In con-

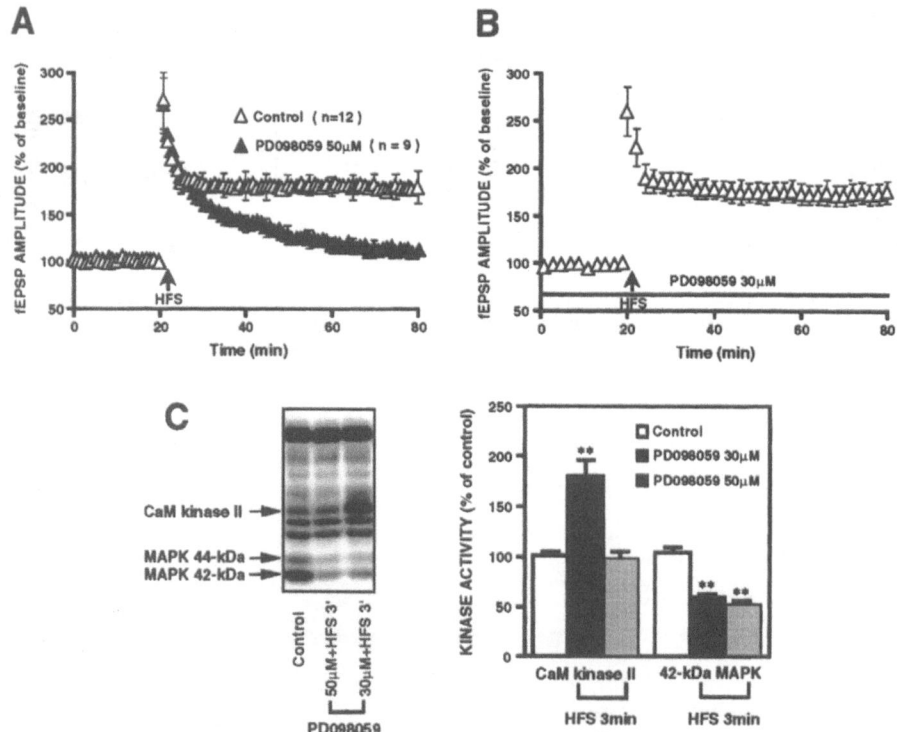

Figure 2. PD098059 caused concentration-dependent inhibitions of both LTP induction and CaM kinase II activation in CA1 areas [13]. (A) Compared with the control slices (open triangles; n = 12), preincubation of slices with 50 μM PD098059 markedly prevented HFS-sensitive induction of hippocampal LTP (filled triangles; n = 9). (B) At a lower concentration of 30 μM, PD098059 did not affect the induction of LTP (n = 6). (C) Analysis of CaM kinase II and MAP kinase activities in the slices treated with different concentrations of PD098059. At 50 μM (n = 9), PD098059 suppressed both 42- and 44-kDa MAP kinase activity, and prevented HFS-induced CaM kinase II activation. At 30 μM (n = 6) of the drug, only MAP kinase, but not CaM kinase II, was inhibited. Data are expressed as mean ± SE values. **p < 0.01. HFS, high frequency stimulation.

trast, the potentiation of field excitatory postsynaptic potential (fEPSP) following high-frequency stimulation delivery significantly decayed to nearly basal levels after 60 min in slices pretreated with 50 μM PD098059 (Fig. 2A), whereas a stable LTP was maintained for over 1 h in control slices (Fig. 2A). However, at a concentration of 30 μM, the drug preincubation for the same period of 90 min did not attenuate the induction of LTP (Fig. 2B). These results suggest that the inhibitory effect of PD098059 on LTP was concentration-dependent.

Inhibitory effects of PD098059 on activities of CaM kinase II and MAP kinase during LTP induction

As described above, at concentrations of 30 and 50 μM, PD098059 appeared to have different effects on the LTP induction. To determine if the lack of inhibitory effect seen with 30 μM is due to the weak blocking action of the drug on the activation of MAP kinase, we next examined effects of PD098059 on PK activity in the hippocampal slices. Because a maximal increase in MAP kinase activity was observed at 3 min after LTP, we designed an experimental protocol of collecting slices at 3 min after high-frequency stimulation to direct attention to this time point. In PD098059-treated slices, the activities of both 44- and 42-kDa MAP kinase were markedly inhibited by the drug (Fig. 2C). Thus, at both concentrations, PD098059 similarly inhibited the LTP-induced transient activation of 42-kDa MAP kinase. These observations indicate that the effect of 50 μM PD098059 on LTP induction could not be simply explained by inhibitory effects of the drug on MAP kinase.

We then analyzed the changes in CaM kinase II activity detectable in the same gel. With the same protocol for MAP kinase assay, CaM kinase II activity was also examined at a time point of 3 min after LTP. When the drug was applied at 50 μM, high-frequency stimulation did not produce any significant increase in CaM kinase II activity in the slices in which high-frequency stimulation failed to elicit the LTP. In contrast, in slices pretreated with 30 μM PD098059, a brief high-frequency stimulation markedly activated CaM kinase II as well as LTP. Statistical analysis showed that high-frequency stimulation produced an increase in CaM kinase II under the condition of 30 μM drug treatment, whereas no significant increase was detectable in 50 μM drug-treated slices. Thus, in addition to MEK and its downstream target, PD098059 also has inhibitory effects on CaM kinase II activation of hippocampal slices.

A question arose here as to whether this effect occurs by directly inhibiting activity of CaM kinase II. To answer this question, we next tested the effect of PD098059 on the activity of purified CaM kinase II *in vitro*. At concentrations of 10, 50 and 100 μM, PD098059 did not affect the activity of this kinase, suggesting that PD098059 has no direct inhibitory effect on CaM kinase II.

Effects of brain-derived neurotrophic factor (BDNF) and short high-frequency stimulation on LTP and kinase activities

It has been proposed that neurotrophic factors can regulate neuronal activity and synaptic efficacy via the activation of TrK family of tyrosine kinase receptors [14, 15]. One of these factors, brain-derived neurotrophic factor (BDNF), is synthesized predominantly in neurons, and is highly expressed in the hippocampus of the adult species brain. The binding of BDNF to the TrKB receptor triggers the kinase cascade and thereby activates MAP kinase in culture neurons [11, 16]. We then used BDNF as a modulator for MAP kinase to further explore the role of MAP kinase in LTP in hippocampal slices. Application of 50 ng/ml BDNF alone

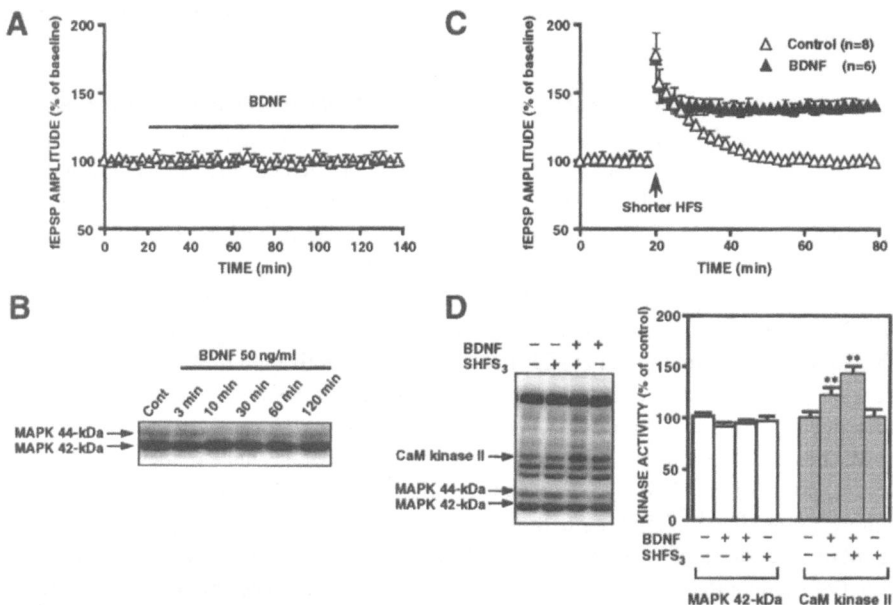

Figure 3. Effects of BDNF and short high-frequency stimulation (SHFS) on LTP induction and activities of MAP kinase and CaM kinase II [13]. *A*, Effects of BDNF (50 ng/ml) on LTP induction. *B*, Representative autoradiograph of MAP kinase activity assay in control (cont) and BDNF-pretreated slices. *C*, Effects of SHFS alone and combination of BDNF and SHFS on LTP induction. *D*, Under these conditions, activities of MAP kinase and CaM kinase II were determined by in-gel kinase assay.

did not affect basal synaptic strength under our experimental conditions (Fig. 3A) and did not increase the protein amounts of 42- and 44-kDa MAP kinase (Fig. 3B).

On the other hand, short high-frequency stimulation (SHFS) was given to hippocampal slices. As 1 s 100 Hz stimulation was usually given, SHFS meant 0.12 s, 100 Hz stimulation. SHFS alone induced only LTP of fEPSP in hippocampal slices (Fig. 3C). However, when slices are preincubated with 50 ng/ml BDNF, the same SHFS induced a typical LTP (Fig. 3C). Under these conditions, activities of MAP kinase and CaM kinase II were determined (Fig. 3D). Even though both BDNF and SHFS were combined, MAP kinase activity did not increase (Fig. 3D). In contrast, CaM kinase II activity increased slightly with BDNF alone and more with both BDNF and SHFS (Fig. 3D). These results suggest that LTP induction is correlated with CaM kinase II activity, but not with MAP kinase activity.

Discussion

In the present study, using both electrophysiological and biochemical approaches, we demonstrated that the induction of LTP is dependent on CaM kinase II activation rather than on MAP kinase activation in CA1 hippocampal slices. The transient activation of 42-kDa MAP kinase in high-frequency stimulation-receiving slices supports the view of activity-dependent regulation of this enzyme. A most interesting finding in our experiments is that we observed a heretofore unknown pharmacological effect of PD098059 that is commonly believed to be a specific inhibitor of MAP kinase kinase. Although earlier studies examined effects of this agent on a variety of PKs and found that only MAP kinase kinase is sensitive to it [17], information about its effect on CaM kinase II was not included. Most recently, one research group suggested that MAP kinase activation is necessary for LTP because they found an inhibitory effect of PD098059 on LTP. In that paper, the authors said that they tested the effect of PD098059 on the activity of CaM kinase II *in vitro* and found no definite actions. This is well consistent with our results. However, when we used it in high-frequency stimulation-induced slices, the data clearly showed that 50 μM PD098059 actually blocked high frequency stimulation-induced activation of CaM kinase II in hippocampal slices. A similar result was also found in cultured hippocampal neurons (data not shown). This evidence suggests that there is an indirect effect of this compound on CaM kinase II, differing from its effect of binding to the inactive form of MEK [17], although the precise blocking mechanisms are unknown. Recently, PD098059 has been found to have a direct inhibitory effect on cyclooxygenase-1 and -2 [18]. Therefore, one must be careful when using this drug as a MEK inhibitor at concentrations of more than 50 μM. Additionally, these findings also suggest that the inhibition of hippocampal LTP induction with PD098059 was due to its effects on CaM kinase II and not on the MEK/MAP kinase pathway. The parallel changes in hippocampal LTP and CaM kinase II activity indicate that CaM kinase II activation plays a critical role in LTP induction.

On the other hand, MAP kinase was certainly activated during LTP, although activation of MAP kinase is not critically correlated to LTP induction. This suggests that activation of MAP kinase is involved in LTP maintenance by stimulation of gene expression. We recently observed that CaM kinase IV is activated during LTP (unpublished observation). These results are summarized in Figure 4. Glutamate is released from presynaptic neurons and mainly stimulates NMDA glutamate receptors. Ca^{2+} enters postsynaptic neurons through NMDA receptor-gated Ca^{2+} channels and activates CaM kinase II and MAP kinase. Activated CaM kinase II induces LTP by phosphorylation of certain substrates such as (S)-α-amino-3-hydroxy-5-methyl-4-isoxazolepropionate (AMPA) glutamate receptors. Activated CaM kinases II and IV and MAP kinase may be involved in stimulation of gene expression by phosphorylation of transcription factors and in maintenance of LTP, which is a model for long-lasting memory.

Several reports recently indicated that LTP promotes formation of spine synapses between axon terminals and dendrites [19]. Thereby structural remodeling of synapses and formation of new synaptic contacts have been implicated in the late phase of LTP [19]. Furthermore, AMPA glutamate receptor-deficient mice showed that the receptors are closely related to hippocampal synaptic plasticity. Redistribution of AMPA receptors was observed after synaptic NMDA glutamate receptor activation [20]. These results suggest that spine formation, structural remodeling of new circuits and redistribution of synapses are necessary to maintain long-lasting memory. To cause these morphological changes, stimulation of gene expression would be indispensable.

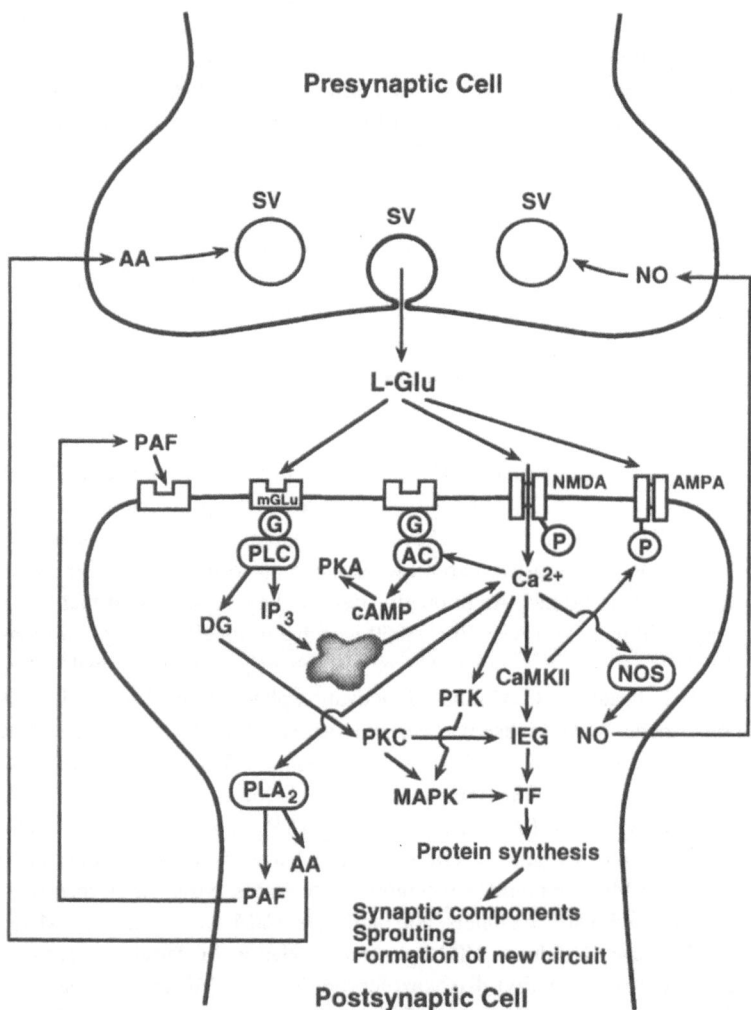

Figure 4. Possible hypothesis for LTP maintenance which is a model for long-term memory. SV, synaptic vesi-cle; PAF, platelet activity factor; G, G protein; PLC, phospholipase C; AC, adenylate cyclase; CaMKII, CaM kinase II; PTK, phosphotyrosine kinase; IEG, immediate early gene; NOs, NO synthase; TF, transcription fac-tor; AA, arachidonic acid.

Acknowledgment
This work was supported in part by Grants-in-Aid for Scientific Research and for Scientific Research on Priority Areas from the Ministry of Education, Science, Sports and Culture of Japan (to K.F. and E.M.), by the Swiss National Research Foundation (FNRS 3100-040815.94 to D.M.), and a research grant from the Human Frontier Science Program (RG30-96 to D.M. and E.M.).

References

1 Bliss TVP, Lømo T (1973) Long-lasting potentiation of synaptic transmission in the dentate area of the anaesthetized rabbit following stimulation of the perforant path. *J Physiol-Lond* 232: 331–356

2 Bliss TVP, Collingridge GL (1993) A synaptic model of memory: long-term potentiation in the hippocampus. *Nature* 361: 31–39

3 Malenka RC, Kauer JA, Perkel DJ, Mauk MD, Kelly PT, Nicoll RA, Waxham MN (1989) An essential role for postsynaptic calmodulin and protein kinase activity in long-term potentiation. *Nature* 340: 554–557

4 Malinow R, Schulman H, Tsien RW (1989) Inhibition of postsynaptic PKC or CaMKII blocks induction but not expression of LTP. *Science* 245: 862–866

5 Sakimura K, Kutsuwada T, Ito I, Manabe T, Takayama C, Kushiya E, Yagi T, Aizawa S, Inoue Y, Sugiyama H et al (1995) Reduced hippocampal LTP and spatial learning in mice lacking NMDA receptor ε1 subunit. *Nature* 373: 151–155

6 Abeliovich A, Chen C, Goda Y, Silva AJ, Stevens CF, Tonegawa S (1993) Modified hippocampal long-term potentiation in PKCγ mutant mice. *Cell* 75: 1253–1262

7 Silva AJ, Stevens CF, Tonegawa S, Wang Y (1992b) Deficient hippocampal long-term potentiation in α-calcium-calmodulin kinase II mutant mice. *Science* 257: 201–206

8 Fukunaga K, Stoppini L, Miyamoto E, Muller D (1993) Long-term potentiation is associated with an increased activity of Ca^{2+}/calmodulin-dependent protein kinase II. *J Biol Chem* 268: 7863–7867

9 Fukunaga K, Muller D, Miyamoto E (1995) Increased phosphorylation of Ca^{2+}/calmodulin-dependent protein kinase II and its endogenous substrates in the induction of long term potentiation. *J Biol Chem* 270: 6119–6124

10 Kurino M, Fukunaga K, Ushio Y, Miyamoto E (1995) Activation of mitogen-activated protein kinase in cultured rat hippocampal neurons by glutamate receptors. *J Neurochem* 65: 1282–1289

11 Fukunaga K, Miyamoto E (1998) Role of MAP kinase in neurons. *Mol Neurobiol* 16: 79–95

12 Atkins CM, Selcher JC, Petraitis JJ, Trzaskos JM, Sweatt D (1998) The MAPK cascade is required for mammalian associative learning. *Nat Neurosci* 1: 602–609

13 Liu J, Fukunaga K, Yamamoto H, Nishi K, Miyamoto E (1999) Differential roles of Ca^{2+}/calmodulin-dependent protein kinase II and mitogen-activated protein kinase activation in hippocampal long-term potentiation. *J Neurosci* 19: 8292–8299

14 Figurov A, Pozzo-Miller LD, Olafsson P, Wang T, Lu B (1996) Regulation of synaptic responses to high-frequency stimulation and LTP by neurotrophins in the hippocampus. *Nature* 381: 706–709

15 Akaneya Y, Tsumoto T, Kinoshita S, Hatanaka H (1997) Brain-derived neurotrophic factor enhances long-term potentiation in rat visual cortex. *J Neurosci* 17: 6707–6716

16 Finkbeiner S, Tavazoie SF, Maloratsky A, Jacobs KM, Harris KM, Greenberg ME (1997) CREB: a major mediator of neuronal neurotrophin responses. *Neuron* 19: 1031–1047

17 Alessi DR, Cuenda A, Cohen P, Dudley DT, Saltiel AR (1995) PD 098059 is a specific inhibitor of the activation of mitogen-activated protein kinase kinase *in vitro* and *in vivo*. *J Biol Chem* 270: 27489–27494

18 Borsch-Haubold AG, Pasquet S, Watson SP (1998) Direct inhibition of cyclooxygenase-1 and -2 by the kinase inhibitors SB 203580 and PD 98059. SB 203580 also inhibits thromboxane synthase. *J Biol Chem* 273: 28766–28772

19 Engert F, Bonhoeffer T (1999) Dendritic spine changes associated with hippocampal long-term synaptic plasticity. *Nature* 399: 66–70

20 Shi S-H, Hayashi Y, Petralia RS, Zaman SH, Wenthold RJ, Svoboda K, Malinow R (1999) Rapid spine delivery and redistribution of AMPA receptors after synaptic NMDA receptor activation. *Science* 284: 1811–1816

Pathogenesis of dementia – tau

Neuroscientific Basis of Dementia
C. Tanaka, P.L. McGeer, Y. Ihara (eds)
© 2001 Birkhäuser Verlag Basel/Switzerland

Transgenic mice overexpressing the shortest human tau isoform develop a progressive tauopathy

Takeshi Ishihara, Ming Hong, Bin Zhang, John Q. Trojanowski and Virginia M.-Y. Lee

The Center for Neurodegenerative Disease Research, Department of Pathology and Laboratory Medicine, The University of Pennsylvania School of Medicine, Philadelphia, PA, USA

Introduction

Tau is an abundant microtubule-associated protein in the CNS that is implicated in the pathogenesis of frontotemporal dementia with parkinsonism linked to chromosome 17 (FTDP-17), progressive supranuclear palsy (PSP), amyotrophic lateral sclerosis/parkinsonism-dementia complex (ALS/PDC) of Guam, Alzheimer's disease (AD) and a number of other neurodegenerative diseases known as tauopathies [1–4]. Tauopathies are characterized neuropathologically by numerous inclusions formed by aggregated paired helical filaments (PHFs) and/or straight filaments composed of aberrantly phosphorylated tau proteins (PHF-tau) in selectively vulnerable neurons and glial cells throughout widespread regions of the CNS [1–4].

Six alternatively spliced tau isoforms are expressed primarily by neurons in the adult human CNS, and are localized predominantly in axons, although glial cells also contain small amounts of tau [5–7]. Tau proteins bind to microtubules (MTs), promote the assembly of MTs and stabilize MTs in the polymerized state [8, 9], but the formation of PHF-tau results in a loss of these important functions [10, 11]. Moreover, unlike normal tau, PHF-tau is insoluble, accumulates in the somatodendritic domain of neurons and assembles into filaments [12] that aggregate as neurofibrillary tangles (NFTs) in tauopathies [1–4]. In addition to PHF-tau, other cytoskeletal proteins, i.e., neurofilament (NF) subunits, are also found in many NFTs, as is ubiquitin [1, 13, 14].

The massive degeneration of neurons and extensive gliosis associated with the progressive accumulation of PHF-tau lesions provided circumstantial evidence implicating filamentous tau pathology in the onset/progression of neurodegenerative disease. However, the discovery of multiple pathogenic mutations in the *tau* gene of many distinct FTDP-17 families demonstrated directly and unequivocally that tau abnormalities cause neurodegenerative disease. These pathogenic FTDP-17 mutations are located at topographically distinct sites in exons and introns of the *tau* gene and they include exonic missense substitutions, an in-frame deletion and intronic substitutions [3, 15–21]. Significantly, emerging evidence suggests that topographically separate mutations cause FTDP-17 by differential mechanisms that specifically alter the functions or levels of tau isoforms in the CNS [16, 20, 22].

Recently, we tested the hypothesis that neurodegenerative disease can result from altered expression levels of normal tau isoforms by generating Tg mice that overexpressed the shortest human brain tau isoform (T44, also known as "fetal tau") in CNS neurons, and we reported that these Tg mice develop progressive age-dependent accumulations of intraneuronal filamentous inclusions accompanied by neurodegeneration, gliosis and tau protein abnormalities. Because overexpression of normal tau in these Tg mice causes a neurodegenerative disease that partially recapitulates human tauopathies, these mice will be useful in studies to elucidate mechanisms of brain degeneration in tauopathies.

A progressive tauopathy in tau Tg mice

Generation of Tg mice that overexpress the shortest human tau isoform

To generate Tg mice expressing human tau, a cDNA corresponding to the shortest human brain tau isoform ("fetal tau") was cloned into an expression plasmid MoPrP.Xho. This vector was used here because it enables relatively high levels of transgene expression in CNS neurons [24]. After screening potential Tg mouse lines by Southern blots, we identified 3 stable Tg lines that were shown by Western blot analysis to variably overexpress human tau. Using a polyclonal antibody (17026) that recognizes human and mouse tau in quantitative Western blot studies, we showed that the heterozygous Tg mouse lines 7, 43 and 27 overexpressed tau proteins at approximately 5-, 10- and 15-fold higher levels, respectively, than endogenous mouse tau. Since Tg line 27 mice were not viable beyond 3 months, and homozygous Tg mice generated from each of these 3 lines of Tg mice died *in utero* or within 3 months postnatal, the observations summarized below come from studies conducted on 1–12-month-old heterozygous tau Tg mice from lines 7 and 43.

Tau Tg mice acquire CNS tau inclusions with advancing age similar to human tauopathies

Tg mice and their wild-type (WT) littermates from lines 7 and 43 between 1–12 months of age were subjected to histological studies that revealed a widespread expression of human tau in neurons and their processes throughout the CNS of the Tg, but not the WT mice. In Tg mouse spinal cords, T14, a human tau-specific monoclonal antibody (MAb) stained spheroidal intraneuronal inclusions that were observed initially at 1 month, and the size and number of these inclusions increased up to 6–9 months, but they decreased in abundance by 12 months. Notably, many vacuolar lesions that were the same size or larger than the inclusions also were observed in the older Tg mice, which may reflect degeneration of affected axons. The inclusions were about the size of medium-to-large spinal cord neurons, and some appeared to arise within proximal axons of spinal cord neurons. Although they occurred in gray and white matter at all spinal cord levels, these tau-positive inclusions were most frequent at the gray-white junction.

Spinal cord sections were probed with a panel of antibodies to tau and other neuronal cytoskeletal proteins, and the inclusions were immunostained by a MAb, Alz50, commonly used to detect tau protein found in PHFs and by additional antibodies specific for other hyperphosporylated PHF-tau epitopes, including PHF1 (phosphoserine 396 and 404, numbering according to the largest human brain tau), PHF6 (phosphothreonine 231), T3P (phosphoserine 396), AT8 (phosphoserine 202 and 205), AT270 (phosphothreonine 181) and 12E8 (phosphoserine 262) [23]. Therefore, these lesions contain hyperphosphorylated tau similar to PHF-tau. Significantly, these inclusions were also stained strongly with anti-neurofilament (NF) protein antibodies specific to the low (NFL), middle (NFM), and high (NFH) molecular weight NF proteins. Both phosphorylated and non-phosphorylated NFM and NFH were observed in these lesions. Indirect immunofluorescence double labeling confirmed the co-localization of tau and NFs in these inclusions. In addition, anti-tubulin antibodies immunostained these inclusions.

In the brainstem and cortex of the Tg mice, tau-positive intraneuronal aggregates were also detected, but they were smaller and appeared later than the spinal cord inclusions. They were first seen in the pontine neurons of 1-month-old animals, and emerged in the cerebral cortex at about 6 months of age. The immunohistochemical profile of these brain aggregates was similar to that of the spinal cord lesions. However, the morphological features of these inclusions indicate that some are variants of the spinal cord axonal lesions, but others occur in the somatodendritic compartment of cortical neurons and resemble NFTs and dystrophic neurites. Notably, the brain and the spinal cord inclusions were positively stained by the Bodian silver method, similar to human NFTs. However, they were thioflavine S negative, and they were not stained by antibodies to α-internexin, peripherin, ubiquitin and synucleins. Tg mouse line 43 expressed higher levels of human tau than line 7 and similar tau-rich inclusions were also observed to accumulate in the spinal cord and brain of Tg line 43 in an age-dependent manner, but they were larger and more abundant than in line 7. These results indicate that the accumulation of these tau-rich lesions in the tau Tg mice is transgene dose-dependent as well as age-dependent.

Transmission electron microscopy (EM) studies of these inclusions revealed masses of tightly packed aggregates of randomly arranged 10–20 nm straight filaments in myelinated spinal cord axons of Tg, but not WT mice. These aggregates were found in ≈30% of myelinated and unmyelinated axons, ranging from 200 nm to 20 μm in diameter and some inclusions almost completely filled the axon. Mitochondria were trapped within occasional aggregates. Immuno-EM studies showed that the filaments were immunolabeled by antibodies to tau, NFs and tubulin.

Because the argyrophilic filamentous lesions of these tau Tg mice were concentrated in the spinal cord and brain stem and because tau-positive NFTs and dystrophic neurites are found in the brainstem and spinal cord of patients of ALS/PDC, PSP and FTDP-17, we directly compared the tau inclusions in the ALS/PDC spinal cord as well as in the brainstem of PSP and FDTP-17 cases with those in our Tg mice, and we found that they were similar. Significantly, in the ALS/PDC spinal cord, NF immunoreactivity co-localized with tau in many of the inclusions. These data, taken together with the findings described above, suggest that these tau Tg mice develop a neurodegenerative disease that recapitulates the hall-mark lesions of human tauopathies, especially those characteristic of ALS/PDC.

Insoluble tau protein progressively accumulates in the CNS of tau Tg mice

To determine whether tau became insoluble in the tau Tg mice with age and disease pro-gression as in human tauopathies, we analyzed the solubility of tau protein by extracting tau from brain and spinal cord samples, using buffers with increasing extraction strengths. In the WT mice, about 90% of endogenous mouse tau was largely re-assembly buffer (RAB)-sol-uble and no tau immunoreactivity was detected in the 70% formic acid (FA)-soluble fraction. Although the RAB-soluble tau from the Tg mice remained relatively constant at around 75–80% with increasing age, RAB-insoluble tau represented by the RIPA (50 mM Tris, 150 mM NaCl, 1% NP40, 5 mM EDTA, 0.5% sodium deoxycholate and 0.1% SDS, pH 8.0) and FA fractions progressively accumulated in both the brain and the spinal cord of the Tg mice. The accumulated RAB-insoluble tau was mainly fetal human tau expressed from the transgene, and the time course of accumulation correlated with the emergence of the tau inclusions in the Tg mice. In addition, RAB-insoluble tau, especially that in the FA fraction, was more pronounced in the spinal cord than the brain, consistent with more abundant tau aggregates in the spinal cord.

The phosphorylation of tau in the tau Tg mice recapitulates that in human tauopathies

Western blot studies of soluble and insoluble tau extracted from the cerebral cortex of Tg mice were performed using MAb T1 (specific for a non-phosphorylated tau epitope that is located within amino acids 189–207), and this antibody did not recognize PHF-tau, but was immunoreactive with human adult normal tau, fetal tau, and both soluble and insoluble Tg tau. This indicates that tau from the Tg mice is partially dephosphorylated at the T1 epitope. However, several phosphorylation-dependent antibodies, which reacted with PHF-tau and fetal tau, but not with normal adult tau, also recognized both soluble and insoluble tau from the Tg mice. These antibodies include PHF1 (phosphoserine 396 and 404), T3P (phospho-serine 396), PHF6 (phosphothreonine 231), AT8 (phosphoserine 202 and 205), AT270 (phos-phothreonine 181), and 12E8 (phosphoserine 262). Therefore the phosphorylation state of Tg tau recapitulates that of PHF-tau found in human tauopathies including ALS/PDC, PSP, FTDP-17 and AD.

Tau Tg mice develop gliosis, axon degeneration and reduced fast axonal transport linked to progressive motor weakness

To detect astrocytosis, we used a MAb to glial fibrillary acidic protein which stained numer-ous reactive astrocytes in brain and spinal cord of Tg but not WT mice indicating the presence of profound gliosis in regions with neuronal damage. Furthermore, the astrocytosis was almost undetectable at 1 month of age but progressed thereafter with age. Because inclusions in the proximal axons of affected neurons could cause disease by damaging axons, we examined the morphology of spinal cord ventral root axons. In semi-thin sections, the normal L5 ventral root of WT mice contained many large and small myelinated axons, but the ventral root of a six-month-old Tg mouse primarily contained irregularly shaped axons, and at 12 months of age, the endoneurial space appeared to increase, consistent with the removal of degenerated axons in these nerves. Evidence of axonal degeneration also came from comparing axon numbers in L5 ventral roots of Tg and WT mice based on studies of photomicrographs of semi-thin sections. A 20% decrease in the number of axons was seen in 12-month-old Tg mice relative to their WT counterparts. We also showed that, despite a significant reduction of MT density in the 12-month-old Tg mice, the NF density remained unchanged when compared with age-matched WT mice. This finding correlated with the biochemical analysis of β-tubulin and NF subunits in the proximal sciatic nerve, which showed a progressive decrease in β-tubulin levels in the Tg mice and relatively constant levels of NF subunits.

To assess whether or not the neuropathology in ventral roots of 12-month-old tau Tg mice compromised axonal transport, we measured radiolabeled proteins transported in the fast component of axonal transport following microinjection of [^{35}S]methionine into the L5 ventral horn of tau Tg and age-matched WT mice. Significantly, these studies showed that the fast axonal transport of radiolabeled proteins was retarded in the tau TG mice. Finally, in addition to acquiring the spinal cord pathologies described above, the tau Tg mice also developed progressive motor weakness, as demonstrated by their impaired ability to stand on a slanted surface and by retraction of their hind limbs when lifted by their tails. These impairments may explain the fact that the Tg mice weighed about 30%–40% less than age-matched WT littermates.

Discussion

As summarized here, the studies by Ishihara et al. [23] provide compelling evidence that the overexpression of normal human fetal tau protein in Tg mice causes a CNS neurodegenerative tauopathy that recapitulates key aspects of human tauopathies. For example, we observed a progressive, age-dependent accumulation of argyrophilic, tau immunoreactive inclusions in neurons of spinal cord, brainstem and neocortex similar to human tauopathies. Because the inclusions in the Tg mice were most abundant in spinal cord neurons, the tauopathy in these mice most closely resembles ALS/PDC wherein tangles are abundant in the spinal cord [1, 4, 25, 26]. Significantly, ALS/PDC patients who present with motor weaknesses do so about a decade earlier than those who present with parkinsonism and dementia. Thus, the accumulation of tau aggregates in the brains of our Tg mice later than in the spinal cord mirrors disease progression in ALS/PDC patients who present with motor weakness. Moreover, as shown here and reported earlier, these tau tangles also are immunostained by antibodies to NF proteins and tubulin [27], as are the inclusions in our Tg mice.

Finally, ALS/PDC is associated with a progressive motor weakness similar to that observed in the Tg mice.

Although the distribution of the tau pathology in our mice most closely resembles that found in ALS/PDC, PSP and some FTDP-17 syndromes, these filamentous tau aggregates share many characteristics with authentic NFTs in AD and other tauopathies. First, like highly insoluble PHF-tau in AD NFTs [28], a substantial fraction of tau proteins from the Tg mice is extracted only with RIPA and FA despite the fact that normal tau is an extremely soluble protein. Second, the amount of insoluble tau protein progressively accumulates with age and disease progression in the Tg mice similar to AD and other tauopathies. Third, PHF-tau proteins in human NFTs are hyperphosphorylated and so are soluble and insoluble tau proteins recovered from the Tg mice [12, 29]. Fourth, although AD NFTs contain mostly PHFs, straight filaments similar to those found in the inclusions of the Tg mice are also present [1–4]. Finally, in addition to ALS/PDC, NF protein immunoreactivity also occurs in NFTs of AD, PSP and other human tauopathies [13, 14, 26, 27, 30–32

Despite similarities between the tau aggregates in our Tg mice and those of other human tauopathies, the tau inclusions in the Tg mice are not identical to AD NFTs because NFTs contain all 6 tau isoforms, while the tau pathology in the Tg mice does not. In this regard, the filamentous Tg tau inclusions in these Tg mice resemble Pick bodies in the brains of patients with typical Pick's disease, because these pathological bodies are comprised almost exclusively of tau isoforms with 3 MT binding repeats (3R-tau). One notable difference, however, is that all 3 brain 3R-tau isoforms are found in Pick bodies, whereas only the fetal 3R-tau isoform is present in the tau aggregates in the Tg mice. Nevertheless, it is noteworthy that insoluble tau tangles comprised of all 6 tau isoforms (as seen in AD and ALS/PDC), or predominantly the 3 tau isoforms with 4 MT (4-R tau) binding repeats (as detected in PSP), or mostly 3R-tau (as found in Pick's disease) have been reported in human tauopathies, and recent studies of FTDP-17 and other familial tauopathies suggest that tau tangles comprised of different ratios of the 6 alternatively spliced tau isoforms exist in different tauopathies [16, 19, 20, 22].

Because the accumulation of filamentous tau inclusions in spinal cord neurons was associated with the degeneration of ventral root axons in the tau Tg mice, we hypothesize that this reflects a gain of toxic function by the overexpressed tau and several lines of evidence support this hypothesis. First, previously described tau Tg mice that expressed lower levels (<2-fold) of tau protein did not develop filamentous tau inclusions or neurodegeneration [33, 34]. Second, we observed a Tg tau dose-dependent increase in the size and number of tau aggregates in our 2 lines of Tg mice. Thus, one plausible explanation to account for the axonal degeneration in these Tg mice is a gain of toxic function by the excess tau proteins that cannot bind MTs, aggregate in the neuronal cytoplasm, block axonal transport and lead to the degeneration of affected axons.

The reduced numbers of MTs and the reduced levels of tubulin, but not NFs or NF proteins, in the remaining axons of the degenerating ventral roots also imply a loss of the MT-stabilizing function of tau. Because overexpressed human tau could aggregate with endogenous mouse tau leading to progressive insolubility and hyperphosphorylation of both human and mouse tau in the Tg mice, this could impair the ability of endogenous mouse tau to perform a MT-stabilizing function. Indeed, the observed reduction in fast axonal transport in 12-month-old Tg mice is consistent with a loss of MT function, although the loss of axons in the Tg mice may contribute to this. Based on indirect evidence from studies of human tauopathies, we and others have proposed that both gains of toxic functions and losses of normal tau functions could be involved mechanistically in causing neurodegenerative disease [16, 20, 22], and the data presented here support both of these mechanistic hypotheses.

Although a dose effect of the transgene is observed in 2 different Tg mouse lines, the distribution of the tau-rich lesions within the CNS is not completely dependent upon the expression levels across the different regions. For example, spinal cord expresses less Tg tau than brain, but more abundant and larger inclusions accumulate earlier in spinal cord neurons than in the other brain regions. This could be explained by the metabolic differences among diverse types of neurons and excess tau may aggregate at a lower concentration in spinal cord neurons under the influence of local factors or pathological chaperones such as high concentrations of NF proteins. Similarly, the selective distribution of tau pathology in different human tauopathies is likely to be due to other, as yet unidentified, local vulnerability factors. Indeed, the findings described here parallel the well known, but enigmatic "selective vulnerability" that is a constant feature of most human neurodegenerative diseases and Tg mouse models thereof [35]. Moreover, because the Tg tau mice described here exhibit the key neuropathological features of human tauopathies, they will be useful in studies designed to further elucidate mechanisms leading to the formation of tau pathology and the selective degeneration of neurons in FTDP-17, PSP, ALS/PDC, AD, Pick's disease and related tauopathies.

Acknowledgements
We thank Dr. A. Hirano for providing tissue blocks of ALS/PDC, Dr. M.K. Lee for collaboration in the generation of the tau Tg mice, the Biomedical Imaging Core Facility of University of Pennsylvania for assistance in the EM studies, Dr. M. L. Schmidt and Dr. Y.Nakagawa for help with some of these studies, and N. Shah, E. Heatherby, K. H. Szymczyk and Grace Kim for technical assistance. Supported by grants from the NIA.

References

1 Ginsberg SD, Schmidt ML, Crino PB, Eberwine JH, Lee VM-Y, Trojanowski JQ (1999) Molecular pathology of Alzheimer's disease and related disorders. *In*: A Peters, JH Morrison (eds): *Cerebral Cortex: Neurodegenerative And Age-Related Changes In Structure And Function Of Cerebral Cortex*. Kluwer Academic/Plenum Publishers, New York, 14: 603–654

2 Hof PR, Bouras C, Morrison JH (2000) Cortical neuropathology in aging and dementing disorders: Neuronal typology, connectivity, and selective vulnerability. *In*: A Peters, JH Morrison (eds): *Cerebral Cortex: Neurodegenerative And Age-Related Changes In Structure And Function Of Cerebral Cortex*. Kluwer Academic/Plenum Publishers, New York, 14: 175–312

3 Hong M, Trojanowski JQ, Lee VM-Y (2000) Tau-based neurofibrillary lesions. *In*: CM Clark, JQ Trojanowski (eds): *Neurodegenerative Dementias: Clinical Features And Pathological Mechanisms*. McGraw-Hill, New York, 161–175

4 Nakano I, Hirano A (1999) Ultrastructural changes in dementing illnesses. *In*: A Peters, JH Morrison (eds): *Cerebral Cortex: Neurodegenerative And Age-Related Changes In Structure And Function Of Cerebral Cortex*. Kluwer Academic/Plenum Publishers, New York, 14: 399–432

5 Goedert M, Spillantini MG, Jakes R, Rutherford D, Crowther RA (1989) Multiple isoforms of human microtubule-associated protein tau: sequences and localization in neurofibrillary tangles of Alzheimer's disease. *Neuron* 3: 519–526

6 Andreadis A, Brown WM, Kosik KS (1992) Structure and novel exons of the human tau gene. *Biochemistry* 31: 10626–10633

7 Binder LI, Frankfurter A, Rebhun LI (1985) The distribution of tau in the mammalian central nervous system. *J Cell Biol* 101: 1371–1378

8 Weingarten MD, Lockwood AH, Hwo SY, Kirschner MW (1975) A protein factor essential for microtubule assembly. *Proc Natl Acad Sci USA* 72: 1858–1862

9 Drechsel DN, Hyman AA, Cobb MH, Kirschner MW (1992) Modulation of the dynamic instability of tubulin assembly by the microtubule-associated protein tau. *Mol Biol Cell* 3: 1141–1154

10 Bramblett GT, Goedert M, Jakes R, Merrick SE, Trojanowski JQ, Lee VM-Y (1993) Abnormal tau phosphorylation at Ser396 in Alzheimer's disease recapitulates development and contributes to reduced microtubule binding. *Neuron* 10: 1089–1099

11 Yoshida H, Ihara Y (1993) Tau in paired helical filaments is functionally distinct from fetal tau: assembly incompetence of paired helical filament-tau. *J Neurochem* 61: 1183–1186

12 Lee VM-Y, Balin BJ, Otvos L Jr, Trojanowski JQ (1991) A68: a major subunit of paired helical filaments and derivatized forms of normal Tau. *Science* 251: 675–678

13 Perry G, Rizzuto N, Autilio-Gambetti L, Gambetti P (1985) Paired helical filaments from Alzhemier's disease patients contain cytoskeletal components. *Proc Natl Acad Sci USA* 82: 3916–3920

14 Schmidt ML, Lee VM-Y, Trojanowski JQ (1990) Relative abundance of tau and neurofilament epitopes in hippocampal neurofibrillary tangles. *Amer J Pathol* 136: 1069–1075

15 Clark LN, Poorkaj P, Wszolek ZK, Geschwind DH, Nasreddine ZS, Miller B, Payami H, Awert F, Markopoulou K, D'Souza I et al (1998) Pathogenic implications of mutations in the tau gene in pallido-ponto-nigral degeneration and related chromosome 17-linked neurodegenerative disorders. *Proc Natl Acad Sci USA* 95: 13103–13107

16 Hutton M, Lendon CL, Rizzu P, Baker M, Froelich S, Houlden H, Pickering-Brown S, Chakraverty S, Isaacs A, Grover A et al (1998) Association of missense and 5'-splice-site-mutations in tau with the inherited dementia FTDP-17. *Nature* 393: 702–705

17 Iijima M, Tabira T, Poorkaj P, Schellenberg GD, Trojanowski JQ, Lee VM-Y, Schmidt ML, Takahashi K, Nabika T, Matsumoto T et al (1999) A distinct familial presenile dementia with a novel missense mutation in the tau gene. *NeuroReport* 10: 497–501

18 Poorkaj P, Bird TD, Wijsman E, Nemens E, Garruto RM, Anderson L, Andreadis A, Wiederholt WC, Raskind M, Schellenberg GD (1998) Tau is a candidate gene for chromosome 17 frontotemporal dementia. *Ann Neurol* 43: 815–825

19 Spillantini MG, Murrell JR, Goedert M, Farlow MR, Klug A, Ghetti B (1998) Mutation in the tau gene in familial multiple system tauopathy with presenile dementia. *Proc Natl Acad Sci USA* 95: 7737–7741

20 D'Souza I, Poorkaj P, Hong M, Nochlin D, Lee VM-Y, Bird TD, Schellenberg GD (1999) Missense and silent mutations in tau cause FTDP-17 by altering alternative splicing. *Proc Natl Acad Sci USA* 96: 5598–5603

21 Rizzu P, Van Swieten JC, Joosse M, Hasegawa M, Stevens M, Tibben A, Niermeijer MF, Hillebrand M, Ravid R, Oostra BA et al (1999) High prevalence of mutations in the microtubule-associated protein tau in a popolation study of frontotemporal dementia in the Netherlands. *Amer J Hum Genet* 64: 414–421

22 Hong M, Zhukareva V, Vogelsberg-Ragaglia V, Wszolek Z, Reed L, Miller BI, Geschwind DH, Bird TD, McKeel D, Goate A et al (1998) Mutation-specific functional impairments in distinct tau isoforms of hereditary FTDP-17. *Science* 282: 1914–1917

23 Ishihara T, Hong M, Zhang B, Nakagawa Y, Lee MK, Trojanowski JQ, Lee VM-Y (1999) Age-dependent emergence and progression of a tauopathy in transgenic mice overexpressing the shortest human tau isoform. *Neuron* 24: 751–762

24 Borchelt DR, Davis J, Fischer M, Lee MK, Slunt HH, Ratovitsky T, Regard J, Copeland NG, Jenkins NA, Sisodia SS et al (1996) A vector for expressing foreign genes in the brains and hearts of transgenic mice. *Genet Anal* 13: 159–163

25 Hirano A, Malamud N, Kurland LT (1961) Parkinsonism dementia complex: an endemic disease on the island of Guam-pathological features. *Brain* 84: 642–661

26 Matsumoto S, Hirano A, Goto S (1990) Spinal cord neurofibrillary tangles of Guamanian amyotrophic lateral sclerosis and parkinsonism-dementia complex: an immunohistochemical study. *Neurology* 40: 975–979

27 Shankar SK, Yanagihara R, Garruto RM, Grundke-Iqbal I, Kosik KS, Gajdusek DC (1989) Immunocytochemical characterization of neurofibrillary tangles in amyotrophic lateral sclerosis and parkinsonism-dementia of Guam. *Ann Neurol* 25: 146–151

28 Bramblett GT, Trojanowski JQ, Lee VM-Y (1992) Regions with abundant neurofibrillary pathology in human brain exhibit a selective reduction in levels of binding-competent tau and accumulation of abnormal tau-isoforms (A68 proteins). *Lab Invest* 66: 212–222

29 Matsuo ES, Shin RW, Billingsley ML, Van deVoorde A, O'Connor M, Trojanowski JQ, Lee VM-Y (1994) Biopsy-derived adult human brain tau is phosphorylated at many of the same sites as Alzheimer's disease paired helical filament tau. *Neuron* 13: 989–1002

30 Kato T, Hirano A, Weinberg MN, Jacobs AK (1986) Spinal cord lesions in progressive supranuclear palsy: some new observations. *Acta Neuropathol* 71: 11–14

31 Nakazato Y, Sasaki A, Hirato J, Ishida T (1984) Immunohistochemical localization of neurofilament protein in neuronal degenerations. *Acta Neuropathol* 64: 30–36

32 Schmidt ML, Lee VM-Y, Trojanowski JQ (1991) Comparative epitope analysis of neuronal cytoskeletal proteins in Alzheimer's disease senile plaque neurites and neuropil threads. *Lab Invest* 64: 352–357

33 Brion J-P, Tremp G, Octave J-N (1999) Transgenic expression of the shortest human tau affects its compartmentalization and its phosphorylation as in the pretangle stage of Alzheimer's disease. *Amer J Pathol* 154: 255–270

34 Goetz J, Probst A, Spillantini MG, Schaefer T, Jakes R, Buerki K, Goedert M (1995) Somatodendritic localization and hyperphosphorylation of tau protein in transgenic mice expressing the longest human brain tau isoform. *EMBO J* 14: 1304–1313

35 Tu P-H, Gurney ME, Julien J-P, Lee VM-Y, Trojanowski JQ (1997) Oxidative stress, mutant SOD1, and neurofilament pathology in transgenic mouse models of human motor neuron disease. *Lab Invest* 76: 441–456

Neuroscientific Basis of Dementia
C. Tanaka, P.L. McGeer, Y. Ihara (eds)
© 2001 Birkhäuser Verlag Basel/Switzerland

Tau and neurodegenerative disease: genetics and pathogenetic mechanisms

Gerard D. Schellenberg[1-4], Ian D'Souza[1,2], Parvoneh Poorkaj[1,2] and Thomas D. Bird[1-3]

[1] *Geriatric Research Education and Clinical Center, Veterans Affairs Puget Sound Health Care System, Seattle, WA, USA*
[2] *Divisions of Gerontology and Geriatric Medicine, and Medical Genetics, Department of Medicine, University of Washington, Seattle, WA, USA*
[3] *Department of Neurology,* [4] *Department of Pharmacology, University of Washington, Seattle, WA, USA*

Introduction

Neurodegenerative disease is often accompanied by abnormal tau aggregates in the form of neurofibrillary tangles (NFTs) and other intracellular structures. Diseases where tau pathology is found include Alzheimer's disease (AD) [1], Down syndrome, frontotemporal dementia with parkinsonism-chromosome 17 type (FTDP-17), progressive supra-nuclear palsy (PSP) [2], Gerstmann-Straussler Scheinker disease [3], Pick's disease [4], amyotrophic lateral sclerosis/parkinsonism dementia complex of Guam [5], Niemann-Pick type-C disease [6] and to a lesser degree in normal aging. In some disorders such as AD, Down syndrome and some forms of FTDP-17, tau aggregates are exclusively found in neuronal cell bodies as NFTs and as "ghost" tangles, the presumed remains of a dead neuron. In other disorders, such as in some forms of FTDP-17, tau aggregates are also in glial cells. The abnormal tau in these aggregates is highly phosphorylated.

The observation of tau neuropathology in a large number of different diseases suggested that tau aggregates were a consequence of neuronal death but not the cause. However, recently, mutations were identified in the gene (*TAU*) encoding tau protein in subjects from families with autosomal dominant FTDP-17 [7–9]. Thus altered *TAU* gene regulation and altered tau protein function can cause neurodegeneration and the process results in NFTs that in some cases are identical to those seen in AD. Thus, in disorders other than FTDP-17, tau must be considered to be an integral part of neurodegeneration and not just a by-product.

Structure and regulation of *TAU*

Tau is one of a number of homologous proteins belonging a group of proteins referred to as microtubule-associated proteins (MAPs). Though the normal function of tau is not completely understood, this protein is thought to promote microtubule (MT) polymerization and stabilize formed MTs, and possibly promote neurite extension and maintain neuronal integrity [10, 11]. In the central nervous system (CNS), *TAU* is expressed in high levels in neurons but is also present in oligodendrocytes [12] and astrocytes. *TAU* is also expressed in non-neuronal tissues [13, 14].

TAU consists of 16 exons (E0, 1–4, 4a, 5–14; note E0 has also been called E-1) [15–19]. The structure of the gene is known and spans over 100 kb. Multiple protein products are produced and are the result of alternative splicing of exons 2, 3, 4A, 6, 8 and 10 (Fig. 1). The expression patterns of various tau transcripts are complex and depend on the developmental state, the organ, the cell type and the species examined [14]. In rodent and human fetal CNS, only a single transcript is produced (3R0, Fig. 1) and no alternatively spliced exons are included [17, 20]. In the rat, at 7 days after birth, E2+ and E2+/E3+ isoforms appear. E3 is only included when E2 is present, though transcripts with only E2 are observed. E2 and E3 are differentially regulated and appear at different but overlapping times in development [20]. In the human brain, 6 isoforms formed using E2, E3 and E10 (Fig. 1) are observed. In other organisms including rodents, brain isoforms containing E4a are found.

In the C-terminal end of tau there are either 3 or 4 repeated sequences, depending on the isoform, that are MT-binding domains. Three of these repeats are encoded by constitutively included exons. The fourth repeat is specified by alternatively spliced E10. Thus tau protein from transcripts that include E10 have 4 repeats and are called 4R tau, protein from transcripts that do not include E10 are 3-repeat or 3R tau.

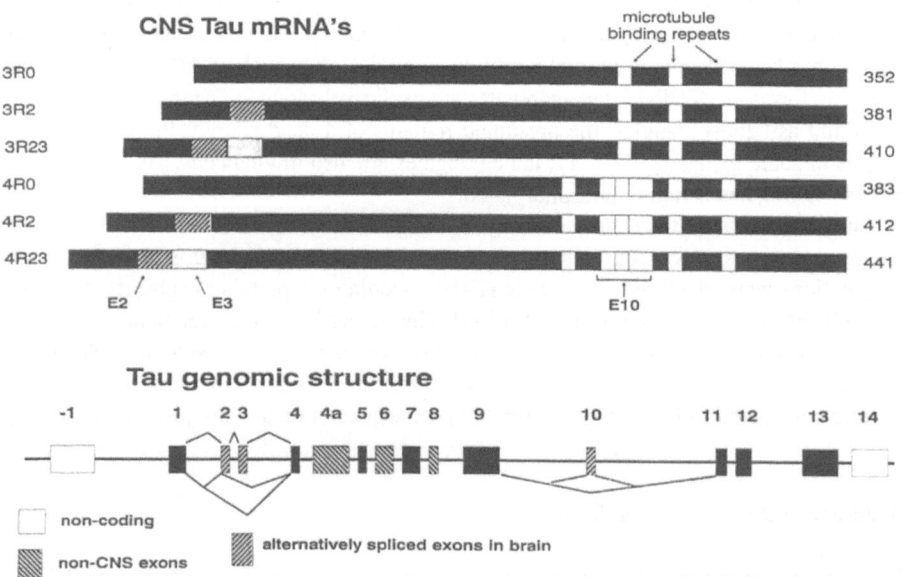

Figure 1. *TAU* gene and cDNA structure (A) Adult human brain tau isoforms. Abbreviations on the left are the exon content for alternatively spliced exons E2, E3 and E10. 4R tau includes E10 and 3R tau does not. (B) Genomic structure of the *TAU* gene. Exons are indicated by boxes and splicing by lines connecting the boxes. The alternative splicing shown is for adult human brain.

FTDP-17 and *TAU*

Tau-induced neurodegeneration is best understood in FTDP-17. This disease is a group of autosomal dominant syndromes with overlapping behavioral, cognitive and motor features [21–27]. FTDP-17 is characterized by frontotemporal atrophy and may include atrophy of the basal ganglia, substantia nigra and amygdala. The hippocampus is left relatively intact. Neuronal loss and gliosis occurs and Pick bodies or ballooned neurons are seen in some cases. Some but not all FTDP-17 families have AD-like NFTs [26, 28–31] and probably have some form of tau pathology. Although *TAU* mutational analyses were initially negative [32, 33], later work resulted in identification of a number of *TAU* mutations responsible for FTDP-17 [7–9]. Subsequently, over 20 pathogenic *TAU* mutations have been identified (Tab. 1) [34–37].

Four types of *TAU* mutations have now been identified. These are: (1) missense mutations that change the biochemical properties of the tau protein (2) missense and silent mutations that alter the regulation of E10 splicing (3) a 3 bp in-frame deletion that alters E10 splicing (4) intronic mutations in intron 10 (I10) that also alter the regulation of E10 splicing. Over 12 missense *TAU* FTDP-17 mutations are known. Most alter the protein properties of tau (e.g., $^P301^L$ and $^V337^M$).

Table 1. Inter-repeat (IR) regions are sequences between the MT binding domains. N-ter and C-ter refer to the N-terminal and C-terminal regions. The stemloop region is E10+1 to E10+21

Mutation	Location in *TAU* gene	Location in Tau protein
$^P189^A$	E9	N-ter
$^G272^V$	E9	R1
E9+33	I9	intron
N279^K	E10	IR1-2
Δ280^K	E10	IR1-2
$^L284^L$	E10	IR1-2
$^P301^L$	E10	R2
$^P301^S$	E10	R2
$^S305^N$	E10	IR2-3/5'splice site
$^V337^M$	E12	IR3-4
$^G389^R$	E13	C-ter
$^R406^W$	E13	C-ter
E10+3	I10	5' splice site
E10+12	I10	stemloop region
E10+13	I10	stemloop region
E10+14	I10	stemloop region
E10+16	I10	stemloop region

Mutations in intron 10 (I10) are found immediately after the 3' end of the exon and are in the first 16 nucleotides of the intron [7, 9, 34, 38]. These mutations (E10+3, E10+12, E10+13, E10+14 and E10+16) increase inclusion of E10 in exon splicing assays. These mutations possibly destabilize an RNA stemloop that may inhibit inclusion of E10 (Fig. 2). Though a stemloop structure does form *in vitro* [39], it is not clear whether this stemloop forms *in vivo* and whether this structure is important for E10 splicing regulation. Figure 2A shows a stemloop ($\Delta G = -9.2$ kcal/mol) that is more extensive than structures proposed by

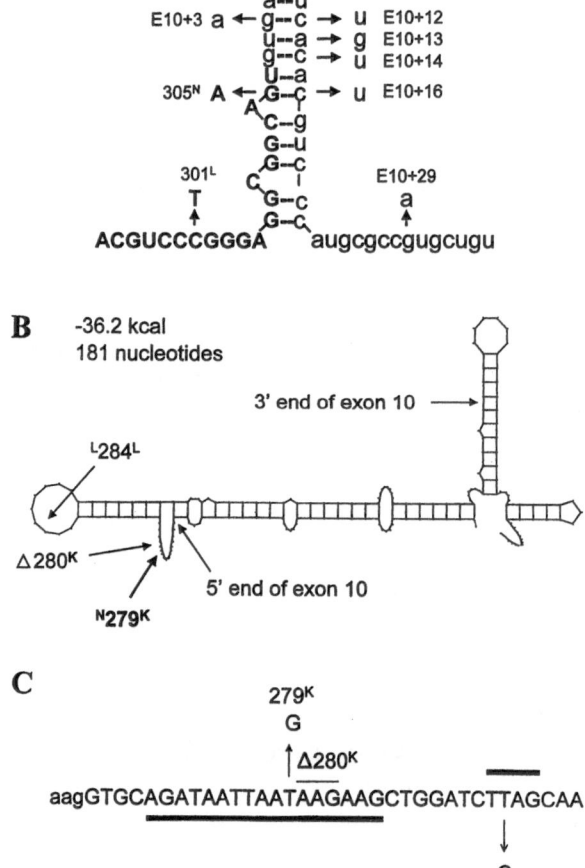

Figure 2. E10/I10 *TAU* mutations and secondary structure. (A) RNA sequences for the E10/I10 junction. Capital letters are E10, and lower case letters are I10. (B) Potential secondary structure generated from 181 nucleotide of tau E10/E10. (C) Part of E10 RNA sequence showing FTDP-17 mutations. The bar below the E10 sequence spans the proposed purine-rich ESE. The bar above the sequence spans the proposed ESS sequence.

others. Both a structure with only the top portion [9] and an intermediate structure containing the top 2 stems ($\Delta G = -8.4$ kcal/mol) [7] have been proposed. Even more extensive structures are possible when additional nucleotides are included (Fig. 2B). Clearly this region of *TAU* has an inhibitory function in *TAU* E10 splicing. However, the linear sequence and not the stemloop may be critical for this inhibition.

I10 mutations cause FTDP-17 by altering splicing of E10 so that transcripts carrying these mutations make almost exclusively E10+ processed messages. At the protein level, the result is an increase in the 4R/3R tau isoform ratio [7, 9]. This 4R/3R ratio is normally approximately 1 in normal and in AD brain [40]. FTDP-17 is an autosomal dominant disease, and affected subjects carry one mutant and one normal copy of *TAU*. If intronic mutations result in complete E10 inclusion, the 4R/3R ratio is expected to change from 1 to 3, at most. This small change in the ratio, even though the amino acid sequence is normal, causes severe neurodegenerative disease in mid-life.

Missense mutations and the single deletion mutation identified to date ($\Delta 280^K$) appear to act by 2 different mechanisms. Mutations $^V337^M$, $^R406^W$, $^P301^L$, $\Delta 280^K$ alter tau binding to MTs [40] and alter the ability of tau to promote MT polymerization [35, 36, 40, 41]. Missense mutations also cause FTDP-17 by altering E10 splicing, with a result similar to the I10 mutations described above. The prototype of this class of mutations is $^N279^K$. Recombinant tau containing this mutation, binds to tubulin normally, and promotes *in vitro* MT polymerization with the same efficiency as normal *TAU* [37, 42]. In splicing assays, the $^N279^K$ mutation, which is located in E10, increases E10 inclusion. Consistent with this observation is that in brain samples from subjects with this mutation, the tau protein 4R/3R ratios are increased to approximately 2 [43]. The functional consequence of this mutation is to strengthen the effects of an exon-splicing enhancer (ESE) element in *TAU* E10 (Fig. 2C). The activity of this ESE is completely abolished by the FTDP-17 mutation $\Delta 280^K$ [35], which is directly adjacent to the $^N279^K$ mutation. The result in *in vitro* splicing assays is to completely exclude inclusion of this exon in *TAU* transcripts. The $\Delta 280^K$ mutation also affects tau protein function and is the most severe in terms of tau-MT binding and its effects on tau-dependent MT polymerization. The splicing data suggest that no E10+ transcript is made and thus no protein with this mutation would be produced. However, this has not been confirmed *in vivo* as no autopsy material is available from subjects with this mutation. Assuming that no E10+ is made from the $\Delta 280^K$ gene, the result would be a reduced tau 4R/3R ratio, a situation that could lead to excess 3R protein in tau aggregates, analogous to Pick's disease [44].

A silent *TAU* E10 mutation $^L284^L$ was recently identified which changes the normal sequence TTAG to TCAG [37]. This is the site of a potential exon splicing silencer (ESS) motif TTAG which has been defined in HIV-1 *tat* exon 3 [37, 45]. This mutation, like the $^N279^K$ mutation, also increases inclusion of E10. Whether this is an ESS or an ESE is currently under investigation. A single autopsy of a subject with the $^L284^L$ mutation has been obtained and the results are unique in that both tau aggregates and amyloid plaques are present. Plaque distribution and density were similar to those seen in AD [37]. The clinical symptoms were consistent with frontotemporal dementia. The autopsied subject died at age 61, at an age when AD is still rare. Aβ deposition is typically absent in FTDP-17. The subject may have had both AD and FTDP-17, though at age 61 both are relatively rare. Alternatively, disruption of the ESS element may alter *TAU* E10 regulation in a cell/region-specific manner that leads to subsequent Aβ production and deposition.

The results from both mutations that alter the biochemical properties of tau and mutations that alter E10 splicing suggest a common pathogenic mechanism. Missense mutations that reduce the affinity of tau for MTs could result in excess free tau not bound to MTs. This free tau may then be available for hyperphosphorylation, neurotoxic effects and aggregation. Likewise, missense mutations and I10 mutations that alter E10 splicing and consequently 4R/3R ratios, may work by the same mechanism. If the ratio of 4R and 3R binding sites is close to 1, and both must bind simultaneously, excess of one isoform could result in excess free tau. The results from the $\Delta280^K$ mutation suggest that in terms of developing FTDP-17, it is not important whether the excess free tau is 3R or 4R.

Another possible consequence of work to date is that different mutations may result in different neuropathologies. Neuropathologies found in 2 FTDP-17 families are interesting because they present an interesting contrast between the Seattle BK (or A) family [8, 26, 29, 46] with mutation $^V337^M$, and the multiple system tauopathy with presenile dementia (MSTD) family with the E10+3 mutation [7, 24, 47]. NFTs are observed in the BK family ($^V337^M$) in the neocortex, amygdala and parahippocampal gyrus but are sparse in the hippocampus itself [46]. NFTs are restricted to neurons and no glial cell tau pathology is found. In all respects studied to date, the abnormal tau is indistinguishable from that found in AD. The NFTs, as in AD, occupy the cell soma and apical dendrites. The filaments have a diameter of 11–20 μm and a periodicity of 80 nm [29]. The 4R/3R tau ratio in the NFT fraction is ≅ 1, which is the expected result because amino acid 337 is in E12 and is present in all isoforms [34]. The soluble 4R/3R tau ratio is also ≅ 1 as seen in AD and normal subjects [40]. In contrast, in the MSTD family, tau pathology is in the neocortex, hippocampus, substantia nigra, multiple brainstem regions and in the spinal cord [24, 28]. Tau deposits are in both neuronal and glial cells (primarily oligodendrocytes), as is seen in the pallido-ponto-nigral degeneration (PPND) [48], disinhibition-dementia-Parkinsonism-amyotrophy complex (DDPAC) [49] and other FTDP-17 families [24, 38, 50]. NFTs are in the cell soma but not in the dendrites, and are globose in shape [38, 50, 51]. In the MSTD family, the tau filaments have a diameter of 6–20 nm and a periodicity of 140–130 nm and are predominantly 4R tau [7, 24]. The difference between these 2 families may be because in the BK family ($^V337^M$), the mutation is present in all 6 isoforms while in the MSTD family (E10+3), the amino acid sequence is normal but the 4R/3R ratio is altered. Another possible explanation for the differences is that the E10+3 mutation selectively alters E10-splicing in specific brain regions and cell types but regulation in other brain regions is normal. Similarly, other E10-splicing regulatory mutations may yield different phenotypes due to differential regulation controlled by the different ESE and ESS control elements in E10.

Conclusions

TAU mutations cause FTDP-17 by multiple different mechanisms. Some missense mutations, a single deletion, and intronic mutations alter the regulation of *TAU* E10 splicing.

Three different *cis*-acting regulatory elements are affected by these regulatory mutations; an ESE element, an ESS element, and an inhibitory element spanning the 5' splice site of E10. Other E10 splicing regulatory elements remain to be characterized. Other *TAU* missense mutations affect the ability of tau to bind to MTs and to facilitate MT polymerization, thus altering the biochemical properties of the tau protein. FTDP-17 phenotypes, which vary depending on the mutation involved, presumably are caused by the multiple ways in which *TAU* mutations affect tau protein and gene function. The variability of the different phenotypes may be the result of cell-specific and brain region-specific regulatory patterns of tau as influenced by different mutations. Identification of the *trans*-acting protein factors that bind to these *cis* elements should provide information on the pathogenetic mechanism of *TAU* mutations.

References

1 Schmidt ML, Martin JA, Lee VMY, Trojanowski JQ (1996) Convergence of Lewy bodies and neurofibrillary tangles in amygdala neurons of Alzheimer's disease and Lewy body disorders. *Acta Neuropathol* 91: 475–481

2 Schmidt ML, Huang R, Martin JA, Henley J, Mawaldewan M, Hurtig HI, Lee VMY, Trojanowski JQ (1996) Neurofibrillary tangles in progressive supranuclear palsy contain the same tau epitopes identified in Alzheimer's disease PHF tau. *J Neuropathol Exp Neurol* 55: 534–539

3 Tagliavini F, Giaccone G, Prelli F, Verga L, Porro M, Trojanowski JQ, Farlow MR, Frangione B, Ghetti B, Bugiani O (1993) A68 is a component of paired helical filaments of GerstmannStraeusslerScheinker Disease, Indiana kindred. *Brain Res* 616: 325–328

4 Lieberman AP, Trojanowski JQ, Lee VMY, Balin BJ, Ding XS, Grossman M (1998) Cognitive, neuroimaging, and pathological studies in a patient with Pick's disease. *Ann Neurol* 43: 259–65

5 Mawal-Dewan M, Schmidt ML, Balin B, Perl DP, Lee VMY, Trojanowski JQ (1996) Identification of phosphorylation sites in PHF-TAU from patients with Guam amyotrophic lateral sclerosis/parkinsonism-dementia complex. *J Neuropathol Exp Neurol* 55: 1051–1059

6 Auer IA, Schmidt ML, Lee VMY, Curry B, Suzuki K, Shin RW, Pentchev PG, Carstea ED, Trojanowski JQ (1995) Paired helical filament tau (PHFtau) in Niemann-Pick type C disease is similar to PHFtau in Alzheimer's disease. *Acta Neuropathol* 90: 547–551

7 Spillantini MG, Murrell JR, Goedert M, Farlow MR, Klug A, Ghetti B (1998) Mutation in the tau gene in familial multiple system tauopathy with presenile dementia. *Proc Natl Acad Sci USA* 95: 7737–7741

8 Poorkaj P, Bird TD, Wijsman E, Nemens E, Garruto RM, Anderson L, Andreadis A, Wiederholt WC, Raskind M, Schellenberg GD (1998) Tau is a candidate gene for chromosome 17 frontotemporal dementia. *Ann Neurol* 43: 815–825

9 Hutton M, Lendon CL, Rizzu P, Baker M, Froelich S, Houlden H, Pickeringbrown S, Chakraverty S, Isaacs A, Grover A et al (1998) Association of missense and 5'-splice-site mutations in tau with the inherited dementia FTDP-17. *Nature* 393: 702–705

10 Caceres A, Kosik KS (1990) Inhibition or neurite polarity by tau antisense oligonucleotides in primary cerebellar neurons. *Nature* 343: 461–463

11 Caceres A, Potrebic S, Kosik KS (1991) The effect of tau antisense oligonucleotides on neurite formation of cultured cerebellar macroneurons. *J Neurosci* 11: 1515–1523

12 LoPresti P, Szuchet S, Papasozomenos SC, Zinkowski RP, Binder LI (1995) Functional implications for the microtubule-associated protein tau: localization in oligodendrocytes. *Proc Natl Acad Sci USA* 92: 10369–10373

13 Vanier MT, Neuville P, Michalik L, Launay JF (1998) Expression of specific tau exons in normal and tumoral pancreatic acinar cells. *J Cell Sci* 111: 1419–1432

14 Gu YJ, Oyama F, Ihara Y (1996) Tau is widely expressed in rat tissues. *J Neurochem* 67: 1235–1244

15 Himmler A (1989) Structure of the bovine tau gene: Alternatively spliced transcripts generate a protein family. *Mol Cell Biol* 9: 1389–1396

16 Himmler A, Drechsel D, Kirschner MW, Martin DW (1989) Tau consists of a set of proteins with repeat-

ed C-terminal microtubule-binding domains and variable N-terminal domains. *Mol Cell Biol* 9: 1381–1388

17 Goedert M, Spillantini MG, Rotier MC, Ulrich J, Crowther RA (1989) Cloning and sequencing of the cDNA encoding an isoform of microtubule-associated protein tau containing four tandem repeats: differential expression of tau protein mRNA's in human brain. *EMBO J* 8: 393–399

18 Goedert M, Spillantini MG, Jakes R, Rutherford D, Crowther RA (1989) Multiple isoforms of human microtubule-associated protein tau: sequences and localization in neurofibrillary tangles of Alzheimer's disease. *Neuron* 3: 519–526

19 Andreadis A, Brown WM, Kosik KS (1992) Structure and novel exons of the human-tau gene. *Biochemistry* 31: 10626–10633

20 Collet J, Fehrat L, Pollard H, Depouplana LR, Charton G, Bernard A, Moreau J, Benari Y, Khrestchatisky M (1997) Developmentally regulated alternative splicing of mRNAs encoding N-terminal tau variants in the rat hippocampus: Structural and functional implications. *Eur J Neurosci* 9: 2723–2733

21 Wijker M, Wszolek ZK, Wolters ECH, Rooimans MA, Pals G, Pfeiffer RF, Lynch T, Rodnitzky RL, Wilhelmsen KC, Arwert F (1996) Localization of the gene for rapidly progressive autosomal dominant parkinsonism and dementia with pallido-ponto-nigral degeneration to chromosome 17q21. *Hum Mol Genet* 5: 151–154

22 Wilhelmsen KC, Lynch T, Pavlou E, Higgens M, Nygaard TG (1994) Localization of disinhibition-dementia-Parkinsonism-Amyotrophy complex to 17q21-22. *Amer J Hum Genet* 55: 1159–1165

23 Yamaoka LH, Welshbohmer KA, Hulette CM, Gaskell PC, Murray M, Rimmler JL, Helms BR, Guerra M, Roses AD, Schmechel DE et al (1996) Linkage of frontotemporal dementia to chromosome 17: Clinical and neuropathological characterization of phenotype. *Amer J Hum Genet* 59: 1306–1312

24 Spillantini MG, Goedert M, Crowther RA, Murrell JR, Farlow MR, Ghetti B (1997) Familial multiple system tauopathy with presenile dementia: A disease with abundant neuronal and glial tau filaments. *Proc Natl Acad Sci USA* 94: 4113–4118

25 Heutink P, Stevens M, Rizzu P, Bakker E, Kros JM, Tibben A, Niermeijer MF, Vanduijn CM, Oostra BA, Vanswieten JC (1997) Hereditary frontotemporal dementia is linked to chromosome 17q21-q22: A genetic and clinicopathological study of three Dutch families. *Ann Neurol* 41: 150–159

26 Bird TD, Wijsman EM, Nochlin D, Leehey M, Sumi SM, Payami H, Poorkaj P, Nemens E, Raskind M, Schellenberg GD (1997) Chromosome 17 and hereditary dementia: Linkage studies in three non-Alzheimer families and kindreds with late-onset FAD. *Neurology* 48: 949–954

27 Foster NL, Wilhelmsen K, Sima AAF, Jones MZ, D'Amato CJ, Gilman S, Conference Participants (1997) Frontotemporal dementia and parkinsonsim linked to chromosome 17: A consensus conference. *Ann Neurol* 41: 706–715

28 Spillantini MG, Bird TD, Ghetti B (1998) Frontotemporal Dementia and Parkinsonism linked to chromosome 17: A new group of tauopathies. *Brain Pathol* 8: 387–402

29 Spillantini MG, Crowther RA, Goedert M (1996) Comparison of the neurofibrillary pathology in Alzheimer's disease and familial presenile dementia with tangles. *Acta Neuropathol* 92: 42–48

30 Reed LA, Grabowski TJ, Schmidt ML, Morris JC, Goate A, Solodkin A, Vanhoesen GW, Schelper RL, Talbot CJ, Wragg MA et al (1997) Autosomal dominant dementia with widespread neurofibrillary tangles. *Ann Neurol* 42: 564–572

31 Reed LA, Schmidt ML, Wszolek ZK, Balin BJ, Soontornniyomkij V, Lee VMY, Trojanowski JQ, Schelper RL (1998) The neuropathology of a chromosome 17-linked autosomal dominant parkinsonism and dementia ("pallido-ponto-nigral degeneration"). *J Neuropathol Exp Neurol* 57: 588–601

32 Froelich S, Basun H, Forsell C, Lilius L, Axelman K, Andreadis A, Lannfelt L (1997) Mapping of a disease locus for familial rapidly progressive frontotemporal dementia to chromosome 17q12-21. *Amer J Med Genet* 74: 380–385

33 Baker M, Kwok JBJ, Kucera S, Crook R, Farrer M, Houlden H, Isaacs A, Lincoln S, Onstead L, Hardy J et al (1997) Localization of frontotemporal dementia with parkinsonism in an Australian kindred to chromosome 17q21-22. *Ann Neurol* 42: 794–798

34 Kitamura Y, Shimohama S, Koike H, Kakimura J, Matsuoka Y, Nomura Y, Gebickehaerter PJ, Taniguchi T (1999) Increased expression of cyclooxygenases and peroxisome proliferator-activated receptor-gamma in Alzheimer's disease brains. *Biochem Biophys Res Commun* 254: 582–586

35 Rizzu P, Van Swieten JC, Joosse M, Hasegawa M, Stevens M, Tibben A, Niermeijer MF, Hillebrand M, Ravid R, Oostra BA et al (1999) High prevalence of mutations in the microtubule-associated protein tau in a population study of fronto-temporal dementia in the Netherlands. *Amer J Hum Genet* 64: 414–421

36 Dumanchin C, Camuzat A, Campion D, Verpillat P, Hannequin D, Dubois B, Saugierveber P, Martin C,

Penet C, Charbonnier F et al (1998) Segregation of a missense mutation in the microtubule-associated protein tau gene with familial frontotemporal dementia and parkinsonism. *Hum Mol Genet* 7: 1825–1829

37 D'Souza I, Poorkaj P, Hong M, Nochlin D, Lee VMY, Bird TD, Schellenberg GD (1999) Missense and silent tau gene mutations cause front temporal dementia with parkinsonism-chromosome 17 type by affecting multiple alternative RNA splicing regulatory elements. *Proc Natl Acad Sci USA* 96: 5598–5603

38 Yasuda M, Kawamata T, Komure O, Poorkaj P, Tanimukai S, Yamamoto Y, Hasegawa H, Sasahara M, Hazama F, Schellenberg GD et al. A mutation in the microtubule associated protein tau in a family with pallido-nigro-luysian degeneration. *Neurology* 53: 864–868

39 Varani L, Hasegawa M, Spillantini MG, Smith MJ, Murrell JR, Ghetti B, Klug A, Goedert M, Varani G (1999) Structure of tau exon 10 splicing regulatory element RNA and destabilization by mutations of frontotemporal dementia and parkinsonism linked to chromosome 17. *Proc Natl Acad Sci USA* 96: 8229–8234

40 Hong M, Zhukareva V, Vogelsberg-Ragaglia V, Wszolek Z, Reed L, Miller BI, Geschwind DH, Bird TD, Mckeel D, Goate A et al (1998) Mutation-specific functional impairments in distinct Tau isoforms of hereditary FTDP-17. *Science* 282: 1914–1917

41 Hasegawa M, Smith MJ, Goedert M (1998) Tau proteins with FTDP-17 mutations have a reduced ability to promote microtubule assembly. *FEBS Lett* 437: 207–210

42 Hasegawa M, Smith MJ, Iijima M, Tabira T, Goedert M (1999) FTDP-17 mutations N279K and S305N in tau produce increased splicing of exon 10. *FEBS Lett* 443: 93–96

43 Clark LN, Poorkaj P, Wszolek Z, Geschwind DH, Nasreddine ZS, Miller B, Li D, Payami H, Awert F, Markopoulou K et al (1998) Pathogenic implications of mutations in the tau gene in pallido-ponto-nigral degeneration and related neurodegenerative disorders linked to chromosome 17. *Proc Natl Acad Sci USA* 95: 13103–13107

44 Bueescherrer V, Hof PR, Buee L, Leveugle B, Vermersch P, Perl DP, Olanow CW, Delacourte A (1996) Hyperphosphorylated tau proteins differentiate corticobasal degeneration and Pick's disease. *Acta Neuropathol* 91: 351–359

45 Si Z-H, Rauch D, Stoltzus M (1998) The exon splicing silencer in human immunodificiency virus type 1 tat exon 3 is bipartite and acts early in splicesome assembly. *Mol Cell Biol* 18: 5404–5413

46 Sumi SM, Bird TD, Nochlin D, Raskind MA (1992) Familial presenile dementia with psychosis associated with cortical neurofibrillary tangles and neurodegeneration of the amygdala. *Neurology* 42: 120–127

47 Murrell JR, Koller D, Foroud T, Goedert M, Spillantini MG, Edenberg HJ, Farlow MR, Ghetti B (1997) Familial multiple-system tauopathy with presenile dementia is localized to chromosome 17. *Amer J Hum Genet* 61: 1131–1138

48 Yamada T, McGeer EG, Schelper RL, Wszolek ZK, McGeer PL, Pfeiffer RF, Rodnitzky RL (1993) Histological and biochemical pathology in a family with autosomal dominant parkinsonism and dementia. *Neurol Psychiatr Brain Res* 2: 26–35

49 Sima AAF, Defendini R, Keohane C, D'Amato C, Foster NL, Parchi P, Gambetti P, Lynch T, Wilhelmsen KC (1996) The neuropathology of chromosome 17-linked dementia. *Ann Neurol* 39: 734–743

50 Petersen RB, Tabaton M, Chen SG, Monari L, Richardson SL, Lynches T, Manetto V, Lanska DJ, Markesbery WR, Currier RD et al (1995) Familial progressive subcortical gliosis: Presence of prions and linkage to chromosome 17. *Neurology* 45: 1062–1067

51 Spillantini MG, Roses AD, Yamaoka LH, Gaskell PC, Welsh-Bohmer KA, Pericak-Vance MA, Hulette CM (1997) Neuropathological features of frontotemporal dementia and parkinsonism linked to chromosome 17q21-22 (FTDP-17): Duke family 1684. *Brain Pathol* 7: 1149

Neuroscientific Basis of Dementia
C. Tanaka, P.L. McGeer, Y. Ihara (eds)
© 2001 Birkhäuser Verlag Basel/Switzerland

Tau mutations altering splicing of tau exon 10 in japanese frontotemporal dementia

Minoru Yasuda[1], Junichi Takamatsu[2], Osamu Komure[3], Sadako Kuno[3], Ian D'Souza[4], Toshio Kawamata[1], Masato Hasegawa[6], Takeshi Iwatubo[5], Parvoneh Poorkaj[4], Michel Goedert[6], Gerard D. Schellenberg[4] and Chikako Tanaka[1]

[1] *Hyogo Institute for Aging Brain and Cognitive Disorders, Himeji, Japan*
[2] *Division of Clinical Research, Kikuchi National Hospital, Kumamoto, Japan*
[3] *Department of Neurology, Utano National Hospital, Kyoto, Japan*
[4] *Geriatric Research Education and Clinical Center, Veterans Affairs Puget Sound Health Care System, Seattle, WA, USA*
[5] *Graduate School of Pharmaceutical Sciences, University of Tokyo, Tokyo, Japan*
[6] *MRC Laboratory of Molecular Biology, Cambridge, UK*

Introduction

Recent studies have shown that mutations in the tau gene cause familial frontotempotal dementia and parkinsonism linked to chromosome 17 (FTDP-17) [1–5]. Known tau mutations are either intronic mutations located close to the splice-donor site of the intron following exon 10 or missense, deletion or silent mutations in the coding region. Here we describe two pathogenic mutations in the tau gene of Japanese patients with frontotemporal dementia and present the ensuing biochemical and morphological abnormalities.

A Japanese case with pallido-nigro-luysian degeneration (PNLD) was a member of a family with parkinsonism and dementia [6]. The kindred family, called frontotemporal dementia-Kumamoto (FTD-K), also has multiple patients with parkinsonism and dementia [7]. The characteristics of clinical and neuropathological findings in patients with these mutations are summarized in Table 1. FTD-K seems to be clinically and pathologically typical of FTDP-17. On the other hand, the neuropathology of PNLD was distinct from the dominant pathologies of other FTDP-17 cases reported, because PNLD exhibited prominent changes in the motor cortex, motor nuclei of the brainstem, amygdala and entorhinal cortex with lesser alterations in the frontal and temporal cortices.

Genomic DNA was extracted from the temporal cortex of the brains. Tau exons and flanking intronic sequences were amplified and sequenced on both strands, as described [5].

Total RNA was extracted from 200 mg of the proband's frontal cortex and the frontal cortex of 10 control individuals. Tau exons 9–11 and 9–13 were amplified, as described [1]. After the optimal condition was determined, the resulting bands corresponding to exon 10-containing or -lacking tau were quantified. Exon trapping analysis was performed as described [3].

The FTD-K mutation is located at position +12 of the intron following exon 10 of the tau gene. It is located within a stem-loop structure that is believed to be important for the regu-

Table 1. Clinicopathological features of FTD-Kumamoto and PNLD pedigrees

Pedigree	FTD-Kumamoto [7, 12]	PNLD [5, 6]
Mutation	exon 10 splice donor +12 site	N279K
Age of onset	Mean 53 years	35–41 years
Duration of illness	Mean 7 years	6–8 years
Initial symptoms	Inactiveness, amnesia	Parkinsonism
Other symptoms	Frontal lobe dementia aphasia, disinhibition aggressive reaction rigidity, increased tendon reflex, forced grasping	Dementia aggressiveness, vertical gaze palsy, nuchal stiffness blepharospasm, pyramidal signs pallilalia
Main affected regions	Frontal and temporal lobes	Globus pallidus, luysian body, substantia nigra
Other affected regions	Insular cortex	Frontal and temporal lobes
Main pathologic findings	Neuronal loss and gliosis	Neuronal loss and gliosis
	Ballooned neurons	Ballooned neurons
Abnormal inclusion bodies	Abnormally phosphorylated tau-containing inclusions	Abnormally phosphorylated tau-containing inclusions

lation of alternative mRNA splicing of exon 10 [2–5]. Mutations at positions −1, +3, +13, +14 and +16 of the stem-loop structure have been found in FTDP-17 families [2, 3]. The clustering of mutations in the upper part of the stem strongly suggests that this region is crucial for the quantitative regulation of the splicing-in of exon 10 [8, 9].

The mutation in PNLD is exon 10 of tau that causes a substitution at codon 279 (N279K). However, this mutation may not act at the protein level but rather may cause disease by altering splicing of tau exon 10. At the DNA sequence level, the 279 mutation occurs in a potential exon-splicing enhancer sequence which controls alternative splicing at exon 10. The 279 mutation is a T to G change that creates a GAR repeat motif (where R is a purine) which promotes incorporation of alternative spliced exons.

By exon trapping, the +12 mutation and the N279K mutation both led to a marked increase in the proportion of exon 10-containing transcripts, when compared with the wild-type construct. Similar results have been obtained for other mutations that increase splicing-in of exon 10 [3–5, 8, 9].

By RT-PCR, a large increase in the proportion of exon 10-containing transcripts was found in FTD-K and PNLD brains, whereas in controls, tau mRNA without exon 10 was more abundant. Immunoblotting of sarkosyl-insoluble tau extracted from FTD-K brains was performed. They were probed with the phosphorylation-independent anti-tau serum BR133 (1:3,000) and the phosphorylation-dependent anti-tau antibodies AT8, M4, C5 and AP422 and detected using avidin-biotin-peroxidase (Vector). The immunoblot analyses revealed that the tau protein isoforms of the affected brains were mainly 4-repeat tau isoforms in FTD-K brains. Other studies also reported the excess soluble four-repeat tau in brains from subjects with the 279 mutation. Similar findings have been reported for the +3, +14 and +16 mutations [3, 10]. Overproduction of four-repeat tau is the primary effect of the known mutations in the intron following exon [11]. It is believed to lead to the assembly of four-repeat tau into abnormal filaments in both nerve cells and glia cells. As expected, sarkosyl-insoluble tau from FTD-K brains contained wide twisted ribbons made of hyperphosphorylated tau isoforms with four repeats [12]. Wide twisted ribbons made of four-repeat tau have also been described in families with the +3 and +16 intronic mutations [3, 13]. The tau pathology of FTD-K is both neuronal and glial [8], as is that in other families with mutations in the intron after tau exon 10 [11]. These biochemical data indicate that both mutations may cause disease by altering splicing of tau exon 10 and resulting in overproduction of 4-repeat tau isoforms.

Conclusions

In summary, we found two pathogenic mutations in the tau gene of Japanese patients with frontotemporal dementia. One was a missense mutation that causes a substitution at codon 279 (N279K), and the other was a C-to-T transition at the exon 10 splice donor +12 site of the tau gene. An abnormal preponderance of exon 10 containing tau transcripts and the overproduction of tau protein isoforms with 4 microtubule-binding repeats were identified in both patients with the 279 mutation and the +12 mutation. Immunostaining revealed inclusions containing abnormally phosphorylated tau in degenerating neurons and glial cells. Isolated tau filaments had a twisted ribbon-like morphology in both mutations. A major implication that derives from the present study is that an overproduction of four-repeat tau is sufficient to lead to neurodegeneration and to cause dementia. The mechanism by which overproduction of four–repeat tau leads to neurodegeneration remains to be clarified.

References

1 Poorkaj P, Bird TD, Wijsman E, Nemens E, Garruto RM, Anderson L, Andreadis A, Wiederholt WC, Raskind M, Schellenberg GD (1998) Tau is a candidate gene for chromosome 17 frontotemporal dementia. *Ann Neurol* 43: 815–825

2 Hutton M, Lendon CL, Rizzu P, Baker M, Froelich S, Houlden H, Pickering-Brown S, Chakraverty S, Isaacs A, Grover A et al (1998) Association of missense and 5'-splice-site mutations in *tau* with inherited dementia FTDP-17. *Nature* 393: 702–705

3 D'Souza I, Poorkaj P, Hong M, Nochlin D, Lee VMY, Bird TD, Schellenberg GD (1999) Missense and silent tau gene mutations cause frontotemporal dementia with parkinsonism-chromosome 17 type, by affecting multiple alternative RNA splicing regulatory elements. *Proc Natl Acad Sci USA* 96: 5598–5603

4 Delisle MB, Murrell JR, Richardson R, Trofatter JA, Rascol O, Soulages X, Mohr M, Calvas P, Ghetti B (1999) A mutation at codon 279 (N279K) in exon 10 of the Tau gene causes a tauopathy with dementia and supranuclear palsy. *Acta Neuropathol* 98: 62–77

5 Yasuda M, Kawamata T, Komure O, Kuno S, D'Souza I, Poorkaj P, Kawai J, Tanimukai S, Yamamoto Y, Hasegawa H et al (1999) A mutation in the microtubule-associated protein tau in pallido-nigro-luysian degeneration. *Neurology* 53: 864–868

6 Kawai J, Sasahara M, Hazama F, Kuno S, Komure O, Nomura S, Yamaguchi M (1993) Pallidonigroluysian degeneration with iron deposition: a study of three autopsy cases. *Acta Neuropathol* 86: 609–616

7 Takamatsu J, Kondo A, Ikegami K, Kimura T, Fujii H, Mitsuyama Y, Hashizume Y (1998) Selective expression of Ser199/202 phosphorylated tau in a case of frontotemporal dementia. *Dement Geriatr Cogn Disord* 9: 82–89

8 Grover A, Houlden H, Baker M, Adamson J, Lewis J, Prihar G, Pickering-Brown S, Duff K, Hutton M (1999) 5' Splice site mutations in tau associated with the inherited dementia FTDP-17 affect a stem-loop structure that regulates alternative splicing of exon 10. *J Biol Chem* 274: 15134–15143

9 Varani L, Hasegawa M, Spillantini MG, Smith MJ, Murrell JR, Ghetti B, Klug A, Goedert M, Varani G (1999) Structure of tau exon 10 splicing regulatory element RNA and destabilization by mutations of frontotemporal dementia and parkinsonism linked to chromosome 17. *Proc Natl Acad Sci USA* 96: 8229–8234

10 Hong M, Zhukareva V, Vogelsberg-Ragaglia V, Wszolek Z, Reed L, Miller BI, Geschwind DH, Bird TD, McKeel D, Goate A et al (1998) Mutation-specific functional impairments in distinct tau isoforms of hereditary FTDP-17. *Science* 282: 1914–1917

11 Goedert M, Crowther RA, Spillantini MG (1998) Tau mutations cause frontotemporal dementias. *Neuron* 21: 955–958

12 Yasuda M, Takamatsu J, D'Souza I, Crowther RA, Kawamata T, Hasegawa M, Hasegawa H, Spillantini MG, Tanimukai S, Poorkaj P et al (2000) A novel mutation at position +12 in the intron following exon 10 of the tau gene in familial frontotemporal dementia (FTD-Kumamoto). *Ann Neurol* 47: 422–429

13 Spillantini MG, Goedert M, Crowther RA, Murrell JR, Farlow MR, Ghetti B (1997) Familial multiple system tauopathy with presenile dementia: A disease with abundant neuronal and glial tau filaments. *Proc Natl Acad Sci USA* 94: 4113–4118

Neuroscientific Basis of Dementia
C. Tanaka, P.L. McGeer, Y. Ihara (eds)
© 2001 Birkhäuser Verlag Basel/Switzerland

Amyotrophic lateral sclerosis/parkinsonism-dementia complex of the Kii peninsula of Japan (Kii ALS/PDC) may be a familial tauopathy. Epidemiological trends, clinical features, neuropathology and molecular genetics

Shigeki Kuzuhara, Ryogen Sasaki, Yasumasa Kokubo and Yugo Narita

Department of Neurology Mie University School of Medicine, Tsu, Japan

Introduction

The Kii peninsula of Japan and the island of Guam in the western Pacific are two major foci of very high prevalence rates of amyotrophic lateral sclerosis (ALS) in the world [1, 2]. ALS accumulated in some villages on Guam and in the Kii peninsula at rates 50–100 times of those in the other areas of the United States or Japan. Another peculiar endemic disease parkinsonism-dementia complex (PDC), also accumulates at very high prevalence rates in the same areas. PDC is characterized by progressive parkinsonism and dementia clinically and by loss of nerve cells and marked neurofibrillary degeneration without accompanying senile plaques neuropathologically [3]. Many neurofibrillary tangles (NFTs) are seen in the temporal lobe cortex, substantia nigra, and brainstem nuclei, and a few in the cerebellum and spinal cord in ALS and PDC on Guam [4] and ALS in the Kii peninsula foci [5]. ALS and PDC in the western Pacific have thus been regarded as being two different clinical phenotypes of a nosologically single disease.

Intensive investigations to clarify the causes and pathogenic mechanism of the accumulation of ALS/PDC were done on genetics, infections, toxins, nutrition and deficiencies [1]. The environmental factors, especially deficiencies of such minerals as calcium and magnesium and excessive intake of heavy metals such as manganese and aluminum were suspected as the cause of the accumulation of this peculiar neurodegenerative disease, [6] but this hypothesis was not proved clinically and experimentally. In the meantime, it was reported that the incidence rates of ALS/PDC had markedly declined during 1970, not only on Guam [7] but also in the Kii peninsula [8]. The cause of the marked decline remained unknown; a hypothesis that changes in environmental factors, especially food and water, might have caused the reduction of ALS/PDC was proposed [7], although it was not substantiated.

In this review, we show the continuously high rates of prevalence and incidence of Kii ALS/PDC, high rates of familial occurrence, and accumulation of tau proteins in the central nervous system neuropathologically, suggesting that Kii ALS/PDC may be a familial tauopathy.

Geography

The mountainous areas of the southern coast of the Kii peninsula of Japan have been called the Muro district; here a high prevalence of ALS was known to exist. Two villages in this district, Hohara and Kozagawa, were reported to have extremely high incidence rates, 50–100 times of those in the other areas of Japan [2].

Epidemiological studies

In 1993, several years after the report of disappearance of new cases of ALS from Hohara and Kozagawa, one of the authors (SK) saw a new patient with ALS from Hohara, and started the epidemiological study to clarify if Kii ALS was really gone. During several years, the authors [9] found continuing high incidence rates and frequent familial occurrence of ALS/PDC in Hohara. In order to investigate the accurate prevalence rate, we conducted a house-to-house survey during 1997 and 1998 in Hohara, which had approximately 400 homes and a population of nearly 1,500 in 1997.

Prevalence rates of ALS and PDC

Table 1 shows the findings of the authors' survey [10] and those of the survey of Yase and his colleagues [2] done in 1969 in Hohara. The present survey disclosed clinically 2 patients with ALS alone, 6 patients with PDC plus some features of motor neuron disease (MND) and 5 patients with PDC alone, while the survey of Yase and his colleagues [2] had disclosed 4 patients with ALS and 4 patients with PDC including some with MND signs. Six patients out of a population of 2,059 in 1969 had clinical features of ALS, including ALS alone and PDC plus MND signs; the prevalence rate of ALS and ALS plus was 291 per 100,000 population. In 1997, eight patients out of 1,476 population had clinical features of ALS, and the

Table 1. The number and prevalence rates of ALS and PDC in the Hohara village during 1969–1970 and during 1997–1998

Population	1969[*] 2,059		1997 1,476	
Patients	Number	Prevalence rate per 100,000[**]	Number	Prevalence rate per 100,000[**]
ALS	4	194	2	136
PDC plus MND signs[***]	2	97	6	406
PDC	2	97	5	339
ALS/PDC total	8	389	13	881

[*] Data during 1969 are from Shiraki and Yase [4]; [**] not age-adjusted; [***] MND signs include muscle atrophy of the hand, tongue or limbs, hyperactive deep tendon reflexes and Babinski sign.
ALS: amyotrophic lateral sclerosis; PDC: parkinsonism-dementia complex; MND: motor neuron disease.

prevalence rate of ALS and ALS plus was 542 per 100,000 population. The prevalence rate of ALS and ALS plus as well as that of ALS/PDC total seems to have doubled in 1997 in comparison with those in 1969.

Annual incidence rates of ALS

The annual incidence of patients with ALS in Hohara had been investigated by Yase and his colleagues [11] during 1960 and 1984; the authors [10] added the data on new patients during 1985 and 1997. The 5-year average annual incidence rates of ALS per 100,000 population was 14–49 during 1960 and 1984, while it was 12–87 during 1985 and 1997.

Higher rates of incidence and prevalence in recent years probably reflect that our survey was done more thoroughly than the previous ones. The high incidence rates of ALS in the Hohara focus thus remain unchanged since 1960, despite the marked westernization of food and water supply to the residents of the village. These findings seem to stand against the hypothesis [7] that changes in dietary habits and local water supplies, and much less dependence on locally grown food as a consequence of increased westernization of life caused a dramatic decline in the high incidence rates of ALS/PDC of Guam and Kii peninsula.

Clinical features

The authors [10] found 21 patients with ALS/PDC in whom the onset of the disease was between 1985 and 1998. Out of them, 16 were natives of Hohara, while two were women who had come into Hohara from the neighboring villages when they had gotten married. Four patients were from those who had moved out of Hohara in their late teens, and their disease had developed 20 to 60 years later. The sex ratio showed female dominancy.

The patients were grouped into three major clinical subtypes. Seven of them showed MND features alone from the onset to the terminal stage of the disease, and a diagnosis of ALS of either classical type, progressive bulbar type, or spinal muscular atrophy type was made. The second group included 7 patients in whom either parkinsonism or dementia was noted at the onset and both parkinsonism and dementia were manifested within a few years, and a diagnosis of PDC was made. The third group included 7 patients in whom dementia with or without parkinsonism occurred first, followed by MND signs such as muscular atrophy of hands and limbs, bulbar signs, fasciculations, spasticity, ankle clonus, hyperactive deep tendon reflexes, positive Babinski sign and forced crying or laughing, and a diagnosis of "PDC plus ALS" was made.

Family history and genetics

A family history of either ALS or PDC, or both, in the close relatives was confirmed in 15 of the 21 patients (71%). A similar high frequency of familial occurrence of ALS was reported in Hohara by the previous studies in 1969 [2]. ALS and PDC frequently affected members of the same family. These findings suggest that ALS and PDC are genetically a single disease which may be transmitted in either autosomal dominant fashion with low penetration or autosomal recessive fashion which occurs on the background of a high frequency of gene carriers in consequence of consanguineous marriage in a small community.

Brain-imaging

Computed tomography (CT) or magnetic resonance imaging (MRI) of the brain of patients with PDC or ALS with dementia showed frontal and temporal lobe-dominant atrophy. In contrast, there was no remarkable atrophy in patients with ALS features alone. There was a marked decrease in regional cerebral blood flow in the frontotemporal areas by N-iso-propyl-p-[^{123}I]-iodoamphetamine (IMP) single photon emission computed tomography (SPECT) done in some of the patients.

Neuropathological findings

Approximately 10 autopsy cases of Kii ALS, mostly from the Kozagawa focus or its vicinities, had been reported, [5] [12] but no neuropathologically verified case of Kii PDC had been reported until the authors [13] first reported the neuropathological findings of Kii PDC. The authors [10] have so far examined neuropathologically 6 cases of ALS/PDC from the

Table 2 Neuropathological findings of 6 autopsy cases (reproduced from Kuzuhara [10])

Case number	1	2	3	4	5	6
Age of death (years)	66	49	63	66	77	70
Sex	F	M	M	F	M	F
Clinical Diagnosis	ALS	ALS	PDC	PDC +ALS	PDC +ALS	PDC
Neuropathologic findings						
ALS changes						
Lower motor neuron	+++	+++	N.A.	+	++	++
Spinal pyramidal tract	+++	++	N.A.	N.A.	+	+++
Atrophy and NFTs						
Temporal cortex	+	++	+++	+++	+++	+++
Substantia nigra	+	+++	+++	N.A.	++	+++
Brainstem nuclei	+	++	+++	+++	+++	+++
Senile plaques	–	–	++	–	–	–
Lewy bodies	–	–	++	+	–	–

F: female; M: male; ALS: amyotrophic lateral sclerosis; PDC: parkinsonism-dementia complex; PDC+ALS: PDC plus motor neuron disease features; N.A.: Preparations are not available.
Severity of lesions: none (–), slight (+), moderate (++), severe (+++)
Cases 2 and 6, and cases 3 and 4 are siblings of unrelated families

Hohara focus as shown in Table 2. The neuropathological findings consisted of two major changes: first, marked Alzheimer neurofibrillary degeneration with marked loss of nerve cells and presence of many NFTs, and second, classical ALS pathology consisting of upper and lower motor neuron involvement. NFTs were seen most abundantly in the temporal lobe cortex (Fig. 1a), limbic system, substantia nigra (Fig. 1b) and brainstem nuclei, a few were seen in the dentate nucleus of the cerebellum and the gray matter of the spinal cord. The neuropathological findings were common to all the 6 cases in spite of the different clinical manifestations of individual cases. The clinical manifestation was different even in the members of the same family. Case 1 (ALS) and case 6 (PDC) were siblings. Case 4 (ALS and PDC simultaneously) and case 5 (PDC followed by ALS) were also siblings, and case 3 was one of their cousins. These findings seem to support the hypothesis that ALS and PDC of the high-incidence focus of the Kii peninsula are different phenotypes of, nosologically, a single disease entity.

As we report here, brains of PDC, with or without features of MND, were affected by many NFTs and nerve cell loss without many senile plaques. These findings are very similar to those of PDC on Guam, especially PDC with features of MND [3, 4]. These similarities between Kii ALS/PDC and Guamanian ALS/PDC imply that the two conditions may belong to the same disease entity.

Tau protein analysis

Tau protein is composed of 6 isoforms which are controlled by alternative splicing of the tau gene on chromosome 17 [14]. The insoluble PHF-tau protein extracted from the fresh brains of two autopsy cases of Kii ALS/PDC were analyzed by western blotting and immunostained with AT8, the phosphorylation-dependent monoclonal anti-tau antibodies. Three bands of 60, 64, and 68 kD were revealed, which resembled those of Alzheimer's disease. Immunoblots of dephosphorylated PHF-tau with BR 133, the antibodies which recognize the amino terminus of all tau isoforms, revealed 6 isoforms of tau protein in almost equal quantity (unpublished observations). The immunoblotting pattern was similar to that of the normal brain, and none of the isoforms showed abnormal increase or decrease. Similar findings were reported in Guamanian ALS/PDC also [15].

Molecular genetics

The high frequency of familial occurrence of Kii ALS/PDC and the manner of transmission of the disease in the family pedigrees suggest a possibility that Kii ALS/PDC may be a heredodegenerative disorder related to or produced by abnormal gene(s). Therefore, the authors performed gene analyses, using the DNA extracted from the peripheral blood cells. Some of the known candidate genes related to ALS, Alzheimer's disease and frontotemporal dementia were investigated in 12 cases of Kii ALS/PDC from Hohara.

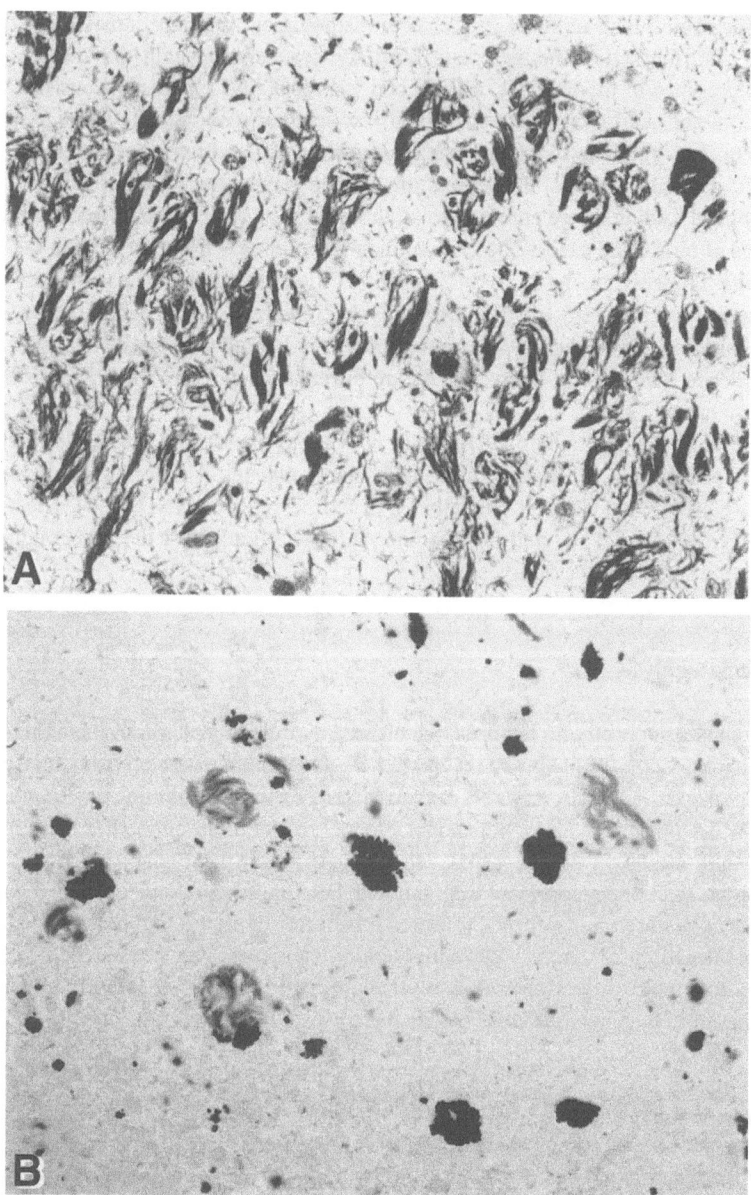

Figure 1. Neurofibrillar degeneration. (A) Hippocampus. There are many NFTs with loss of nerve cells (× 300, Bielschowsky stain). (B) Substantia nigra. The pigmented nerve cells are almost totally lost with free neuromelanin, and there is a moderate number of NFTs (× 480, Gallyas stain).

Copper-zinc superoxide dismutase (SOD1) gene

Abnormalities of the SOD1 gene on chromosome 21 are the cause of 15–20% of the autosomal-dominant familial ALS cases [16] However, none of the Kii ALS/PDC patients had mutations of the SOD1 gene. (Sasaki et al., in preparation)

Apolipoprotein E (apoE) polymorphism

Apo E4 is one of the genetic risk factors of sporadic Alzheimer's disease. The frequency of the apo E4 gene was not significantly high in the Kii ALS/PDC patients.

Tau gene

Tau protein is one of the microtubule-associated proteins, and has important roles in the cytoskeletons of the nerve cells and axonal transport in neurons. In recent years, mutations of the gene on chromosome 17 were found to cause neuronal degeneration with tau accumulation in some families of autosomal dominant form of frontotemporal dementia with parkinsonism linked to chromosome 17 (FTDP-17) and some of the familial tauopathies [17]. None of the present Kii ALS/PDC patients had mutations of the tau gene (Sasaki et al., in preparation).

The gene analysis study by the authors for abnormalities or polymorphism which may be related to Kii ALS/PDC either primarily or secondarily has so far failed to disclose candidate genes. Similar studies done on ALS/PDC on Guam also failed to reveal any abnormalities of the SOD1 gene, [18] apoE polymorphism [19] and tau gene [15, 17]. However, McGeer and his colleagues [20] reported the familial nature and continuing morbidity of ALS/PDC on Guam, and implied the existence of the role of genetic factors on the pathogenesis of Guamanian ALS/PDC. More investigations are necessary to find gene abnormalities in ALS/PDC of Kii and Guam.

Conclusion

We reported the clinical, neuropathological and epidemiological features of the ALS/PDC patients from Hohara, one of the high-incidence ALS foci of the Kii peninsula of Japan. We confirmed the neuropathological findings of Kii PDC, which was characterized by abundant NFTs widely in the central nervous system without a marked presence of senile plaques. More than 70% of the patients had a positive family history of ALS/PDC, and the rates of incidence and prevalence remain continuously very high, despite the dramatic changes in environmental factors, including food and water supply, by westernization of daily life. Kii ALS/PDC may thus be included in the familial tauopathies without amyloid accumulation. These findings indicate that Kii ALS/PDC may be a condition influenced by genetic factors rather than environmental factors, although abnormalities of the known candidate genes such as the SOD1 gene mutation, apoE4 polymorphism and tau gene mutation were not found in the affected patients.

Acknowledgments
This study was partly supported by a Grant-in-Aid from the Research Committee of CNS Degenerative Diseases, the Ministry of Welfare, Japan, and by a Grant-in-Aid for the Scientific Research from the Ministry of Education, Science, Sports, and Culture, Japan.
The tau protein analysis was done by Masato Hasegawa, PhD and Takeshi Iwatsubo, MD, PhD, Department of Neuropathology and Neuroscience, Graduate School of Pharmaceutical Sciences, University of Tokyo. For his invaluable comments and advice we thank Yasuo Ihara, MD, PhD, Department of Neuropathology, Graduate School of Medicine, University of Tokyo, and Shoji Tsuji, MD, PhD, Institute of Brain Research, Niigata University. We thank Ms. Asami Akatsuka for her technical assistance in neuropathological studies and Ms. Kayoko Kihira for her clerical assistance.

References

1 Garruto RM, Yanagihara R (1991) Amyotrophic lateral sclerosis in the Mariana Islands. *In*: PJ Vinken, GW Bruyn, HL Klawans (eds): *Handbook of clinical neurology. vol 15 (59), Diseases of the motor system.* North Holland Publishing, Amsterdam, 253–271

2 Shiraki H, Yase Y (1975) Amyotrophic lateral sclerosis in Japan. *In*: PJ Vinken, GW Bruyn (eds): *Handbook of clinical neurology, vol 22, System disorders and atrophies, Part II.* North Holland Publishing, Amsterdam, 353–419

3 Hirano A, Malamud N, Kurland LT (1961) Parkinsonism-dementia complex, an endemic disease on the island of Guam. II. Pathological features. *Brain* 84: 662–679

4 Hirano A, Malamud N, Elizan TS, Bethesda Kurland LT (1966) Amyotrophic lateral sclerosis and parkinsonism-dementia complex on Guam. Further pathologic studies. *Arch Neurol* 15: 35–51

5 Yase Y, Matsumoto N, Azuma K, Nakai Y, Shiraki H (1972) Amyotrophic lateral sclerosis. Association with schizophrenic symptoms and showing Alzheimer's tangles. *Arch Neurol* 27: 118–128

6 Yase Y (1972) The pathogenesis of amyotrophic lateral sclerosis. *Lancet* 2: 292–296

7 Garruto RM, Yanagihara R, Gajdusek DC (1985) Disappearance of high-incidence amyotrophic lateral sclerosis and parkinsonism-dementia on Guam. *Neurology* 35: 193–198

8 Yoshida S (1991) Environmental factors in western Pacific foci of ALS and possible pathogenetic role of aluminum (Al) in motor neuron degeneration. *Clin Neurol (Tokyo)* 31: 1310–1312

9 Kuzuhara S, Kokubo Y, Narita Sasaki R (1998) Continuing high incidence rates and frequent familial occurrence of amyotrophic lateral sclerosis and parkinsonism-dementia complex of the Kii peninsula of Japan (abs). *Neurology* 50:A173

10 Kuzuhara S (1999) Familial ALS-parkinsonism-dementia in the Kii peninsula. *Neurol Med (Tokyo)* 50: 137–145

11 Uebayashi Y (1988) Amyotrophic lateral sclerosis in the western Pacific: a new aspect for the progressive disease process based on the changing epidemiological pattern. *In*: T Tsubaki, Y Yase (eds): *Amyotrophic lateral sclerosis.* Elsevier Science Publishers, Amsterdam, 17–23

12 Kihira T, Yoshida S, Mitani K, Yasui M, Yase Y (1993) ALS in the Kii peninsula of Japan, with special reference to neurofibrillary tangles and aluminum. *Neuropathology* 13: 125–136

13 Kuzuhara S, Kokubo Y, Narita Y (1996) Parkinsonism-dementia complex – An endemic neurodegenerative dementia syndrome in Guam island, west New Guinea and Kii peninsula. *Dementia (Tokyo)* 10: 378–390

14 Goedert M, Spillantini MG, Cairns NJ, Crowther RA (1992) Tau proteins of Alzheimer paired helical filaments: abnormal phosphorylation of all six brain isforms. *Neuron* 8: 159–168

15 Perz-Tur J, Buee L, Morris HR, Waring SC, Onstead L, Wavrant-De Vrieze Crook R, Buee-Scherrer V, Hof PR, Petersen RC et al (1999) Neurodegenerative diseases of Guam: analysis of TAU. *Neurology* 53: 411–413

16 Rosen DR, Siddique T, Patterson D, Figlewicz DA, Sapp P, Hentati A, Donaldson D, Goto J, O'Regan JP, Deng H-X et al (1993) Mutations in Cu/Zn superoxide dismutase gene are associated with familial amyotrophic lateral sclerosis. *Nature* 362: 59–62

17 Poorkaj P, Bird TD, Wijsman E, Nemens E, Garruto RM, Anderson M, Andreadis A, Wiederholt WC, Raskind M, Schellenberg GD (1998) Tau is a candidate gene for chromosome 17 frontotemporal dementia. *Ann Neurol* 43: 815–825
18 Figlewicz DA, Garruto RM, Krizus A, Yanagihara R, Rouleau GA (1994) The Cu/Zn superoxide dismutase gene in ALS and parkinsonism-dementia of Guam. *NeuroReport* 5: 557–560
19 Chen X, Xia Y, Gresham LS, Molgaard CA, Thomas RG, Galasko D, Wiederholt WC, Saito T (1996) ApoE and CYP2D6 polymorphism with and without parkinsonism-dementia complex in the people of Chamorro, Guam. *Neurology* 47: 779–784
20 McGeer PL, Schwab C, McGeer EG, Haddock RL, Steele JC (1997) Familial nature and continuing morbidity of the amyotrophic lateral sclerosis-parkinsonism dementia complex of Guam. *Neurology* 49: 400–409

Neuroscientific Basis of Dementia
C. Tanaka, P.L. McGeer, Y. Ihara (eds)
© 2001 Birkhäuser Verlag Basel/Switzerland

Senile dementia of the neurofibrillary tangle type (SD-NFT): a clinical, neuropathological and molecular genetic study

Masahito Yamada[1, 2], Yoshinori Itoh[3], Nobuyuki Sodeyama[2], Naomi Suematsu[4], Eiichi Otomo[3], Masaaki Matsushita[5] and Hirdehiro Mizusawa[2]

[1] *Department of Neurology, Kanazawa University School of Medicine, Kanazawa, Japan*
[2] *Department of Neurology, Tokyo Medical and Dental University, Tokyo, Japan*
Departments of [3] *Internal Medicine and* [4] *Pathology, Yokufukai Geriatric Hospital, Tokyo, Japan*
[5] *Department of Neuropathology, Tokyo Metropolitan Institute of Psychiatry, Tokyo, Japan*

Introduction

A subset of senile dementia is characterized by numerous neurofibrillary tangles (NFTs) in the hippocampal region, similar to Alzheimer's disease (AD), but with absence or scarcity of senile plaques (SPs) throughout the brain [1]. These cases have been reported as an atypical or NFT-predominant form of AD [2]. We studied demented elderly patients with such neuropathological features [3, 4], compared them with age-matched AD patients in clinicopathological and genetic aspects, and investigated whether the senile dementia characterized by NFTs was just a subtype of AD or an independent disease entity [4]. Our results indicated that this neuropathological condition represents a disease entity with a pathogenetic process different from that of AD [4]. Therefore, we have proposed the term, "senile dementia of the NFT type (SD-NFT)", which is distinguished from late-onset AD, i.e., senile dementia of the Alzheimer type [4]. Thereafter, the identical neuropathological condition has been recently reported under terms such as NFT-dominant form of senile dementia [5], senile dementia with tangles [6], and limbic neurofibrillary tangle dementia [7].

We report here the incidence, clinical and neuropathological features, and genetic aspects of SD-NFT in comparison with AD.

Incidence of SD-NFT

We performed neuropathological examinations of the 105 patients who developed dementia at age 70 or later in an autopsy series of elderly Japanese at Yokufukai Geriatric Hospital, Tokyo.

The causes of dementia were AD in 57 patients (54.3%), vascular dementia in 31 (29.5%), SD-NFT in 5 (4.8%), dementia with Lewy bodies in 5 (4.8%) and others [4].

Clinical and neuropathological features of SD-NFT

The clinical features of SD-NFT were compared with AD [4]. The patients with SD-NFT were characterized clinically by very late-onset (average age, 89.4 years) of dementia in which memory disturbance was predominant with relative preservation of the other cognitive functions. Ages at onset and death in SD-NFT were significantly higher than those in AD. All the patients with SD-NFT had been diagnosed as having late-onset AD.

Neuropathological features of SD-NFT were compared with AD [4]. SD-NFT was characterized neuropathologically by abundant NFTs in the hippocampal region and scarcity of SPs throughout the brain. A large number of NFTs with neuropil threads were found in the hippocampal regions, including the hippocampus (CA1 > CA2), subiculum and entorhinal and transentorhinal regions, associated with neuronal loss and gliosis, while NFTs were scarce in the neocortical regions. The distribution of NFTs in SD-NFT corresponds to the Limbic Stages (Stages III and IV) of Braak and Braak's Classification [8]; however, the density of the NFTs in SD-NFT far exceeds the Limbic Stages. Electron-microscopically, the structure of paired helical filaments (PHF) of NFTs in SD-NFT was similar to that in AD [3]. NFTs and neuropil threads in SD-NFT were immunohistochemically labelled with antibodies to phosphorylated (AT8, AT100, AT180 and AT270) and non-phosphorylated epitopes (TAU-2 and anti-human tau) of tau (Fig. 1) and an antibody to ubiquitin [9]. There was no significant difference in the immunoreactivities of NFTs with the anti-tau antibodies between SD-NFT and AD [9].

Aβ deposits (SPs and amyloid angiopathy) were absent or scarce in SD-NFT in contrast with AD [4, 9].

We have proposed criteria for neuropathological diagnosis of SD-NFT (Tab. 1).

Comparisons of SD-NFT with AD and centenarians

To elucidate the mechanism of neurodegeneration in SD-NFT, we morphometrically analyzed the hippocampal lesion of SD-NFT for the atrophy, neuronal cells, NFTs, presynaptic terminals and astrocytic and microglial changes in comparison with AD, using immunohistochemistry with antibodies to synaptophysin, Aβ, tau, glial fibrillary acidic protein (GFAP), and KiM1P (a microglial marker) in addition to routine neuropathological stainings [9].

The densities of hippocampal NFTs in SD-NFT were significantly higher than those in age-matched AD patients; in contrast, the hippocampal atrophy, synaptic loss and proliferation of astrocytes and microglias in SD-NFT were significantly mild compared with AD [9].

The results indicated that the neurodegenerative process with NFT formation of the hippocampal region in SD-NFT would be different from that in AD. We speculate that neuronal death without NFT formation may be important in the disease progression of AD compared with SD-NFT.

Furthermore we compared neuropathological findings of 13 centenarians without obvious dementia [10] with those of AD and SD-NFT. The patterns of NFT distribution, Aβ deposition, synaptic loss and glial changes in the centenarians were different from those of AD

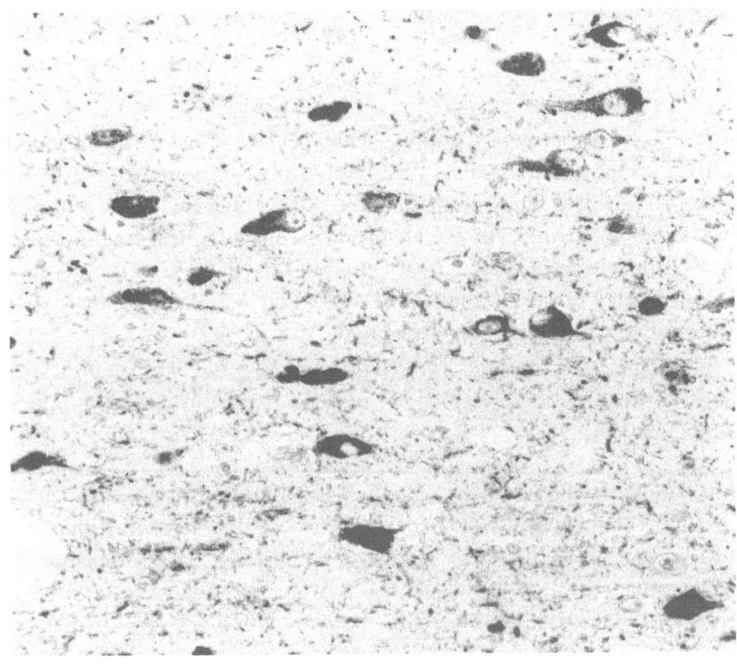

Figure 1. Hippocampal CA1 in SD-NFT. NFTs and neuropil threads are immunostained with a phosphoryla-tion-dependent anti-tau antibody, AT100 (AT100, x175).

[10], and appear similar to those in SD-NFT, although the number of NFTs in the hip-pocampal region of the centenarians appeared less than that in SD-NFT. These observations suggest that SD-NFT, but not AD, may represent a condition of accelerated brain aging.

Table 1. Criteria for the neuropathological diagnosis of SD-NFT

A. Late-onset dementia with the following neuropathological features:

   1. Abundant neurofibrillary tangles (NFTs) in the hippocampal region:
   a large number of neurofibrillary tangles (NFTs) are observed in the hippocampus and parahippocampus (especially, CA1/subiculum/entorhinal and transentorhinal cortex) with neuropil threads and neuronal loss. NFTs are also found in the amygdala, insula and nucleus basalis of Meynert, but rarely in the neocortex.

   2. Absence or scarcity of senile plaques (Aβ deposits) throughout the brain.

B. Exclusion of other dementing disorders with NFTs:
   Alzheimer's disease, progressive supranuclear palsy, diffuse NFTs with calcification, frontotemporal dementia and parkinsonism linked to chromosome 17 (FTDP-17), amyotrophic lateral sclerosis/parkinsonism/dementia complex, etc.

Molecular genetic studies

Apolipoprotein E genotype

In the analyses of apolipoprotein E (ApoE) genotypes, we found no ε4 allele in the patients with SD-NFT; in contrast, the ε4 allele was significantly overpresented in AD, as reported in many studies [4]. Low frequency of the ε4 allele [5] and high frequency of the ε2 allele [11] have been reported in patients with the identical neuropathological condition. Taken together with these results, the frequency of the ε4 allele in SD-NFT is significantly lower than that in AD. Low frequency of the ε4 allele in SD-NFT may be related to low risk for development of AD until very old age, and to scarcity of Aβ deposits.

Analysis of the tau gene

Recently, in familial frontotemporal dementia and parkinsonism linked to chromosome 17 (FTDP-17), different mutations in the tau gene have been reported with different phenotypes. A family with FTDP-17 shows some similarity with SD-NFT, i.e., numerous NFTs in the hippocampal region.

We sequenced the exons 9 to 14 and the flanking intronic regions, which encompassed different mutations identified in FTDP-17, in 3 patients with SD-NFT [9]. We failed to find any mutation in the tau gene from the patients with SD-NFT. However, recent studies suggested that polymorphism in an intron of the tau gene might be associated with risk of progressive supranuclear palsy, another sporadic neurodegenerative disease with characteristic NFT accumulation. Further study is necessary to clarify a possible role of the tau gene polymorphism in the etiology of SD-NFT.

Conclusions

SD-NFT is a cause of dementia among very elderly individuals. Our results suggest that SD-NFT would be an entity with a pathogenetic process different from AD, and may represent a condition of accelerated brain aging without development of AD.

Acknowledgements
The study was supported in part by a Health Science Research Grant (to M.Y.) from the Ministry of Health and Welfare, Japan, by a Grant-in-Aid for Scientific Research (to M.Y.) from the Ministry of Education, Science, Sports and Culture, Japan, and by a Grant (to M.Y.) from Novartis Foundation for Gerontological Research.

References

1 Ulrich J, Spillantini MG, Goedert M, Dukas L, Stähelin HB (1992) Abundant neurofibrillary tangles without senile plaques in a subset of patients with senile dementia. *Neurodegeneration* 1: 257–264
2 Bancher C, Jellinger KA (1994) Neurofibrillary predominant form of senile dementia of Alzheimer type:

a rare subtype in very old subjects. *Acta Neuropathol* 88: 565–570

3   Itoh Y, Yamada M, Yoshida R, Suematsu N, Oka T, Matsushita M, Otomo E (1996) Dementia character-
    ized by abundant neurofibrillary tangles and scarce senile plaques: a quantitative pathological study. *Eur
    Neurol* 36: 94–97

4   Yamada M, Itoh Y, Otomo E, Suematsu N, Matsushita M (1996) Dementia of the Alzheimer type and relat-
    ed dementias in the aged: DAT subgroups and senile dementia of the neurofibrillary tangle type.
    Neuropathology 16: 89–98

5   Bancher C, Egensperger R, Kösel S, Jellinger K, Graeber MB (1997) Low prevalence of apolipoprotein E
    ε4 allele in the neurofibrillary tangle predominant form of senile dementia. *Acta Neuropathol* 94: 403–409

6   Jellinger KA, Bancher C (1998) Senile dementia with tangles (tangle predominant form of senile demen-
    tia). *Brain Pathol* 8: 367–376

7   Kosaka K, Iseki E, Odawara T, Hino H, Kanai A, Katoh M (1997) Limbic neurofibrillary tangle dementia
    (LNTD). *Brain Pathol* 7: 1114. Abstract

8   Braak H, Braak E (1991) Neuropathological stageing of Alzheimer-related changes. *Acta Neuropathol* 82:
    239–259

9   Yamada M, Itoh Y, Sodeyama N, Suematsu N, Otomo E, Matsushita M, Mizusawa H (2000) Senile demen-
    tia of the neurofibrillary tangle type: a comparison with Alzheimer's disease. *Dement Geriat Cog Disord*;
    in press

10 Itoh Y, Yamada M, Suematsu N, Matsushita M, Otomo E (1998) An immunohistochemical study of cente-
    narian brains: a comparison. *J Neurol Sci* 157: 73–81

11 Ikeda K, Akiyama H, Arai T, Sahara N, Mori H, Usami M, Sakata M, Mizutani T, Wakabayashi K,
    Takahashi H (1997) A subset of senile dementia with high incidence of the apolipoprotein E ε2 allele. *Ann
    Neurol* 41: 693–695

Neuroscientific Basis of Dementia
C. Tanaka, P.L. McGeer, Y. Ihara (eds)
© 2001 Birkhäuser Verlag Basel/Switzerland

The dual role of tau in cell polarisation and organelle trafficking

Eva-Maria Mandelkow, Jacek Biernat, Karsten Stamer, Bernhard Trinczek and Eckhard Mandelkow

Max-Planck-Unit for Structural Molecular Biology, Hamburg, Germany

Introduction

Tau is one of the microtubule-associated proteins (MAPs) which copurify with microtubules through cycles of assembly and disassembly [1, 2]. One of its functions is to stabilize micro-tubules [3], although this function can also be assumed by other MAPs so that transgenic mice lacking tau develop almost normally [4]. Other functions include the induction of microtubule bundles during cell process formation and neuritogenesis [5–9], anchoring of cellular enzymes such as kinases, phosphatases, or lipases [10–13], interactions between the cytoskeleton and the plasma membrane [14], and the regulation of intracellular traffic [15, 16]. In Alzheimer's disease (AD), tau aggregates into paired helical filaments (PHFs) which coalesce into neurofibrillary tangles and related deposits. The progression of tau aggregates correlates with the stages of AD and is an early sign of cellular degeneration [17–20]. Similar features occur in other tauopathies, notably FTDP-17 associated with mutations in the tau gene (reviews [21, 22]). Finally, tau is elevated in the cerebrospinal fluid of AD patients which makes it a potential tool for early diagnosis [23].

Tau contains a number of phosphorylation sites which could regulate its functions (Fig. 1). There are up to 17 Ser/Thr-Pro motifs which can be targetted by proline-directed kinases (e.g., MAP kinase, GSK-3, cdk5, cdc2). The repeats in the C-terminal half contain KXGS motifs which can be phosphorylated by MARK (affecting Ser262, 293, 324, 356). PKA phosphorylates Ser214, the KXGS motifs, Ser409 and others, CaMK-II phosphorylates Ser416, and several other sites have been reported (reviews [24–26]). In addition, tau can be phosphorylated at Tyr by the scr-like kinase fyn [27]. Several of the phosphorylated motifs are elevated in AD and are recognized by phosphorylation-sensitive antibodies, which are therefore diagnostic tools in analyzing brain tissue or cell models of AD; this holds particu-larly for the Ser/Thr-Pro motifs. The nature of the sites may shed light on signal transduc-tion pathways affecting tau and microtubule interactions. Some of the sites are particularly potent in detaching tau from microtubules and thus destabilizing microtubules, particularly Ser262 targetted by MARK [28, 29] and Ser214 targetted by PKA [30–32].

While studying cell models for tau's cellular functions, we recently noted novel functions of tau and its phosphorylation sites. Cell processes form in Sf9 cells after transfection of tau [6, 33]. We have now introduced tau mutated at key phosphorylation sites and found that the dependence on phosphorylation is the opposite of what we had intuitively expected: phos-

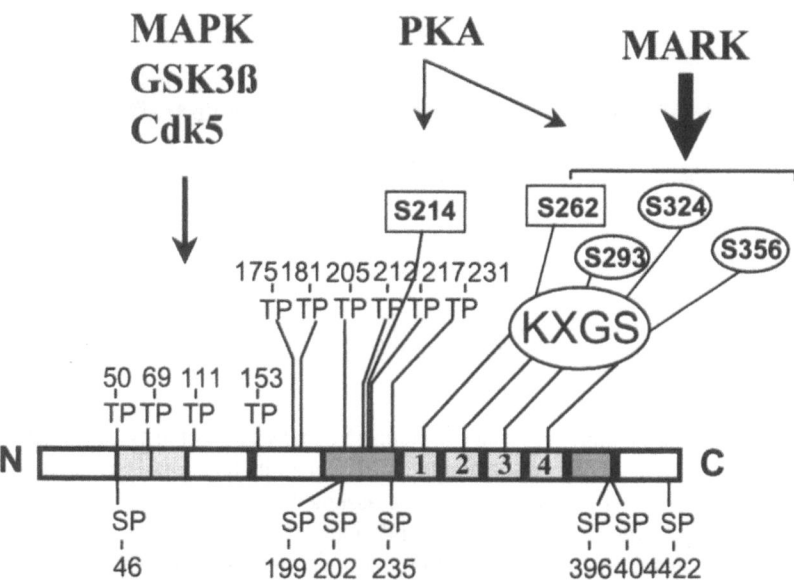

Figure 1. Diagram of tau (isoform htau40, 441 residues), its domains and major phosphorylation sites. One or more of the grey areas may be absent in other isoforms due to alternative splicing. The shortest isoform (htau23, 352 residues) does not contain the grey inserts, it is expressed preferentially in fetal tissue. The C-terminal domain binds to microtubules and contains 3 or 4 repeats of ~31 residues. Most tau mutations in frontotemporal dementias are clustered in the vicinity of the second repeat (coded by exon 10, see other contributions in this volume). Each repeat contains a KXGS motif (with serines 262, 293, 324, 356); they can be phosphorylated by the protein kinase MARK which strongly reduces tau's affinity for microtubules. Microtubule binding is also strongly reduced when Ser214 is phosphorylated by PKA (this kinase can also phosphorylate KXGS motifs to some extent). Tau also contains up to 17 Ser-Pro or Thr-Pro (14 in the fetal isoform htau23) which can be phosphorylated by proline-directed kinases (MAP kinase, GSK-3, cdc2 or cdk5, and others). The phosphorylation mutants used to study cell process formation were derived from the fetal isoform htau23. In the AP-mutant, all Ser-Pro or Thr-Pro motifs were replaced by Ala-Pro, making them inaccessible to proline-directed kinases. In the KXGA mutant, the KXGS motifs in the repeats were replaced by KXGA, making them inaccessible to protein kinases of the MARK family.

phorylation at the KXGS motifs in the repeat domain is essential for process outgrowth, even though this phosphorylation tends to destabilize microtubules. We also transfected tau into mammalian cell cultures (CHO) and found that tau disrupts intracellular traffic and therefore makes cells more vulnerable. Both observations have important implications for the tau-based pathology in AD.

Materials and methods

Methods have been described in detail elsewhere. The infection of Sf9 cells with tau and tau mutants, the analysis of cell process formation and the analysis of tau phosphorylation is described in [9]. The effects of tau on intracellular transport of mitochondria, vesicles and endoplasmic reticulum is described in [15]. The analysis of the interference between tau and motor proteins is given in [16]. The identification of cellular phosphorylation sites of tau by two-dimensional phosphopeptide mapping and related methods is described in [32]. Finally, the effect of microtubule-disrupting kinases operating on tau is analyzed in [29, 34] for MARK, and in [31] for PKA.

Results and discussion

Tau promotes cell process formation in a phosphorylation-dependent manner

The expression of tau is closely linked to the generation of cell processes [5]. This is accompanied by changes in the phosphorylation pattern, but the nature of these changes has been difficult to study because of the low levels of tau and the difficulties of detecting phosphorylation sites. Phosphorylation-sensitive antibodies have been used for cell models, but there are limitations because the antibody reactions are non-linear, minor sites can become exaggerated and major sites may remain undetected. We have therefore searched for a cell system that expresses sufficient tau for biochemical analysis and lends itself to mutational analysis, and for a method that allows quantification of phosphorylation. This is achieved by metabolic labeling of cells, i.e., by adding ^{32}P-labeled phosphoric acid to the medium which is used by the cell to synthesize $[\gamma\text{-}^{32}\text{P}]$ATP, leading to incorporation of ^{32}P into proteins by the cellular protein kinases. This can be analyzed by two-dimensional phosphopeptide mapping, phosphopeptide sequencing and mass spectrometry. With this method one can show that exogenous tau introduced into different cell lines becomes phosphorylated in a fashion that is similar to that of neuronal cells [32]. When tau becomes introduced into Sf9 cells by baculovirus vectors, these cells develop one long, homogeneous cell process reminiscent of a neurite [6]. Thus we generated tau variants containing mutations at crucial phosphorylation sites, transfected them into Sf9 cells, and monitored their behavior and phosphorylation pattern [9] (see Fig. 2, 3). The results can be summarized as follows:

(a) The kinases of Sf9 cells phosphorylate tau at the sites characteristic of Alzheimer sites so that the diagnostic antibodies developed against PHF-tau recognize tau from Sf9 cells as well. This includes the doubly phosphorylated motifs recognized by certain antibodies, such as Ser202/Thr205 (antibody AT-8), Ser396/Ser404 (antibody PHF-1) and in particular the motif Thr212/Ser214 (antibody AT-100). This latter motif is one of the most specific ones in AD-tau [35], presumably because it is generated by a sequential phosphorylation reaction by two different kinases (Thr212 by GSK-3 first, Ser214 by PKA second [31].

(b) Sf9 cells start generating processes about 30 h after transfection (Fig. 2). The rate increases linearly until the cells are harvested (72 h after infection). Because there is only one process per cell, the increase reflects the number of cells carrying one process. When transfecting with human tau23 (the smallest, fetal isoform, 352 residues), about half of

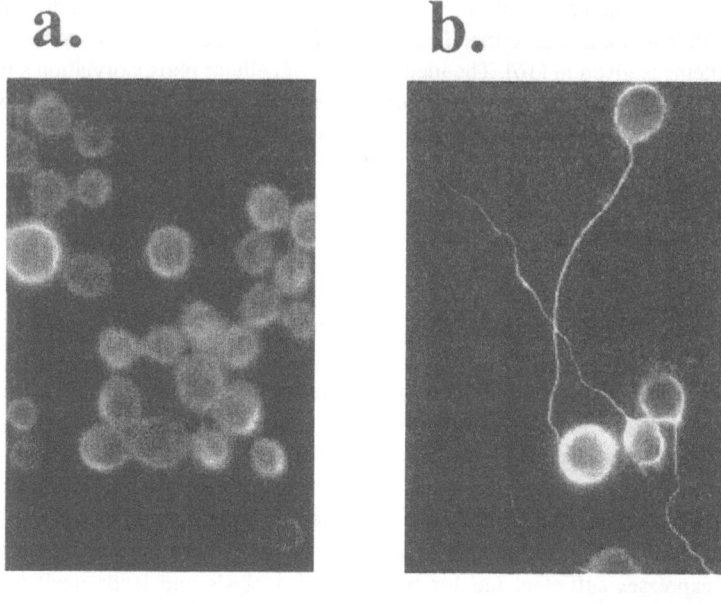

Figure 2. Sf9 cells transfected with tau using baculovirus vectors. Normal Sf9 cells are round (left), but when they express tau they acquire cell processes (right), typically one per cell, with a microtubule organization similar to that of neurites (plus ends pointing to the periphery). Bar = 50μm.

the cells have acquired a process at the end of the experiment.

(c) For distinguishing the effect of phosphorylation at distinct classes of sites we made two constructs (Fig. 1): in "AP-tau" all 14 Ser-Pro (SP) or Thr-Pro (TP) motifs of htau23 were changed into Ala-Pro (AP) so that they are not accessible to proline-directed kinases. In "KXGA-tau", all 3 KXGS motifs in the repeat domain of tau23 were changed into KXGA, thus making them inaccessible to the kinases MARK or PKA (X = I or C). AP-tau led to a moderate increase in process formation (from 55 to 75% of the cells), but KXGA-tau almost completely suppressed process formation (~10% of cells).

(d) The quantification of the phosphorylation sites of tau23 by 2D-phosphopeptide mapping showed that the majority of phosphate (>80%) was incorporated into S/T-P motifs, but only a minor fraction into KXGS motifs (<5%, most of this in Ser262).

These results are summarized in Figure 3. They show that phosphorylatable KXGS motifs are much more important than the SP or TP motifs because process formation is blocked when the KXGA-tau mutants are expressed, even though the phosphorylation at KXGS motifs accounts only for a small percentage of the total phosphate incorporated into tau. This is surprising in the light of the phosphorylation effects on tau-microtubule interac-

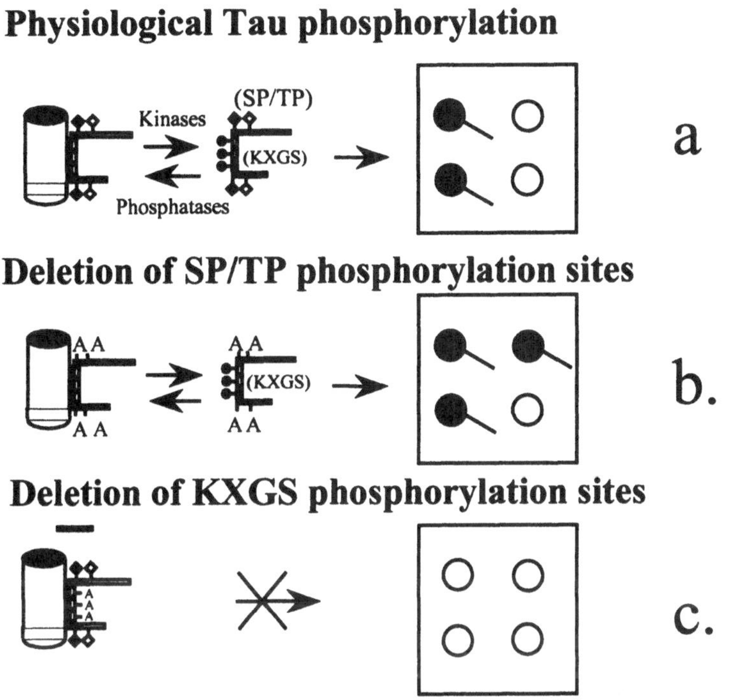

Physiological Tau phosphorylation

Deletion of SP/TP phosphorylation sites

Deletion of KXGS phosphorylation sites

Figure 3. Effects of tau mutants on process formation in Sf9 cells. (a) After three days in culture, normal tau (htau23) induces process in about 50% of the cells. This is indicated on the right by four cells, of which two have processes (b) With AP-tau ~75% of the cells obtain processes (three out of four, right). This indicates that proline-directed kinases are somewhat inhibitory to process formation. (c) With KXGA-tau processes become very rare, showing that some phosphorylation at KXGS sites is necessary to initiate process formation, even though these sites contain only a minor fraction of the total phosphate on tau, and even though this type of phosphorylation disrupts the tau-microtubule interaction [9].

tions. We had shown earlier [28, 36] that phosphorylation of SP or TP motifs had only a modest effect in reducing the tau-microtubule interaction, whereas phosphorylation of KXGS motifs, expecially at S262, strongly inhibited it. One would thus expect that preventing phosphorylation at SP or TP motifs would strengthen the stability of microtubules and microtubule bundles, and this would be consistent with the observed moderate increase in process formation. However, by the same argument one would also expect that preventing phosphorylation at KXGS motifs should strongly increase microtubule stability and process formation, which is the opposite of what is observed.

One interpretation is that microtubules must be destabilized transiently and perhaps only in a limited region (say, at the site of process formation), before processes are stabilized by microtubule bundles. A kinase phosphorylating the KXGS motifs could act as a "gatekeeper", making tubulin polymers become destabilized and small enough to enter the process.

This would be reminiscent of the action of cytochalasin D which softens the cortical actin network and thus allows microtubule bundles to form spikes at the cell surface [37]. The kinase phosphorylating the KXGS motifs of tau most efficiently is MARK, and it is known that this kinase or relatives of it are responsible for cell polarity and differentiation [38]. Future work will have to show whether kinases of the MARK family are indeed activated concomitant with cell process formation in Sf9 cells or other cells, such as neurons.

Tau and intracellular transport

Two main classes of proteins interact with microtubules, MAPs and motor proteins (review [39]). MAPs stabilize microtubules, space them apart from other cell components, or serve as anchors for enzymes, whereas motors are mobile, have transient interactions with microtubules, and generate energy derived from the hydrolysis of ATP. Several authors have asked how MAPs and motors might interact on the microtubule surface, using *in vitro* assays of motor activity in the presence of various MAPs [40–43]. Although inhibition of motors by MAPs could be observed in some conditions, the results remained controversial because the effects were weak, and it remained unclear whether the results were representative of MAP-motor interactions in a cellular environment. In addition, attemps to observe MAP effects on motors by microinjecting MAPs or by transient transfection were complicated by other effects, such as development of cell processes (see above, and [5, 44]). These studies showed, however, a general reduction in random vesicle movements, especially in the presence of large MAPs which have a substantial projection domain such as MAP2 or MAP4 [45, 46].

We noted a directional bias in MAP-motor interactions during studies intended to test the role of tau phosphorylation in cells. CHO cells were stably transfected with tau such that the level of expressed tau remained low in order to avoid the formation of microtubule bundles, typically only 2-3 times the level of endogenous MAP4 [15, 24]. These cells tended to round up and their cell cycle became slower. Staining the cells for mitochondria revealed a conspicuous change. Instead of showing a random distribution (as in normal cells), mitochondria clustered in the region of the Golgi apparatus, around the MTOC (microtubule-organizing center) where microtubules are nucleated before they radiate throughout the cell. Since microtubules are polar structures, the MTOC contains the "minus"-end of microtubules, whereas the "plus"-ends point to the cell periphery (or towards the synapse in the case of axons). This effect becomes most pronounced in cells containing cell processes (Fig. 4).

The Golgi apparatus is centered around the MTOC because it is driven by a minus-end directed microtubule motor, dynein (review [47]). We therefore tested whether the same explanation would be applicable to the clustering of mitochondria around the MTOC in the tau-transfected cells. The key results can be summarized as follows [15]: (i) clustering of mitochondria depends on the polar arrangement of microtubules (ii) the distribution of mitochondria in normal cells depends on a balance between plus- and minus-end directed motors (kinesin or dynein). Clustering around the MTOC requires active dynein; if dynein is inactivated by overexpression of its subunits dynamitin [48], the clustered mitochondria become

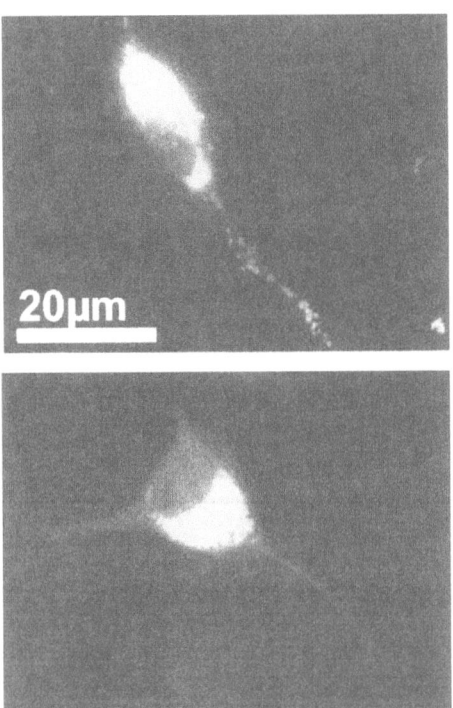

Figure 4. Neuroblastoma cells (N2a), (a) mock transfected with vector only (control), (b) stably transfected with tau (htau40). In the control, mitochondria (stained by MitoTracker red) are distributed throughout the cell body and the neurite. When tau is overexpressed, anterograde transport is inhibited, mitochondria disappear from the neurite and accumulate in the cell body [15].

dispersed again. By the same argument, when kinesin is inhibited by an antibody, mitochondria retract into the cell interior by the action of dynein [49] (iii) the retardation of plus-end directed transport by tau is observed with all microtubule-dependent processes examined so far; e.g., the intermediate filament protein vimentin accumulates in the cell center, the endoplasmatic reticulum retracts from the plasma membrane to the cell interior, peroxisomes become clustered in the cell center, and exocytotic vesicles (carrying transferrin receptors) become retarded.

The mechanism of inhibition was analyzed by observing the movements of fluorescently tagged vesicles or mitochondria in CHO cells [16] (Fig. 5). This revealed that tau did not reduce the speed of instantaneous transport along microtubules in either direction; furthermore, tau reduced the run lengths in both directions to an equal extent. However, the attachment of kinesin-dependent cargoes to microtubules was much more inhibited than the attachment of dynein-dependent cargoes. The net effect is the inhibition of plus-end directed transport or the relative enhancement of minus-end directed transport.

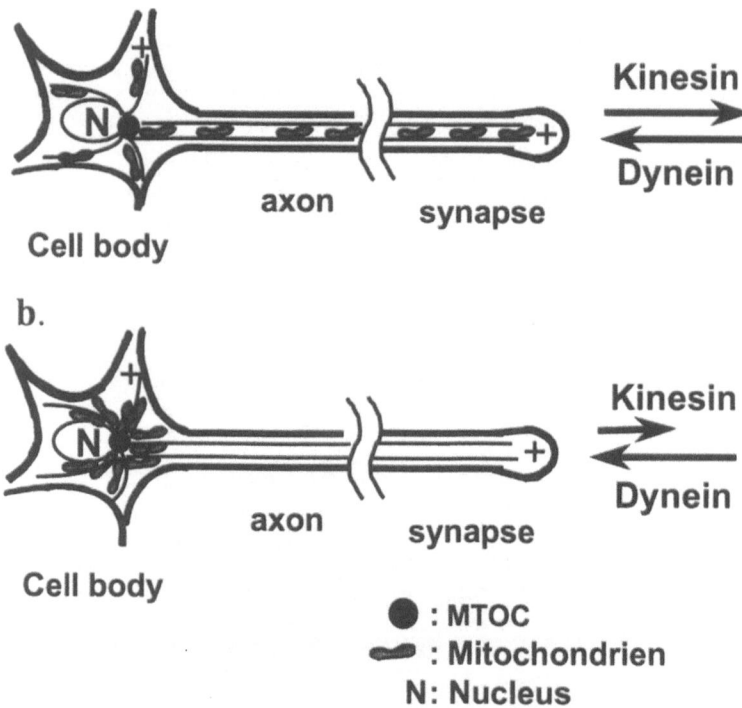

Figure 5. Diagram of the effects of tau on intracellular and axonal transport. Tau represents an obstacle to vesicles or organelles pulled by motor proteins towards the periphery (kinesin) or towards the cell center (dynein). The inhibition of kinesin-dependent movement is more pronounced than that of dynein, leading to a predominance of movements towards the cell center and hence to a depletion of organelles from the periphery.

Implications for neuronal degeneration in Alzheimer's disease

Neurons are asymmetric cells whose supply of nutrients and energy depends on an efficient transport system. Thus the inhibition of plus-end directed transport along the axon would lead to an impaired supply of neurotransmitters, energy and membrane elements, resulting in vulnerability and eventually degeneration. Indeed one observes (Fig. 4) that when tau is elevated by transfection in neuroblastoma cells (N2a), mitochondria are excluded from cell processes and accumulate in the cell body. The cell processes show retarded growth, and when challenged with stressors (e.g., oxidative stress) these cell processes disintegrate quickly. This could be a model of events taking place in AD. In AD, tau is elevated in affected neurons [50]; this leads not only to its precipitation in the form of PHFs, but before that it could impair intracellular transport and make the cells vulnerable.

To defend itself against this reaction a neuron could possibly redistribute tau from the axon to the somatodendritic compartment. The neuron possesses a sorting mechanism that guides tau into the axon which is based on the targeting of tau mRNA to the axon hillock

[51]. However, the sorting breaks down when the system is overloaded, e.g., by microinjection or transient transfection of tau; in that case tau appears throughout the neuronal cell [52]. Another line of defense would be to dissociate the excess tau from microtubules, because only microtubule-bound tau is inhibitory to axonal traffic [15]. This dissociation can be achieved by phosphorylation of critical residues, for example, the KXGS motifs in the repeats by MARK, or Ser214 by PKA [31]. Both types of phosphorylation are elevated in Alzheimer tau [53, 54], possibly because the kinases are activated in response to the increase in tau. Although this may be beneficial to the neuron in the short term, it is detrimental in the long run: soluble tau which is not degraded can aggregate. This problem is particularly serious in aging neurons which can no longer keep their reducing potential high enough. In this situation tau becomes oxidized and forms dimers by disulfide crosslinking (especially at Cys322) which then act as efficient building blocks for the aggregation into PHFs [55]. The aggregation then proceeds through a beta-sheet type interaction involving a short hexapeptide in the microtubule-binding domain of tau [56].

This mechanism differs from earlier descriptions of the tau pathology in neurons where binding of tau to microtubules was considered to be beneficial because it stabilized the tracks of intracellular transport, and consequently the kinases causing tau's detachment from microtubules were seen as detrimental. Now we see tau as a Janus-faced molecule: its attachment to microtubules may be good for microtubules, but too much tau disrupts transport. By analogy, tau kinases may play a positive role because they alleviate the clogging of the transport system. These considerations are oversimplified, but they suggest further experimentation to reveal the physiological and pathological functions of tau.

Acknowledgements
We thank I. Thielke and K. Alm for excellent technical assistance. This work was supported in part by the Deutsche Forschungsgemeinschaft.

References

1 Weingarten MD, Lockwood AH, Hwo SY, Kirschner MW (1975) A protein factor essential for microtubule assembly. *Proc Natl Acad Sci USA* 72: 1858–1862
2 Lee G, Cowan N, Kirschner M (1988) The primary structure and heterogeneity of tau protein from mouse brain. *Science* 239: 285–288
3 Drubin D, Kirschner M (1986) Tau protein function in living cells. *J Cell Biol* 103: 2739–2746
4 Harada A, Oguchi K, Okabe S, Kuno J, Terada S, Ohshima T, Sato-Yoshitake R, Takei Y, Noda T, Hirokawa N (1994) Altered microtubule organization in small-caliber axons of mice lacking tau protein. *Nature* 369: 488–491
5 Kanai Y, Takemura R, Oshima T, Mori H, Ihara Y, Yanagisawa M, Masaki T, Hirokawa N (1989) Expression of multiple tau isoforms and microtubule bundle formation in fibroblasts transfected with a single tau cDNA. *J Cell Biol* 109: 1173–1184
6 Baas PW, Pienkowski TP, Kosik KS (1991) Processes induced by tau expression in Sf9-cells have an axon-like microtubule organization. *J Cell Biol* 115: 1333–1344
7 Esmaeli-Azad B, Mccarty JH, Feinstein SC (1994) Sense and antisense transfection analysis of tau-function: Tau influences net microtubule assembly, neurite outgrowth and neuritic stability. *J Cell Sci* 107: 869–879
8 Matus A (1994) Stiff microtubules and neuronal morphology. *Trends Neurosci* 17: 19–22
9 Biernat J, Mandelkow E-M (1999) The development of cell processes induced by tau protein requires phosphorylation of serine 262 and 356 in the repeat domain and is inhibited by phosphorylation in the pro-

line-rich domains. *Mol Biol Cell* 10: 727–740
10 Liao H, Li YR, Brautigan DL, Gundersen GG (1998) Protein phosphatase-1 is targeted to microtubules by
     the microtubule-associated protein tau. *J Biol Chem* 273: 21901–21908
11 Sontag E, Nunbhakdi-Craig V, Lee G, Bloom GS, Mumby MC (1996) Regulation of the phosphorylation
     state and microtubule-binding activity of tau by protein phosphatase 2a. *Neuron* 17: 1201–1207
12 Sontag E, Nunbhakdi-Craig V, Lee G, Brandt R, Kamibayashi C, Kuret J, White C, Mumby M, Bloom G
     (1999) Molecular interactions among protein phosphatase 2A, tau, and microtubules. Implications for the
     regulation of tau phosphorylation and the development of tauopathies. *J Biol Chem* 274: 25490–25498
13 Jenkins SM, Johnson GVW (1998) Tau complexes with phospholipase C-gamma *in situ*. *NeuroReport* 9:
     67–71
14 Brandt R, Leger J, Lee G (1995) Interaction of tau with the neural plasma-membrane mediated by tau
     amino-terminal projection domain. *J Cell Biol* 131: 1327–1340
15 Ebneth A, Godemann R, Stamer K, Illenberger S, Trinczek B, Mandelkow E-M, Mandelkow E (1998)
     Overexpression of tau protein alters vesicle trafficking, distribution of mitochondria and organization of
     the endoplasmic reticulum in living cells: Implications for Alzheimer's disease. *J Cell Biol* 143: 777–794
16 Trinczek B, Ebneth A, Mandelkow E-M, Mandelkow E (1999) Tau regulates the attachment/detachment
     but not the speed of motors in microtubule-dependent transport of single vesicles and organelles. *J Cell Sci*
     112: 2355–2367
17 Braak H, Braak E (1991) Neuropathological staging of Alzheimer-related changes. *Acta Neuropathol* 82:
     239–259
18 Arriagada PV, Growdon JH, Hedley-Whyte E, Hyman BT (1992) Neurofibrillary tangles but not senile
     plaques parallel duration and severity of Alzheimers disease. *Neurology* 42: 631–639
19 Braak E, Braak H, Mandelkow E-M (1994) A sequence of cytoskeleton changes related to the formation
     of neurofibrillary tangles and neuropil threads. *Acta Neuropathol* 87: 554–567
20 Hyman BT, Trojanowski JQ (1997) Editorial on consensus recommendations for the *post mortem* diagno-
     sis of Alzheimer's disease from the National Institute on Aging and the Reagan Institute working group on
     diagnostic criteria for the neuropathological assessment of Alzheimer's disease. *J Neuropathol Exp Neurol*
     56: 1095–1097
21 Spillantini M, Goedert M (1998) Tau protein pathology in neurodegenerative diseases. Tr. Neurosci. 21:
     428–433
22 Wilhelmsen K C (1999) The tangled biology of tau. *Proc Natl Acad Sci USA* 96: 7120–7121
23 Blennow K, Wallin A, Agren H, Spenger C, Siegfried J, Vanmechelen E (1995) Tau protein in cere-
     brospinal fluid: A biochemical marker for axonal degeneration in Alzheimer disease? *Mol Chem
     Neuropathol* 26: 231–245
24 Imahori K, Uchida T (1997) Physiology and pathology of tau protein kinases in relation to Alzheimer's
     disease. *J Biochem (Tokyo)* 121: 179–188
25 Johnson GV, Hartigan JA (1998) Tau protein in normal and Alzheimer's disease brain – an update. *Alz Dis
     Rev* 3: 125–141
26 Friedhoff P, Mandelkow E (1999) Tau Protein. *In*: T Kreis, R Vale (eds.): *Guidebook to the Cytoskeletal
     and Motor Proteins*. Oxford University Press, pp 230–236
27 Lee G, Newman S, Gard D, Band H, Panchamoorthy G (1998) Tau interacts with src-family non-receptor
     tyrosine kinases. *J Cell Sci* 111: 3167–3177
28 Biernat J, Gustke N, Drewes G, Mandelkow E-M, Mandelkow E (1993) Phosphorylation of serine 262
     strongly reduces the binding of tau protein to microtubules: Distinction between PHF-like immunoreac-
     tivity andmicrotubule binding. *Neuron* 11: 153–163
29 Drewes G, Ebneth A, Preuss U, Mandelkow E-M, Mandelkow E (1997) MARK – a novel family of pro-
     tein kinases that phosphorylate microtubule-associated proteins and trigger microtubule disruption. *Cell*
     89: 297–308
30 Brandt R, Lee G, Teplow D, Shalloway D, Abdelghany M (1994) Differential effect of phosphorylation
     and substrate modulation on tau's ability to promote microtubule growth and nucleation. *J Biol Chem* 269:
     11776–11782
31 Zheng-Fischhöfer Q, Biernat J, Mandelkow E-M, Illenberger S, Godemann R, Mandelkow E (1998)
     Sequential phosphorylation of tau-protein by GSK-3β and protein kinase A at Thr212 and Ser214 gener-
     ates the Alzheimer-specific epitope of antibody AT-100 and requires a paired helical filament-like confor-
     mation. *Eur J Biochem* 252: 542–552
32 Illenberger S, Zheng-Fischhöfer Q, Preuss U, Stamer K, Godemann R, Baumann K, Mandelkow E-M,
     Mandelkow E (1998) The endogenous and cell-cycle dependent phosphorylation of the microtubule-asso-

ciated protein tau in neuroblastoma and CHO cells: Implications for protein kinases cdc2 and PKA. *Mol Biol Cell* 9,1495–1512

33 Kosik KS, McConlogue L (1994) Microtubule-associated protein function: Lessons from expression in Spodoptera frugiperda cells. *Cell Motility Cytoskel* 28: 195–198

34 Ebneth A, Drewes G, Mandelkow E-M, Mandelkow E (1999) Phosphorylation of MAP2c and MAP4 by MARK kinases leads to the destabilization of microtubules in cells. *Cell Motility Cytoskel* 44: 209–224

35 Matsuo ES, Shin RW, Billingsley ML, Vandevoorde A, O'connor M, Trojanowski JQ, Lee VMY (1994) Biopsy-derived adult human brain tau is phosphorylated at many of the same sites as Alzheimer's disease paired helical filament tau. *Neuron* 13: 989–1002

36 Trinczek B, Biernat J, Baumann K, Mandelkow E-M, Mandelkow E (1995) Domains of tau protein, differential phosphorylation, and dynamic instability of microtubules. *Mol Biol Cell* 6: 1887–1902

37 Edson K, Weisshaar B, Matus A (1993) Actin depolymerization induces process formation on MAP2-transfected nonneuronal cells. *Development* 117: 689–700

38 Drewes G, Ebneth A, Mandelkow E-M (1998) MAPs, MARKs, and microtubule dynamics. *Trends Biochem Sci* 23: 307–311

39 Hirokawa N (1998) Kinesin and dynein superfamily proteins and the mechanism of organelle tranport. *Science* 279: 519–526

40 Paschal BM, Obar RA, Vallee RB (1989) Interaction of brain cytoplasmic dynein and MAP2 with a common sequence at the C-terminus of tubulin. *Nature* 342: 569–572

41 Von Massow A, Mandelkow E-M, Mandelkow E (1989) Interaction between kinesin, microtubules, and microtubule-associated protein-2. *Cell Motility Cytoskel* 14: 562–571

42 Hagiwara H, Yorifuji H, Sato-Yoshitake R, Hirokawa N (1994) Competition between motor molecules (kinesin and cytoplasmic dynein) and fibrous microtubule-associated proteins in binding to microtubules. *J Biol Chem* 269: 3581–3589

43 Lopez LA, Sheetz MP (1993) Steric inhibition of cytoplasmic dynein and kinesin motility by MAP2. *Cell Motility Cytoskel* 24: 1–16

44 Weisshaar B, Doll T, Matus A (1992) Reorganization of the microtubular cytoskeleton by embryonic microtubule-associated protein 2 (MAP2c). *Development* 116: 1151–1161

45 Sato-Harada R, Okabe S, Umeyama T, Kanai Y, Hirokawa N (1996) Microtubule-associated proteins regulate microtubule function as the track for intracellular membrane organelle transports. *Cell Struct Funct* 21: 283–295

46 Bulinski JC, McGraw TE, Gruber D, Nguyen H-L, Sheetz MP (1997) Overexpression of MAP4 inhibits organelle motility and trafficking *in vivo*. *J Cell Sci* 110: 3055–3064

47 Lippincott-Schwartz J (1998) Cytoskeletal proteins and Golgi dynamics. *Curr Opin Cell Biol* 10: 52–59

48 Burkhardt JK, Echeverri CJ, Nilsson T, Vallee R B (1997) Overexpression of the dynamitin (p50) subunit of the dynaction complex disrupts dynein-dependent maintenance of membrane organelle distribution. *J Cell Biol* 139: 469–484

49 Rodionov VI, Gyoeva FK, Tanaka E, Bershadsky AD, Vasiliev JM, Gelfand V I (1993) Microtubule-dependent control of cell-shape and pseudopodial activity is inhibited by the antibody to kinesin motor domain. *J Cell Biol* 123: 1811–1820

50 Khatoon S, Grundke-Iqbal I, Iqbal K (1992) Brain levels of microtubule-associated protein tau are elevated in Alzheimer's disease: A radioimmune slot blot assay for nanograms of proteins. *J Neurochem* 59: 750–753

51 Litman P, Barg J, Ginzburg I (1994) Microtubules are involved in the localization of tau messenger-RNA in primary neuronal cell cultures. *Neuron* 13: 1463–1474

52 Kanai Y, Hirokawa N (1995) Sorting mechanisms of tau and MAP2 in neurons: Suppressed axonal transit of MAP2 and locally regulated microtubule binding. *Neuron* 14: 421–432

53 Hasegawa M, Morishima-Kawashima M, Takio K, Suzuki M, Titani K, Ihara Y (1992) Protein sequence and mass spectrometric analyses of tau in the Alzheimer's disease brain. *J Biol Chem* 267: 17047–17054

54 Morishima-Kawashima M, Hasegawa M, Takio K, Suzuki M, Yoshida H, Titani K, Ihara Y (1995) Proline-directed and non-proline-directed phosphorylation of PHF-tau. *J Biol Chem* 270: 823–829

55 Wille H, Drewes G, Biernat J, Mandelkow E-M, Mandelkow E (1992) Alzheimer-like paired helical filaments and antiparallel dimers formed from microtubule-associated protein tau *in vitro*. *J Cell Biol* 118: 573–584

56 von Bergen M, Friedhoff P, Biernat J, Heberle J, Mandelkow E-M, Mandelkow E (2000). Assembly of tau protein into Alzheimer paired helical filaments depends on a local sequence motif (306-YQIVYK-311) forming beta structure. *Proc Natl Acad Sci USA* 97: 5129–5134

Neuroscientific Basis of Dementia
C. Tanaka, P.L. McGeer, Y. Ihara (eds)
© 2001 Birkhäuser Verlag Basel/Switzerland

Rearrangement of microtubule networks by tau bearing missense mutations

Naruhiko Sahara, Takami Tomiyama and Hiroshi Mori

Department of Neuroscience, Osaka City University Medical School, Osaka, Japan

Tau is an axonal microtubule-associated phosphoprotein that promotes tubulin polymerization in the normal brain. Tau is comprised of paired helical filaments (PHF), a characteristic ultrastructural feature of Alzheimer's disease (AD) brain [1]. Tau is also of great interest because of the formation of filamentous tau inclusions in various dementias, such as progressive supranuclear palsy, Pick's disease and corticobasal degeneration [2, 3]. Recently, exonic and intronic mutations in the tau gene have been identified in familial frontotemporal dementia and parkinsonism linked to chromosome 17 (FTDP-17) [4–6].

The tau gene is alternatively spliced to produce six isoforms in adult human brain. These isoforms differ in possessing three or four microtubule-binding domains and one or two amino-terminal inserts of unknown function. Most intronic mutations associated with FTDP-17, close to the exon 10 splice site, lead to excess expression of four-repeat tau. Overproduction of four-repeat tau may result in disruption of microtubule dynamics or lead to increased amounts of cytosolic unbound tau, which then aggregates to form paired helical or straight filaments [6–8]. Most exonic mutations are localized with or close to the microtubule-binding domain, although one mutation (R406W) is in the adjacent C-terminal region. Recently, two groups reported that these exonic mutations result in a reduced ability of tau to promote microtubule assembly [9, 10]. Transfection studies have shown that both wild-type and mutant tau bind to microtubules, but that mutations of tau reduced the length of microtubule extension in transfected cells [11]. In another study, G272V and P301L transfectants showed greater instability of microtubules in the presence of Colcemid, a microtubule-depolymerizing drug [12]. In addition, it has been reported that these mutations have larger effects in 3-repeat than in 4-repeat tau isoforms [10, 11]. However, the pathological effects of exonic mutations are not completely understood. In this study, we morphologically examined the effects of both tau mutations and isoform differences by confocal imaging of COS-7 cells transfected with tau cDNA and found a clear effect on cellular cytoskeletal networks.

The experiments on FTDP-17 mutations (G272V, P301L, V337M or R406W) were first performed using the longest isoforms of tau441 containing exons 2, 3 and 10. Cells transfected with wild-type tau spread well with several processes on culture dishes, as did untransfected cells (Fig. 1a–c). The shapes of COS-7 cells transfected with wild-type and mutant tau were almost identical (Fig. 1c, 1f, 1i, 1 l and 1o). Furthermore, all four mutant taus as well as wild-type tau were found to be associated with cytoplasmic microtubule networks (Fig. 1e, 1h, 1k and 1n). Notably, microtubule organizing centers (MTOC, arrowheads

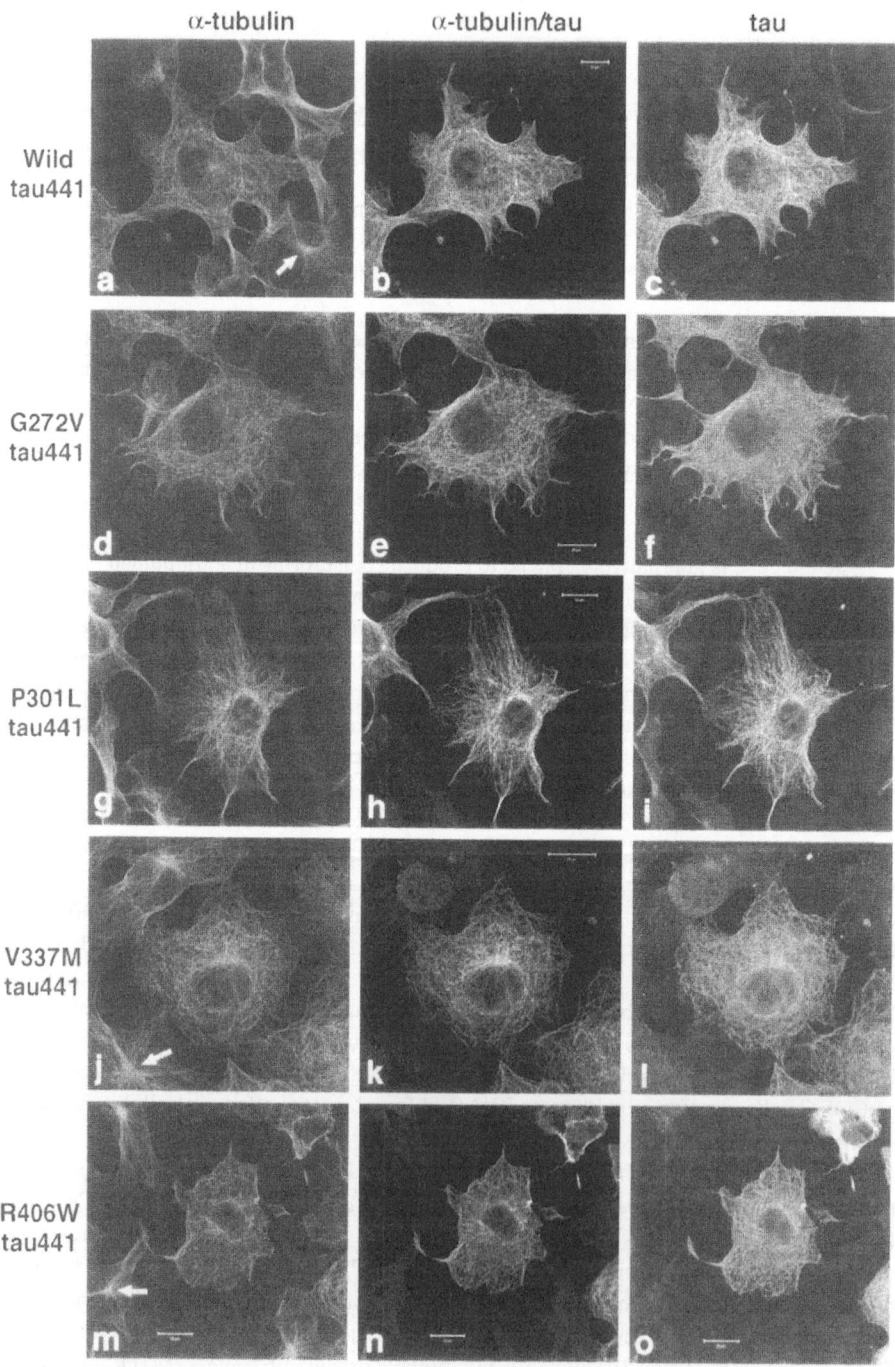

◄ Figure 1. Microtubule networks in COS cells transfected with 4-repeat tau. Confocal images showing the distribution of microtubules (a, d, g, j and m) and tau (c, f, i, l and o) in COS cells transfected with 4-repeat tau. Two antibodies (mouse anti-α-tubulin monoclonal antibody and rabbit anti-human tau antibody) were mixed and used as the primary antibodies. Goat anti-mouse IgG conjugated with rhodamine and goat anti-rabbit IgG conjugated with fluorescein (Jackson ImmunoResearch Lab, West Grove, PA), were used as the secondary antibodies. Figures (b, e, h, k and n) were presented in white and black for overlapping portion of tubulin and tau. The cDNA used were (a–c) wild-type tau441, (d–f) G272V tau441, (g–i) P301L tau441, (j–l) V337M tau441 and (m–o) R406W tau441. Scale bars, 20 μm.

in Fig. 1a, 1j and m) were evident in untransfected COS-7 cells but became vague or had disappeared in COS-7 cells transfected with either wild-type or mutant tau. Thus, tau altered the formation of MTOC which was closely related with cytoplasmic microtubule projections in an unidentified equilibrium condition.

In cells with a high level of tau expression, but not those with a low level of expression, all four mutant taus were found to induce the formation of tau-decorated microtubule bundles, as previously documented for wild-type tau [13, 14]. Microtubule bundle formation may require a sufficient amount of tau expression. There was no clear qualitative difference between wild-type and any of the mutant tau441 in either cellular or cytoskeletal morphology (data not shown).

We then used tau410, which contains amino-terminal inserts of exons 2 and 3 but no C-terminal insert of exon 10, resulting in a 3-repeat tau. We observed a slight effect of wild-type tau410 on cell shape (Fig. 2a–c). Although the effect on shape was most clear in the cells transfected with wild-type tau352, the shortest isoform [15], cell processes were slightly shrunken and cells became round, as shown in Figure 1b and Figure 2b. When the G272V, V337M or R406W mutation was introduced, disruption of microtubule networks was observed with the V337M and R406W mutations. For V337M-mutated tau410 (Fig. 2g), tubulin staining was weaker at the cytoplasmic marginal regions but relatively prominent at the perinuclear regions compared with wild-type tau410 (Fig. 2a). Tau was also found to exist in the cytoplasm where it appeared to encircle the nucleus in a diffuse rather than a well-organized fashion (Fig. 2i). Notably, tubulin and tau around the nucleus did not appear to co-localize (Fig. 2h), in accordance with a previous report [15]. These findings may have been due to mutation-induced reduction in microtubule-binding ability of tau. With R406W-mutated tau, tubulin staining was observed as dotted spots in the cytoplasm rather than as filamentous networks (Fig. 2j), suggesting that microtubules were depolymerized in the cytoplasm. Tau was present diffusely within the cytoplasm (Fig. 2l) and co-localized with tubulin on the dotted spots (Fig. 2k). These findings suggest that the effect of R406W mutation is not reduction in microtubule-binding ability of tau but instead the disruption of microtubule networks. The reduction in ability of tau to promote microtubule assembly induced by the R406W mutation may be explained by conformational change from the reduction in positive charge [9, 10]. In COS-7 cells transfected with R406W-mutated tau410, the binding of tau to microtubules may result in microtubule depolymerization.

It was reported that a single tau cDNA transfected into non-neuronal cells resulted in multiple bands on western blot [11–13]. In our studies, three bands of wild tau441 and tau410 were detected with the anti-tau antibody pool2 (Fig. 3) and with TAU-1. The lowest

α-tubulin   α-tubulin/tau   tau

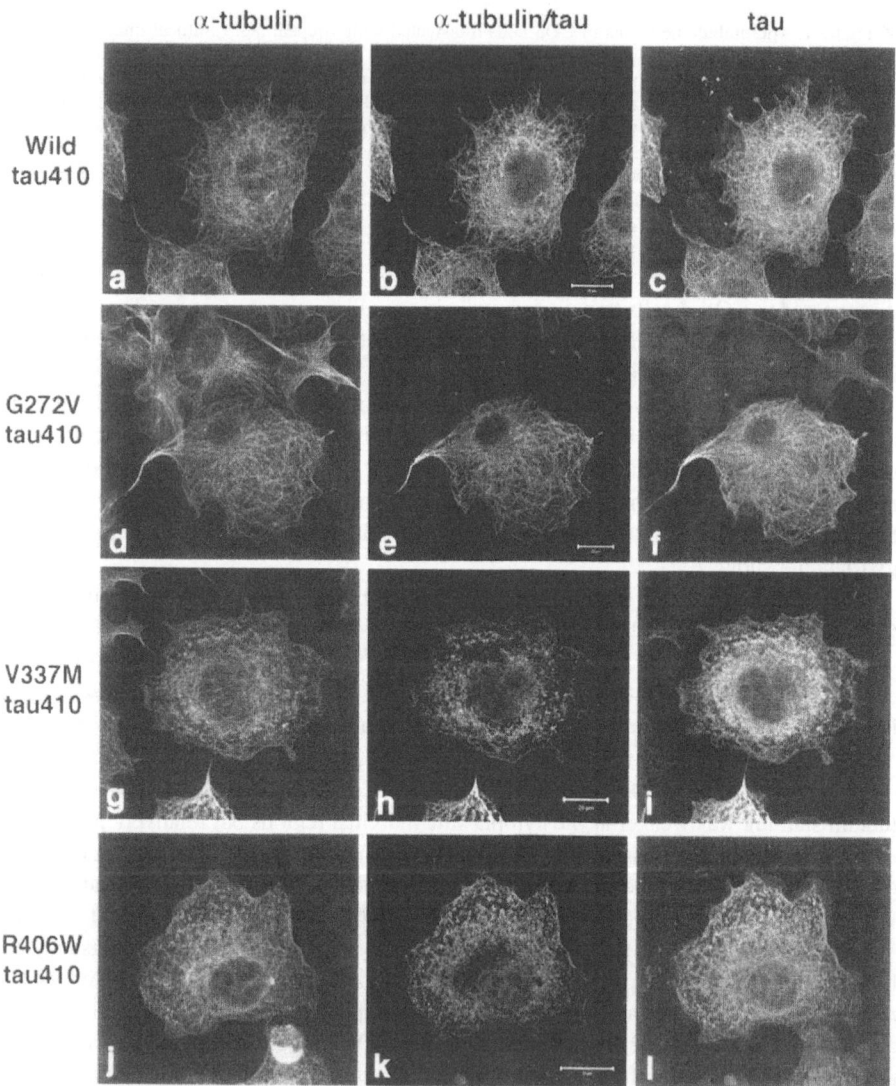

Figure 2. Microtubule networks in COS cells transfected with 3-repeat tau. The similar experimental protocols as Figure 2 were used except tau isoforms of 3-repeat tau. The DNA used were (a–c) wild-type tau410, (d–f) G272V tau410, (g–i) V337M tau410 and (j–l) R406W tau410. Scale bars, 20 μm.

band in Figure 3 was not detected with the antibody Alz-50, which recognized the amino terminus of tau, suggesting that this band was an amino-terminally-truncated product of tau in cells. This was evident from the careful comparison with recombinant tau441 and tau410, as represented in the central lane in Figure 3. TAU-1 antibody, which recognizes the dephos-

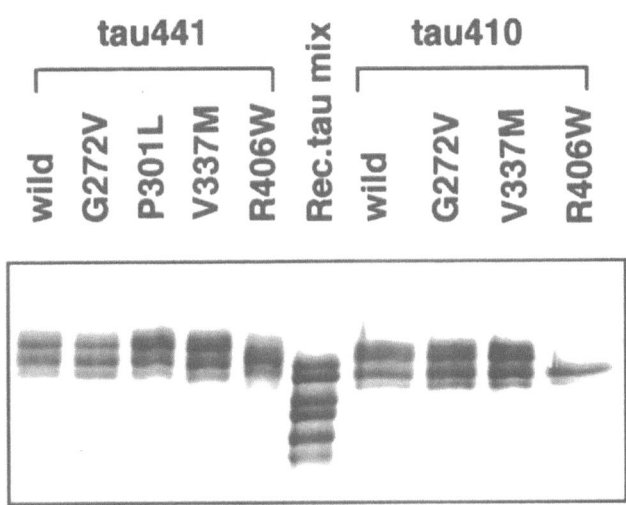

Figure 3. Western blotting analysis of tau expressed in COS-7 cells transfected with tau cDNA. Cultured COS cells were collected in 2 days and electrophoresed on 10% polyacrylamide gel. Western blot analysis was done with antibodies; the phosphorylation-independent antibody, pool2, recognized two isoforms of tau (tau441 and tau410) with or without exonic mutations. Rec. tau mix was six tau isoforms expressed in *E. coli.*

phorylated form of Ser (199th/202th residues), stained the upper bands of wild tau441 and tau 410 (data not shown). Compared with tau proteins in cells transfected with wild-type tau or tau bearing G272V, P301L or V337M mutations, those in cells transfected with tau bearing R406W mutation had a different biochemical profile. In COS transfectants with tau (R406W), only a single band was detected with pool2 antibody (Fig. 3), which had the same electrophoretic mobility as recombinant tau, suggesting that it was not a phosphorylated form. This band was not affected by alkaline phosphatase treatment, indicating that R406W –mutated tau did not undergo abnormal phosphorylation (data not shown).

The effects of tau mutations were also examined on cellular organelles such as the mitochondria, endoplasmic reticula (ER) and lysosomes, which play important roles in cellular function. Specific antibodies Grp75, BiP and LAMP-2 visualized the distribution of mitochondria, ER and lysosomes, respectively. As shown in previous studies [15], effects of tau mutations including G272V, P301L, V337M and R406W on the distributions of Grp75, BiP or LAMP-2 were little (data not shown). Although we expected tau degradation in the lysosomal compartment, LAMP-2 co-localized with neither wild-type tau nor mutant tau. Such little effect of tau mutations on cell organelles might be only in transiently expressed cells. Transgenic mice with tau mutation would then be useful models for explaining the intracellular transport or accumulation of tau inclusion.

In vitro studies have shown that G272V and P301L mutations reduced the ability of tau to promote microtubule assembly [10]. On the other hand, in our assay system, these two mutations had little effect on cellular microtubule networks. One possible explanation for this is that these mutations may reduce the ability of tau to replace endogenous MAPs in COS-7 cells; tau would then have insufficient effect on the formation of microtubule networks (Fig. 4b). Probably tau with these mutations just associate with microtubules with their low affinity and thus appeared to co-localize with tubulin in transfected cells. Tau with the other two mutations, V337M and R406W, will compete sufficiently with endogenous MAPs and cause disruption of microtubule networks (Fig. 4c). The region from repeat domain 1 to repeat domain 2 of tau has been claimed to be of crucial importance for microtubule assembly [16]. Therefore it is likely that G272V mutation in repeat domain 1 and P301L mutation in repeat domain 2 result in decreased binding affinity of tau to microtubules.

It has been reported that phosphorylation regulates the properties of tau, principally tubulin binding and microtubule polymerization [17]. However, our results showed that exonic mutations did not cause abnormal phosphorylation of tau protein. As clearly indicated by the experiment using tau (R406W), abnormal phosphorylation was a secondary event in tauopathy with FTDP-17, at least with R406W. The findings obtained with mutations such as

a. Wild Tau

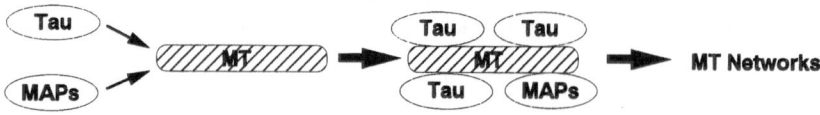

b. G272V (R1) and P301L (R2)

c. V337M (R3) and R406W

Figure 4. A hypothesis for a pathological role of mutated tau. Wild-type or mutant tau (tau*) was introduced in COS-7 cells. The G272V and P301L mutations introduce reduced tubulin-binding activities to tau. Their activities are not strong enough to replace endogenous MAPs. The V337M and R406W mutations are thought to show different effects from the G272V and P301L mutations. The V337 and R406 mutations may cause severe loss of microtubule assembly activity rather than tubulin-binding activity. The hypothesis suggests an occurrence of multiple domains involved in tubulin-binding and filament assembly in tau molecule.

R406W or G389R in FTDP-17 indicate that the non-tubulin-binding domains also play an essential role in microtubule assembly although such non-tubulin-binding domains are yet fully to be characterized.

The issue that the mutation effect changes tau-isoform-dependently was also of great interest because tau pathology is closely related to the distribution of tau isoforms [18]. *In vitro* studies have demonstrated that the ability of 3-repeat tau isoform for microtubule assembly was 2.5–3.0 times lower than that of 4-repeat tau isoform [19], and that exonic mutations have larger effects on 3-repeat than on 4-repeat tau isoforms [10]. Our finding showed that V337M- and R406W-mutated 3-repeat tau isoforms disrupted microtubule networks but that 4-repeat tau isoforms did not. It has been shown *in vivo* that the developmentally regulated transition in expression from 3-repeat to 4-repeat tau is associated with increased microtubule stability and decreased cytoskeletal plasticity [20]. In our study, we considered that 4-repeat tau isoforms bearing missense mutations virtually did not exhibit any effect of missense mutations on microtubule structures, probably because at least three domains in mutant tau were sufficiently capable of promoting polymerization of tubulin (Fig. 4a). It will thus be interesting to examine whether pathological filaments are composed of 3-repeat or 4-repeat isoforms, since both the V337M and R406W mutations lead to tau pathology characterized by PHF. The present findings demonstrate that the length of tau isoforms and the localization of missense mutation sites have different effects on cytoplasmic microtubule networks. These findings may explain the wide variation in clinicopathological changes associated with tauopathies.

Acknowledgements
This work was supported by Grants-in-Aid for Scientific Research from the Ministry of Education, Science and Culture, Japan.

References

1 Wischik CM, Novak M, Edwards PC, Klug A, Tichelaar W, Crowther RA (1988) Structural characterization of the core of the paired helical filament of Alzheimer disease. *Proc Natl Acad Sci USA* 85(13): 4884–8

2 Hof PR, Delacourte A, Bouras C (1992) Distribution of cortical neurofibrillary tangles in progressive supranuclear palsy: a quantitative analysis of six cases. *Acta Neuropathol* 84(1): 45–51

3 Feany MB, Mattiace LA, Dickson DW (1996) Neuropathologic overlap of progressive supranuclear palsy, Pick's disease and corticobasal degeneration. *J Neuropathol Exp Neurol* 55(1): 53–67

4 Poorkaj P, Bird TD, Wijsman E, Nemens E, Garruto RM, Anderson L, Andreadis A, Wiederholt WC, Raskind M, Schellenberg GD (1998) Tau is a candidate gene for chromosome 17 frontotemporal dementia. *Ann Neurol* 43(6): 815–25

5 Hutton M, Lendon CL, Rizzu P, Baker M, Froelich S, Houlden H, Pickering-Brown S, Chakraverty S, Isaacs A, Grover A et al (1998) Association of missense and 5'-splice-site mutations in tau with the inherited dementia FTDP-17. *Nature* 18;393(6686): 702–5

6 Clark LN, Poorkaj P, Wszolek Z, Geschwind DH, Nasreddine ZS, Miller B, Li D, Payami H, Awert F, Markopoulou K et al (1998) Pathogenic implications of mutations in the tau gene in pallido-ponto-nigral degeneration, related neurodegenerative disorders linked to chromosome 17. *Proc Natl Acad Sci USA* 95(22): 13103–7

7 Hasegawa M, Smith MJ, Iijima M, Tabira T, Goedert M (1999) FTDP-17 mutations N279K and S305N in tau produce increased splicing of exon 10. *FEBS Lett* 443(2): 93–6

8 D'Souza I, Poorkaj P, Hong M, Nochlin D, Lee VM, Bird TD, Schellenberg GD (1999) Missense and silent

tau gene mutations cause frontotemporal dementia with parkinsonism-chromosome 17 type, by affecting multiple alternative RNA splicing regulatory elements. *Proc Natl Acad Sci USA* 96(10): 5598–603

9 Hong M, Zhukareva V, Vogelsberg-Ragaglia V, Wszolek Z, Reed L, Miller BI, Geschwind DH, Bird TD, McKeel D, Goate A et al (1998) Mutation-specific functional impairments in distinct tau isoforms of hereditary FTDP-17. *Science* 282(5395): 1914–7

10 Hasegawa M, Smith MJ, Goedert M (1998) Tau proteins with FTDP-17 mutations have a reduced ability to promote microtubule assembly. *FEBS Lett* 437(3): 207–10

11 Dayanandan R, Van Slegtenhorst M, Mack TG, Ko L, Yen SH, Leroy K, Brion JP, Anderton BH, Hutton M, Lovestone S (1999) Mutations in tau reduce its microtubule-binding properties in intact cells and affect its phosphorylation. *FEBS Lett* 446(2–3): 228–32

12 Matsumura N, Yamazaki T, Ihara Y (1999) Stable expression in Chinese hamster ovary cells of mutated tau genes causing frontotemporal dementia and parkinsonism linked to chromosome 17 (FTDP-17). *Amer J Pathol* 154(6): 1649–56

13 Kanai Y, Takemura R, Oshima T, Mori H, Ihara Y, Yanagisawa M, Masaki T, Hirokawa N (1989) Expression of multiple tau isoforms and microtubule bundle formation in fibroblasts transfected with a single tau cDNA. *J Cell Biol* 109(3): 1173–84

14 Lee G, Rook SL (1992) Expression of tau protein in non-neuronal cells: microtubule binding and stabilization. *J Cell Sci* 102: 227–237

15 Arawaka S, Usami M, Sahara N, Schellenberg GD, Lee G, Mori H (1999) The tau mutation (val337met) disrupts cytoskeletal networks of microtubules. *Neuroreport* 10(5): 993–7

16 Goode BL, Feinstein SC (1994) Identification of a novel microtubule binding and assembly domain in the developmentally regulated inter-repeat region of tau. *J Cell Biol* 124: 769–782

17 Utton MA, Vandecandelaere A, Wagner U, Reynolds CH, Gibb GM, Miller CC, Bayley PM, Anderton BH (1997) Phosphorylation of tau by glycogen synthase kinase 3beta affects the ability of tau to promote microtubule self-assembly. *Biochem J* 323 (Pt 3): 741–7

18 Delacourte A, Sergeant N, Wattez A, Gauvreau D, Robitaille Y (1998) Vulnerable neuronal subsets in Alzheimer's and Pick's disease are distinguished by their tau isoform distribution and phosphorylation. *Ann Neurol* 43(2): 193–204

19 Goedert M, Jakes R (1990) Expression of separate isoforms of human tau protein: correlation with the tau pattern in brain and effects on tubulin polymerization. *EMBO J* 9(13): 4225–30

20 Goedert M, Spillantini MG, Cairns NJ, Crowther RA (1992) Tau proteins of Alzheimer paired helical filaments: abnormal phosphorylation of all six brain isoforms. *Neuron* 8: 159–168

Neuroscientific Basis of Dementia
C. Tanaka, P.L. McGeer, Y. Ihara (eds)
© 2001 Birkhäuser Verlag Basel/Switzerland

Possible role of tau phosphorylation on ER membrane in Alzheimer pathology

Toshio Kawamata[1], Taizo Taniguchi[1], Hideyuki Mukai[2], Takeshi Hashimoto[1], Hiroshi Hasegawa[1], Niu San-Yu[1], Akira Terashima[1], Masamichi Nakai[1], Minoru Yasuda[1], Kiyoshi Maeda[1], Yoshitaka Ono[2], Koho Miyoshi[1] and Chikako Tanaka[1]

[1] *Hyogo Institute for Aging Brain and Cognitive Disorders, Himeji, Japan*
[2] *Department of Biology, Faculty of Science, Kobe University, Kobe, Japan*

Introduction

The pathology of Alzheimer's disease (AD) is characterized by intracellular neurofibrillary tangles (NFTs), extracellular senile plaques (SPs), neuronal loss and a chronic inflammatory state. The number of NFTs correlates directly with the severity of dementia [1]. NFTs are composed of straight or paired helical filaments (PHFs) with a major component being an aberrantly hyperphosphorylated form of the microtubule-associated protein (MAP) tau [2]. Such abnormal filaments accumulate in the cell bodies of degenerating neurons, as well as in the neuropil threads and dystrophic neurites in SPs. In the normal brain, tau promotes the assembly and stabilization of microtubules [3]. The ability of tau to bind to microtubules is, however, down-regulated after phosphorylation, especially after local phosphorylation in the C-terminal repeats of the microtubule-binding domain [4, 5]. Tau can be phosphorylated at multiple sites by many kinases. Although abnormal tau aggregation into PHFs has long been attributed to abnormal phosphorylation of tau [2], fetal and biopsy-derived tau proteins were found to be phosphorylated at almost the same sites as PHF-tau [6]. Therefore, dysfunctions of protein phosphatases such as calcineurin or PP2A were recently postulated in AD brain tissues.

We have studied the effects on tau phosphorylation of a 120 kDa lipid/Rho-activated serine/threonine kinase, PKN, and a Ca^{2+}/calmodulin-dependent serine/threonine phosphatase, calcineurin (CaN, PP2B), both of which are enriched in the brain [7–9]. PKN consists of a regulatory N-terminal region containing unique repeats of a leucine zipper-like motif and a C-terminal catalytic domain highly homologous to that of the protein kinase C (PKC) family [9–11]. PKN interacts with and phosphorylates α-actinin, crosslinking actin and with intermediate filament proteins, leading to an inhibitory regulation of their polymerization and fibril formation [12–14]. CaN comprises a family of enzymes, and the holoenzyme is a heterodimer, consisting of the 61 kDa catalytic CaN-A and the 19 kDa regulatory CaN-B subunits. Two distinct isoforms of the CaN-A subunit, designated CaN-Aα and CaN-Aβ, were found in the central nervous system [15], although their specificities and activities have been reported to be same. PKN directly phosphorylates tau and disrupts microtubule net-

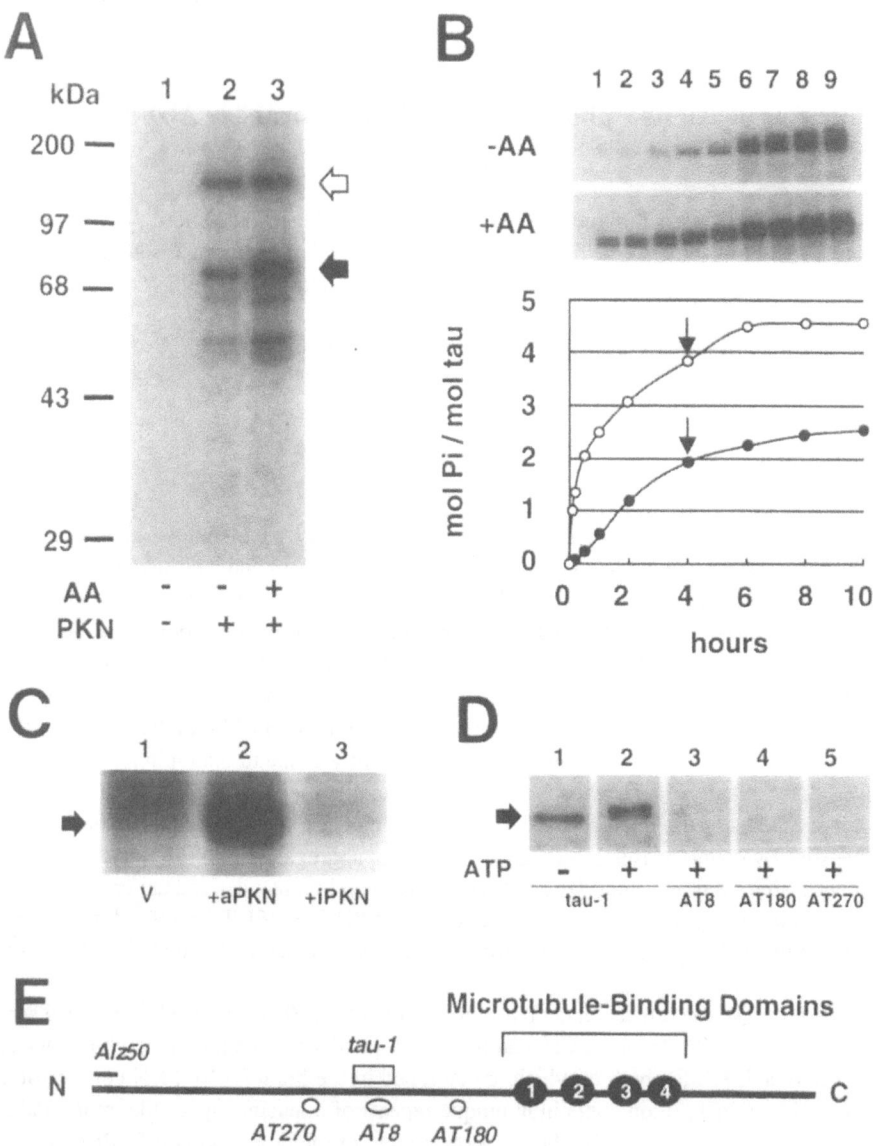

works *in vivo* [16], while CaN is known to dephosphorylate tau which is aberrantly phosphorylated in AD brain tissues [19].

To elucidate the intracellular compartment where abnormal tau phosphorylation occurs,
we determined the intraneuronal distributions of tau, PKN and CaN in relation to cell
organelles in human control and AD brains. We found the co-localization of the kinase and the

◄ Figure 1. Direct phosphorylation of human tau by PKN. *A*, SDS-PAGE and autoradiography revealed the phosphorylation. The white arrow indicates the molecular weight of autophosphorylated recombinant PKN. Recombinant human tau was incubated in assay mixture without (lane 1) or with (lanes 2 and 3) recombinant PKN in the absence (lanes 1 and 2) or presence (lane 3) of arachidonic acid (AA). The black arrow indicates the position of GST-tau fusion protein. Note the upper band of tau in lane 3, indicating molecular weight shift on more phosphorylation than in lane 2. *B*, Time course of tau phosphorylation by PKN. GST-tau was incubated with PKN in the absence (–AA or closed circles) or presence (+AA or open circles) of arachidonic acid. *C*, *In vivo* phosphorylation of tau in human neuroblastoma SK-N-MC cells transfected with vector only or PKN transgenes. Strong phosphorylation signal was seen in the cells expressing active PKN (+aPKN, lane 2), but not in those expressing no (V, lane 1) or inactive PKN (+iPKN, lane 3). *D*, Immunoblot analysis with tau-1 (lanes 1 and 2), AT8 (lane 3), AT180 (lane 4) and AT270 (lane 5) antibodies of recombinant tau phosphorylated by PKN in the absence (lane 1) or presence (lanes 2–5) of ATP. *E*, Schematic diagram of human tau isoform used in this study.

phosphatase in a subset of endoplasmic reticulum (ER)-derived vesicles in normal neurons, and the redistribution of these vesicles to intracellular NFTs in AD damaged neurons. Here we postulate a significance of tau phosphorylation on ER-derived vesicles in AD pathology.

PKN directly phosphorylates tau

Since some cytoskeletal or associated proteins have been found as the substrates for PKN, we came up with the idea that tau may be another candidate for the substrate. To test this idea, we performed *in vitro* kinase assays using recombinant human tau protein. Recombinant PKN directly phosphorylated tau *in vitro*, even in the absence of an activator, arachidonic acid, and the phosphorylation was significantly potentiated by arachidonic acid (Fig. 1A). The level of tau phosphorylation increased in a time-dependent and in an arachidonate-dependent manner (Fig. 1B). A large molecular weight shift took place concomitantly with the phosphorylation. *In vivo* phosphorylation of tau was also seen in the SK-N-MC human neuroblastoma cells expressing the active form of PKN, but not in the cells expressing only a vector or an inactive variant of PKN (Fig. 1C). These data demonstrate that PKN directly phosphorylates human tau protein [16]. To determine the phosphorylation site by PKN, a subsequent kination assay was done by using 4 recombinant tau fragments. The microtubule-binding domain exhibited the highest level of phosphorylation (Fig. 1E). Thus, these data indicate that the PKN, also known as PKC-related kinase (PRK), is classified as one of the second-messenger-activated tau kinases, including PKA or PKC.

PKN is present in a subset of granular ER in human neurons, and associated with NFTs in Alzheimer brain

Immunocytochemistry at the light and electron microscopic levels revealed that PKN was enriched in a subset of granular ER in human brains [16]. In the AD brain, degenerative neurites were intensely labeled for PKN in senile plaques. PKN was enriched in intracellular tangles, but not in extracellular tangles. The remaining neurons displayed higher immunoreactivity for PKN in AD than in control brains. Its protein level was consistently approximately the same in control and AD on immunoblot analysis. Immunoelectron microscopic study of PKN-positive control neurons revealed that dense labeling was associated with ER, multi-vesicular bodies, microtubules, Golgi bodies, secretory vesicles, late endosomes, synaptic vesicles and ribosomes, whereas weak immunoreaction was seen in the nucleoplasm. In AD tangled neurons, straight or twisted tubules were strongly immunoreactive for PKN and were decorated with numerous small vesicles intensely labeled for PKN (Fig. 2A and 2B) [16].

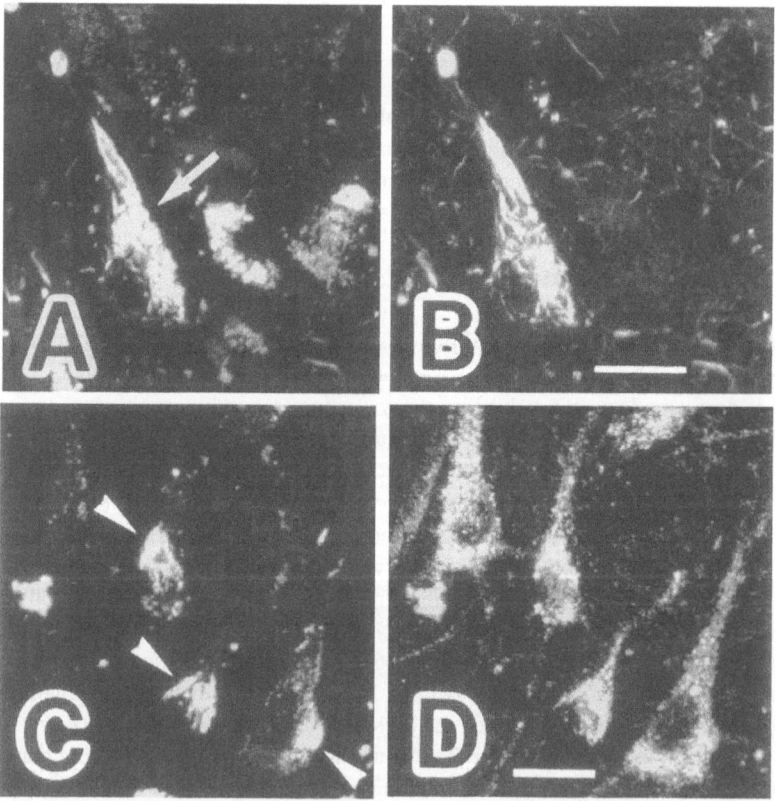

Figure 2. Association of PKN with AD pathology and a specific organelle. A, B, Double-labeling immunoflu-orescence microscopy demonstrating co-localization of PKN (A) and AT8-positive phosphorylated tau protein (B) in tangle-bearing neurons. The AT8 monoclonal antibody (Innogenetics, Zwijndrecht, Belgium, diluted at 1:10000) recognized phosphoepitopes on tau in neuropil threads, degenerative neurites and intracellular NFTs (B), which were decorated with many vesicles labeled for PKN (arrow in A). C, D, Double immunofluores-cent labeling of ERs for PKN (C) and BiP (D). Numerous ERs labeled with the monoclonal antibody to BiP (Stressgen, Victoria, British Columbia) were scattered in the cell bodies and the dendrites (D). A subset of these vesicles was also labeled for PKN (C), and such doubly labeled vesicles were translocated and associated with intracellular tangles (arrowheads in C). Figures A and B, and C and D are at the same magnification, respec-tively. Scale bar in B and D: 25 μm.

Isoform-specific involvement of calcineurin (PP2B) in AD pathology

CaN is a well-characterized protein phosphatase also known as PP2B, one of four principal types of serine/threonine-specific phosphatases and the only phosphatase activated by Ca^{2+} or calmodulin [15, 17, 18]. So far, it has been implicated in crucial neuronal functions, including isoform-specific involvement of CaN-Aα and CaN-Aβ in neuronal reorganization after ischemic brain injury [20] and modulation of the phosphorylation state of tau [19].

Isoform-specific polyclonal antibodies were raised in rabbits against CaN-Aα and CaN-Aβ, and their specificities were confirmed with immunoblot analysis using the recombinant isoforms of the CaN catalytic subunit [20]. Distribution of both CaN-Aα and CaN-Aβ isoforms was localized in control and AD brains with immunohistochemistry. The expression of CaN isoforms was heterogeneous and an isoform-specific localization was observed in the hippocampus. In the AD hippocampus, CaN-Aα immunoreactivity was clearly up-regulated in the remaining neurons and was accumulated in the tortuous neurites. The vast majority of neuronal staining for CaN-Aβ was diminished, but abnormal structures such as NFTs, neuropil threads and degenerative neurites were immunoreactive for CaN-Aβ (Fig. 3A and 3C).

Association of PKN and CaN with NFT

We examined the association of PKN (Fig. 2A and 2B) or CaN-Aβ (Fig. 3A and 3B) with developing tangles in AD hippocampus. In tangle-bearing neurons, PKN or CaN-Aβ was localized in small vesicles clustering in association with tangles (arrows in Fig. 2A and 3A), which is positive for abnormal phosphorylated tau (Fig. 2B and 3B). In addition, PKN was co-localized in tangles with abnormal tau protein, having either phosphoepitopes recognized by AT8/180/270 antibodies or the intramolecular linkage recognized by Alz50 antibody (Fig. 2 A), as well as CaN-Aβ (Fig. 3A). These results indicate the association of PKN or CaN-Aβ with tangles at an early stage of their development.

To characterize the PKN- or CaN-Aβ-positive small vesicles, intraneuronal association of PKN or CaN-Aβ with cell organelles was investigated using markers for ER, Golgi body and lysosome. PKN or CaN-Aβ was localized in a subset of ER-derived vesicles (Fig. 2C and 2D, or 3C and 3D). They appeared in perinuclear or apical compartments of normal neurons. In damaged AD neurons, such ER-derived vesicles were redistributed toward intracellular tangles (Fig. 2C and 3C), while other granular ERs negative for either of the enzymes showed no change in cellular distribution (Fig. 2D and 3D). PKN and CaN-Aβ look much alike, both in their cellular localization in normal neurons and in their redistribution in AD-damaged neurons.

PKN or CaN-Aβ occasionally translocated into nuclei in some AD neurons, which is consistent with the fact of nuclear translocation of PKN in cultured fibroblasts under stress [21] and with the report that CaN is implicated in the transcriptional regulation or the apoptosis [22–24].

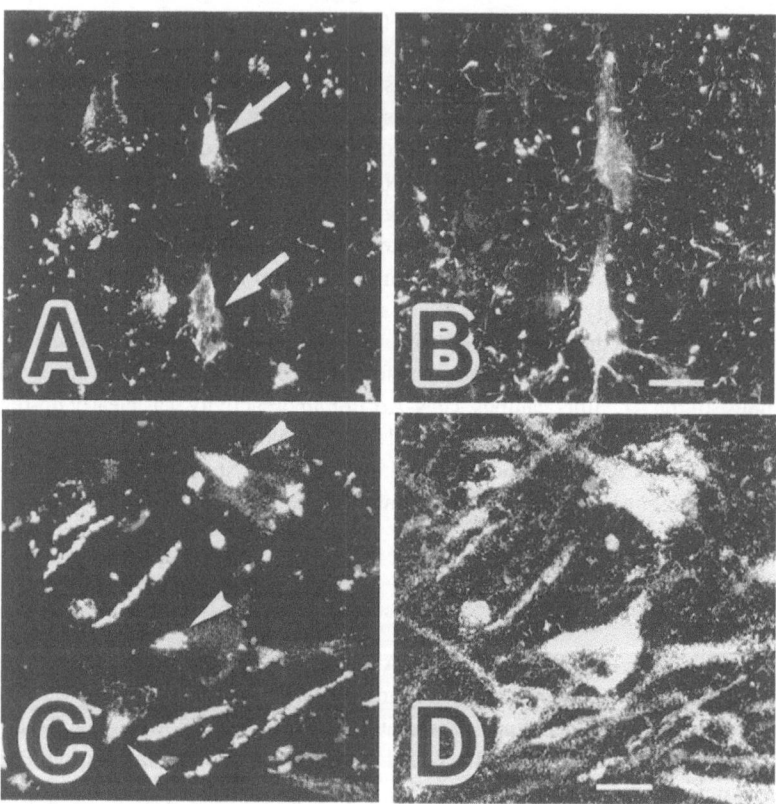

Figure 3. Association of CaN-Aβ with AD pathology and a specific organelle. *A, B*, Double-labeling immuno-fluorescence microscopy illustrating colocalization of CaN-Aβ (A) and AT8-positive phosphorylated tau protein (B) in tangled neurons. Intracellular tangles labeled with AT8 (B) were decorated with CaN-Aβ-positive vesicles (arrows in A). *C, D*, Double immunofluorescent labeling of ERs for CaN-Aβ (C) and BiP (D). Among many ER-derived vesicles scattering in cell bodies and dendrites (D), CaN-Aβ was localized in a subset of the vesicles associated with intracellular NFTs (arrowheads in C). Figures A and B, and C and D are at the same magnification, respectively. Scale bar in B and D: 25 μm.

Conclusion

Our data suggest that a specific subset of granular ERs redistribute to tangles in degenerating AD neurons, and that these ERs scaffold a protein kinase, PKN, and a protein phosphatase, calcineurin-Aβ, both of which work on tau protein. Further investigation of the mechanisms by which the enzymes anchor to these ER-derived vesicles and by which the vesicles distribute within neurons should be needed.

Acknowledgments
We are grateful to Dr. T. Kuno, Dr. H. Mori and Dr. P. Davies for the generous gifts of the expression vectors containing rat calcineurin-Aα, calcineurin-Aβ, and tau cDNAs, and Alz50 antibody, respectively. We also thank Ms. M. Obana, Ms. A. Hori and Ms. M. Sumida for excellent technical assistance. This work was supported in part by a grant from the Sasakawa Research Foundation (Japan).

References

1 Alafuzoff I, Iqbal K, Friden H, Adolfsson R, Winblad B (1987) Histopathological criteria for progressive dementia disorders: clinicopathological correlation and classification by multivariate data analysis. *Acta Neuropathol* 74: 209–225

2 Billingsley ML, Kincaid RL (1997) Regulated phosphorylation and dephosphorylation of tau protein: effects on microtubule interaction, intracellular trafficking and neurodegeneration. *Biochem J* 323: 577–591

3 Drubin DG, Kirschner MW (1986) Tau protein function in living cells. *J Cell Biol* 103: 2739–2746

4 Johnson GV (1992) Differential phosphorylation of tau by cyclic AMP-dependent protein kinase and Ca^{2+}/calmodulin-dependent protein kinase II: metabolic and functional consequences. *J Neurochem* 59: 2056–2062

5 Drewes G, Ebneth A, Preuss U, Mandelkow E-M, Mandelkow E (1997) MARK, a novel family of protein kinases that phosphorylate microtubule-associated proteins and trigger microtubule disruption. *Cell* 89: 297–308

6 Matsuo ES, Shin RW, Billingsley ML, Van deVoorde A, O'Connor M, Trojanowski JQ, Lee VM (1994) Biopsy-derived adult human brain tau is phosphorylated at many of the same sites as Alzheimer's disease paired helical filament tau. *Neuron* 13: 989–1002

7 Goto S, Matsukado Y, Mihara Y, Inoue N, Miyamoto E (1986) The distribution of calcineurin in rat brain by light and electron microscopic immunohistochemistry and enzyme-immunoassay. *Brain Res* 397: 161–172

8 Kincaid RL, Balaban CD, Billingsley ML (1987) Differential localization of calmodulin-dependent enzymes in rat brain: Evidence for selective expression of cyclic nucleotide phosphodiesterase in specific neurons. *Proc Natl Acad Sci USA* 84: 1118–1122

9 Kitagawa M, Mukai H, Shibata H, Ono Y (1995) Purification and characterization of a fatty acid-activated protein kinase (PKN) from rat testis. *Biochem J* 310: 657–664

10 Mukai H, Ono Y (1994) A novel protein kinase with leucine zipper-like sequences: its catalytic domain is highly homologous to that of protein kinase C. *Biochem Biophys Res Commun* 199: 897–904

11 Mukai H, Kitagawa M, Shibata H, Takanaga H, Mori K, Shimakawa M, Miyahara M, Hirao K, Ono Y (1994) Activation of PKN, a novel 120-Kda protein kinase with leucine zipper-like sequences, by unsaturated fatty acids and by limited proteolysis. *Biochem Biophys Res Commun* 204: 348–356

12 Mukai H, Toshimori M, Shibata H, Kitagawa M, Shimakawa M, Miyahara M, Sunakawa H, Ono Y (1996) PKN associates and phosphorylates the head-rod domain of neurofilament protein. *J Biol Chem* 271: 9816–9822

13 Matsuzawa K, Kosako H, Inagaki N, Shibata H, Mukai H, Ono Y, Amano M, Kaibuchi K, Matsuura Y, Azuma I et al (1997) Domain-specific phosphorylation of vimentin and glial fibrillary acidic protein by PKN. *Biochem Biophys Res Commun* 234: 621–625

14 Mukai H, Toshimori M, Shibata H, Takanaga H, Kitagawa M, Miyahara M, Shimakawa M, Ono Y (1997) Interaction of PKN with α-actinin. *J Biol Chem* 272: 4740–4746

15 Cohen P (1989) The structure and regulation of protein phosphatases. *Annu Rev Biochem* 58: 453–508

16 Kawamata T, Taniguchi T, Mukai H, Kitagawa M, Hashimoto T, Maeda K, Ono Y, Tanaka C (1998) A protein kinase, PKN, accumulates in Alzheimer neurofibrillary tangles and associated endoplasmic reticulum-derived vesicles and phosphorylates tau protein. *J Neurosci* 18: 7402–7410

17 Klee CB, Crouch TH, Krinks MH (1979) Calcineurin: a calcium- and calmodulin-binding protein of the nervous system. *Proc Natl Acad Sci USA* 76: 6270–6273

18 Stewart AA, Ingebritsen TS, Manalan A, Klee CB, Cohen P (1982) Discovery of a Ca^{2+}- and calmodulin-dependent protein phosphatase: probable identity with calcineurin (CaM-BP80). *FEBS Lett* 137: 80–84

19 Drewes G, Mandelkow E-M, Baumann K, Goris J, Merlevede W, Mandelkow E (1993) Dephosphorylation of tau protein and Alzheimer paired helical filaments by calcineurin and phosphatase-2A. *FEBS Lett* 336: 425–432

20 Hashimoto T, Kawamata T, Saito N, Sasaki M, Nakai M, Niu S-Y, Taniguchi T, Terashima A, Yasuda M, Maeda K et al (1998) Isoform-specific redistribution of calcineurin Aα and Aβ in the hippocampal CA1 region of gerbils after transient ischemia. *J Neurochem* 70: 1289–1298
21 Mukai H, Miyahara M, Sunakawa H, Shibata H, Toshimori M, Kitagawa M, Shimakawa M, Takanaga H, Ono Y (1996) Translocation of PKN from the cytosol to the nucleus induced by stresses. *Proc Natl Acad Sci USA* 93: 10195–10199
22 Shibasaki F, McKeon F (1995) Calcineurin functions in Ca(2+)-activated cell death in mammalian cells. *J Cell Biol* 131: 735–743
23 Ankarcrona M, Dypbukt JM, Orrenius S, Nicotera P (1996) Calcineurin and mitochondrial function in glutamate-induced neuronal cell death. *FEBS Lett* 394: 321–324
24 Moore AN, Kampfl AW, Zhao X, Hayes RL, Dash PK (1999) Sphingosine-1-phosphate induces apoptosis of cultured hippocampal neurons that requires protein phosphatases and activator protein-1 complexes. *Neuroscience* 94: 405–415

Pathogenesis of dementia – synuclein

Pathogenesis of dementia: updating the role of synuclein pathology in sporadic and hereditary Alzheimer's disease

John E. Duda, Virginia M.-Y. Lee and John Q. Trojanowski,

Center for Neurodegenerative Disease Research, University of Pennsylvania School of Medicine, Philadelphia, PA, USA

Introduction

α-Synuclein (α-syn) is a 140 amino acid highly conserved [1] protein that is expressed in neurons and highly enriched in the presynaptic terminal [2, 3]. The initial suggestion that α-syn may play a role in the neuropathology of dementia came from studies by Ueda and colleagues who demonstrated that 2 amino acid fragments of α-syn co-purified with amyloid from Alzheimer's disease (AD) brain and they called these fragments the non-Aβ component of amyloid plaques or NAC [4]. Upon cloning of a larger precursor protein, referred to as the NAC precursor (NACP), homology was found with synuclein, a protein identified in cholinergic synaptic vesicle preparations of the Torpedo ray [5]. Efforts to understand the contribution of α-syn to neurodegenerative diseases were further stimulated by the discovery of two mutations in the α-syn gene in rare kindreds with autosomal dominant Parkinson's disease (PD) [6, 7]. These efforts led to the identification of α-syn as a major component of the Lewy bodies (LBs) and dystrophic neurites of PD, dementia with Lewy bodies (DLB), and the LB variant of AD (LBVAD) [8–16] as well as the glial and neuronal cytoplasmic inclusions of multiple system atrophy (MSA) [16–25]. Concurrent investigations also postulated a role for α-syn in the regulation of synaptic plasticity with the demonstration of a relationship between the expression of synelfin, the avian homologue of α-syn, and synaptic plasticity in the Zebra finch song learning circuit [2]. Recently, further investigations into the role of α-syn in AD have raised additional questions about synuclein pathologies and neurodegenerative diseases.

Role of α-syn in AD β-amyloid plaques

Our understanding of the role of α-syn in AD neuropathology has been challenged by re-examination of the immunohistochemical profile of α-syn within the brains of AD patients. Previously, a distinction had been made between the immunolabeling of AD brains with antisera that distinguished NAC from full-length α-syn. While antibodies to the N- and C-terminal of α-syn produced a punctate pattern of immunostaining consistent with a synaptic localization and a neuritic pattern at the periphery of Aβ amyloid plaques in AD brains [1, 26, 27], subsequent studies employed 2 rabbit polyclonal antibodies raised to peptides in the

NAC region to immunolabel Aβ-amyloid plaques in AD brains [27]. Masliah and colleagues found that these antibodies immunolabeled mature plaques more frequently than diffuse plaques, and the core more than the periphery of plaques [27]. Recently, two separate groups have reported data that raise questions about the presence of NAC in Aβ-amyloid plaques of AD brains [10, 28]. These studies used novel antibodies generated to disparate peptides within the NAC region to further examine this issue. Bayer and colleagues employed antibodies generated to the N-terminus (amino acids 18–35) and NAC region (amino acids 75–91) of α-syn, and they confirmed a synaptic localization and robust immunostaining of LBs with these antibodies, but they were unable to detect immunostaining of mature Aβ-amyloid plaques in AD brains [10]. In a subsequent study Culvenor and colleagues developed three novel rabbit polyclonal antibodies to amino acids 1–18, 75–91 and 116–131 in α-syn which correspond to the N-terminal region, the NAC region and the C-terminal region of α-syn, respectively [28]. Whereas the authors confirmed synaptic and LB immunoreactivity with each of these antibodies, they did not observe immunoreactivity in mature Aβ-amyloid plaques. Utilizing immunofluorescent double labeling methods, they also were able to demonstrate immunostaining of dystrophic globular neurites at the periphery of Aβ-amyloid plaques in the CA1 region of the hippocampus in patients with mixed DLB and AD pathology, using antisera to the C-terminal and NAC region of α-syn. These findings suggest the presence of full-length α-syn within these neurites, and that the purification of NAC from isolated Aβ-amyloid plaques may have been due to the presence of these neurites in the amyloid isolated from AD brains. However, further studies are needed to substantiate this.

Role of α-syn mutations in AD

Genetic studies concerning α-syn in AD have also led to contradictory results. Early studies found no mutations in α-syn linked to AD [29, 30] after mutations in α-syn (A53T, A30P) were found to be pathogenic for familial PD [6, 7, 31], other studies showed that these mutations were not linked to AD [32]. However, Xia and colleagues suggested that a dinucleotide repeat polymorphism in allele 2 of the α-syn promoter was significantly associated with healthy individuals that carry an apolipoprotein E4 (ApoE ε4) allele, a known risk factor for developing sporadic AD [30]. Thus, these authors proposed that the polymorphism may confer a protective effect on the development of AD. However, a subsequent analysis performed with a similar number of AD and healthy controls who carried the ApoE ε4 allele failed to replicate this finding or any association between AD and other α-syn alleles [33]. However, the low frequency of allele 2 in the general population may require additional studies with larger cohorts to resolve this question.

Role of α-syn in hereditary AD

While the role of α-syn in sporadic AD remains controversial, two recent studies have implicated α-syn in familial AD (FAD) caused by mutations in amyloid precursor protein (APP) and presenilin genes and Down's syndrome [34, 35]. It has been demonstrated previously with anti-ubiquitin immumohistochemistry that LBs are present in some cases of FAD due to mutations in APP or presenilin genes [36]. Lippa et al. [34] extended these observations using antibodies to α-syn and immunohistochemistry by evaluating brain tissue from 74 FAD patients with mutations in either the presenilin or APP genes. Overall, 22% of FAD brains had α-syn-immunopositive LBs. Moreover, of the cases from which the amygdala was available for evaluation, 63% displayed α-syn-immunopositive LBs. In addition, biochemical analysis of FAD amygdala revealed decreased α-syn solubility when compared to control cases. The presence of α-syn-immunopositive LBs was not specifically associated with a particular FAD mutation, raising the possibility of a final common pathway initiated by these genetic influences in AD that leads to the formation of LBs by as yet unknown mechanisms.

Another genetic abnormality that predisposes to the development of AD is trisomy 21 in Down's syndrome [37], and Lippa et al. [35] showed that α-syn-immunopositive LBs and dystrophic neurites occur in 50% of Down's syndrome cases and again these LBs were most prominent in the amygdala, with significantly fewer LBs elsewhere in the brain. The intriguing predilection for LB formation in the amygdala in all of these genetic forms of AD suggests that there might be a selective vulnerability in amygdala neurons for LB formation that warrants further investigation.

Novel axonal synuclein pathology in PD and DLB brains

DLB represents the second most common cause of dementia in the elderly [38, 39, 40], and recently Galvin et al. [41] reported studies of the hippocampal formation from patients with PD, DLB, AD and other diseases using antibodies to α-, β- and γ-synucleins. Notably, β- and γ-synuclein (β- and γ-syn, respectively) are two other members of the synuclein family of proteins [42–45] that had not been implicated previously in neurodegenerative disease [9, 24, 46]. However, Galvin et al. reported novel presynaptic axonal pathology in the hippocampal dentate, hilum and CA2/3 regions that were immunolabeled with antibodies specific to α- and β-syn in PD and DLB brains but not controls. In addition, novel axonal spheroid-like lesions appeared in the molecular layer of the hippocampal dentate immunolabeled with an antibody specific to γ-syn but not α- or β-syn specific antibodies. These lesions co-localized with aggregations of other synaptic proteins, suggesting the possibility of synaptic dysfunction, and presumably, that memory dysfunction might be a consequence of these lesions in the hippocampal formation. While these lesions were not found in pure AD cases, the possibility remains that these lesions may be found in LBVAD, a common subtype of AD, and account for some of the dementia present in this entity.

Future directions

While the bulk of α-syn research focuses on its role in PD, DLB, LBVAD and MSA, collectively referred to as synucleinopathies [47], continued efforts should focus on the role of α-syn and other synucleins in AD and other disorders with prominent synuclein pathologies. Indeed, emerging evidence suggests that abnormal protein-protein interactions may be implicated in the pathogenesis of many different neurodegenerative diseases. Similarly, understanding the role that genetic and epigenetic risk factors play in these interactions also may provide clues to the pathogenesis of neurodegeneration in hereditary as well as common sporadic neurodegenerative diseases.

Acknowledgements
The authors would like to acknowledge the families of the many patients studied by our group and others that make advances in this research possible. The studies summarized here from our laboratory were supported by grants from the National Institute on Aging of the National Institutes of Health and the Alzheimer's Foundation.

References

1 Jakes R, Spillantini MG, Goedert M (1994) Identification of two distinct synucleins from human brain. *FEBS Lett* 345: 27–32
2 George JM, Jin H, Woods WS, Clayton DF (1995) Characterization of a novel protein regulated during the critical period for song learning in the zebra finch. *Neuron* 15: 361–372
3 Iwai A, Masliah E, Yoshimoto M, Ge N, Flanagan L, de Silva HA, Kittel A, Saitoh T (1995) The precursor protein of non-A beta component of Alzheimer's disease amyloid is a presynaptic protein of the central nervous system. *Neuron* 14: 467–475
4 Ueda K, Fukushima H, Masliah E, Xia Y, Iwai A, Yoshimoto M, Otero DA, Kondo J, Ihara Y, Saitoh T (1993) Molecular cloning of cDNA encoding an unrecognized component of amyloid in Alzheimer disease. *Proc Natl Acad Sci USA* 90: 11282–11286
5 Maroteaux L, Campanelli JT, Scheller RH (1988) Synuclein: a neuron-specific protein localized to the nucleus and presynaptic nerve terminal. *J Neurosci* 8: 2804–2815
6 Kruger R, Kuhn W, Muller T, Woitalla D, Graeber M, Kosel S, Przuntek H, Epplen JT, Schols L, Riess O (1998) Ala30Pro mutation in the gene encoding alpha-synuclein in Parkinson's disease. *Nat Genet* 18: 106–108
7 Polymeropoulos MH, Lavedan C, Leroy E, Ide SE, Dehejia A, Dutra A, Pike B, Root H, Rubenstein J, Boyer R et al (1997) Mutation in the alpha-synuclein gene identified in families with Parkinson's disease. *Science* 276: 2045–2047
8 Arima K, Ueda K, Sunohara N, Hirai S, Izumiyama Y, Tonozuka-Uehara H, Kawai M (1998) Immunoelectron-microscopic demonstration of NACP/alpha-synuclein-epitopes on the filamentous component of Lewy bodies in Parkinson's disease and in dementia with Lewy bodies. *Brain Res* 808: 93–100
9 Baba M, Nakajo S, Tu PH, Tomita T, Nakaya K, Lee VM, Trojanowski JQ, Iwatsubo T (1998) Aggregation of alpha-synuclein in Lewy bodies of sporadic Parkinson's disease and dementia with Lewy bodies. *Amer J Pathol* 152: 879–884
10 Bayer TA, Jakala P, Hartmann T, Havas L, McLean C, Culvenor JG, Li QX, Masters CL, Falkai P, Beyreuther K (1999) Alpha-synuclein accumulates in Lewy bodies in Parkinson's disease and dementia with Lewy bodies but not in Alzheimer's disease beta-amyloid plaque cores. *Neurosci Lett* 266: 213–216
11 Irizarry MC, Growdon W, Gomez-Isla T, Newell K, George JM, Clayton DF, Hyman BT (1998) Nigral and cortical Lewy bodies and dystrophic nigral neurites in Parkinson's disease and cortical Lewy body disease contain alpha-synuclein immunoreactivity. *J Neuropathol Exp Neurol* 57: 334–337
12 Mezey E, Dehejia AM, Harta G, Tresser N, Suchy SF, Nussbaum RL, Brownstein MJ, Polymeropoulos MH (1998) Alpha synuclein is present in Lewy bodies in sporadic Parkinson's disease. *Mol Psychiatr* 3: 493–499

13 Spillantini MG, Schmidt ML, Lee VM, Trojanowski JQ, Jakes R, Goedert M (1997) Alpha-synuclein in Lewy bodies. *Nature* 388: 839–840

14 Takeda A, Hashimoto M, Mallory M, Sundsumo M, Hansen L, Sisk A, Masliah E (1998) Abnormal distribution of the non-A beta component of Alzheimer's disease amyloid precursor/alpha-synuclein in Lewy body disease as revealed by proteinase K and formic acid pretreatment. *Lab Invest* 78: 1169–1177

15 Wakabayashi K, Matsumoto K, Takayama K, Yoshimoto M, Takahashi H (1997) NACP, a presynaptic protein, immunoreactivity in Lewy bodies in Parkinson's disease. *Neurosci Lett* 239: 45–4817

16 Wakabayashi K, Hayashi S, Kakita A, Yamada M, Toyoshima Y, Yoshimoto M, Takahashi H (1998) Accumulation of alpha-synuclein/NACP is a cytopathological feature common to Lewy body disease and multiple system atrophy. *Acta Neuropathol* 96: 445–452

17 Arima K, Ueda K, Sunohara N, Arakawa K, Hirai S, Nakamura M, Tonozuka-Uehara H, Kawai M (1998) NACP/alpha-synuclein immunoreactivity in fibrillary components of neuronal and oligodendroglial cytoplasmic inclusions in the pontine nuclei in multiple system atrophy. *Acta Neuropathol* 96: 439–444

18 Dickson DW, Liu W, Hardy J, Farrer M, Mehta N, Uitti R, Mark M, Zimmerman T, Golbe L, Sage J et al (1999) Widespread alterations of alpha-synuclein in multiple system atrophy. *Amer J Pathol* 155: 1241–1251

19 Dickson DW, Lin W, Liu WK, Yen SH (1999) Multiple system atrophy: a sporadic synucleinopathy. *Brain Pathol* 9: 721–732

20 Gai WP, Power JH, Blumbergs PC, Blessing WW (1998) Multiple-system atrophy: a new alpha-synuclein disease? *Lancet* 352: 547–548

21 Gai WP, Power JH, Blumbergs PC, Culvenor JG, Jensen PH (1999) Alpha-synuclein immunoisolation of glial inclusions from multiple system atrophy brain tissue reveals multiprotein components. *J Neurochem* 73: 2093–2100

22 Mezey E, Dehejia A, Harta G, Papp MI, Polymeropoulos MH, Brownstein MJ (1998) Alpha synuclein in neurodegenerative disorders: murderer or accomplice? *Nat Med* 4: 755–757

23 Spillantini MG, Crowther RA, Jakes R, Cairns NJ, Lantos PL, Goedert M (1998) Filamentous alpha-synuclein inclusions link multiple system atrophy with Parkinson's disease and dementia with Lewy bodies. *Neurosci Lett* 251: 205–208

24 Tu PH, Galvin JE, Baba M, Giasson B, Tomita T, Leight S, Nakajo S, Iwatsubo T, Trojanowski JQ, Lee VM (1998) Glial cytoplasmic inclusions in white matter oligodendrocytes of multiple system atrophy brains contain insoluble alpha-synuclein. *Ann Neurol* 44: 415–422

25 Wakabayashi K, Yoshimoto M, Tsuji S, Takahashi H (1998) Alpha-synuclein immunoreactivity in glial cytoplasmic inclusions in multiple system atrophy. *Neurosci Lett* 249: 180–182

26 Irizarry MC, Kim TW, McNamara M, Tanzi RE, George JM, Clayton DF, Hyman BT (1996) Characterization of the precursor protein of the non-A beta component of senile plaques (NACP) in the human central nervous system. *J Neuropathol Exp Neurol* 55: 889–895

27 Masliah E, Iwai A, Mallory M, Ueda K, Saitoh T (1996) Altered presynaptic protein NACP is associated with plaque formation and neurodegeneration in Alzheimer's disease. *Amer J Pathol* 148: 201–210

28 Culvenor JG, McLean CA, Cutt S, Campbell BC, Maher F, Jäkälä P, Hartmann T, Beyreuther K, Masters CL, Li QX (1999) Non-A beta component of Alzheimer's disease amyloid (NAC) revisited: NAC and alpha-synuclein are not associated with a beta amyloid. *Amer J Pathol* 155: 1173–1181

29 Campion D, Martin C, Heilig R, Charbonnier F, Moreau V, Flaman JM, Petit JL, Hannequin D, Brice A, Frebourg T (1995) The NACP/synuclein gene: chromosomal assignment and screening for alterations in Alzheimer disease. *Genomics* 26: 254–257

30 Xia Y, Rohan de Silva HA, Rosi BL, Yamaoka LH, Rimmler JB, Pericak-Vance MA, Roses AD, Chen X, Masliah E, DeTeresa R et al (1996) Genetic studies in Alzheimer's disease with an NACP/alpha-synuclein polymorphism. *Ann Neurol* 40: 207–215

31 Papadimitriou A, Veletza V, Hadjigeorgiou GM, Patrikiou A, Hirano M, Anastasopoulos I (1999) Mutated alpha-synuclein gene in two Greek kindreds with familial PD: incomplete penetrance? *Neurology* 52: 651–654

32 Higuchi S, Arai H, Matsushita S, Matsui T, Kimpara T, Takeda A, Shirakura K (1998) Mutation in the alpha-synuclein gene and sporadic Parkinson's disease, Alzheimer's disease, and dementia with lewy bodies. *Exp Neurol* 153: 164–166

33 Hellman NE, Grant EA, Goate AM (1998) Failure to replicate a protective effect of allele 2 of NACP/alpha-synuclein polymorphism in Alzheimer's disease: an association study. *Ann Neurol* 44: 278–28134

34 Lippa CF, Fujiwara H, Mann DM, Giasson B, Baba M, Schmidt ML, Nee LE, O'Connell B, Pollen DA,

George-Hyslop P et al (1998) Lewy bodies contain altered alpha-synuclein in brains of many familial Alzheimer's disease patients with mutations in presenilin and amyloid precursor protein genes. *Amer J Pathol* 153: 1365–1370

35 Lippa CF, Schmidt ML, Lee VM, Trojanowski JQ (1999) Antibodies to alpha-synuclein detect Lewy bodies in many Down's syndrome brains with Alzheimer's disease. *Ann Neurol* 45: 353–357

36 Lippa CF, Smith TW, Nee L, Robitaille Y, Crain B, Dickson D, Pulaski-Salo D, Pollen DA (1995) Familial Alzheimer's disease and cortical Lewy bodies: is there a genetic susceptibility factor? *Dementia* 6: 191–19437

37 Burger PC, Vogel FS (1973) The development of the pathologic changes of Alzheimer's disease and senile dementia in patients with Down's syndrome. *Amer J Pathol* 73: 457–476

38 Dickson DW, Davies P, Mayeux R, Crystal H, Horoupian DS, Thompson A, Goldman JE (1987) Diffuse Lewy body disease. Neuropathological and biochemical studies of six patients. *Acta Neuropathol* 75: 8–15

39 McKeith IG, Galasko D, Kosaka K, Perry EK, Dickson DW, Hansen LA, Salmon DP, Lowe J, Mirra SS, Byrne EJ et al (1996) Consensus guidelines for the clinical and pathologic diagnosis of dementia with Lewy bodies (DLB): report of the consortium on DLB international workshop. *Neurology* 47: 1113–1124

40 Trojanowski JQ, Schmidt ML, Shin RW, Bramblett GT, Rao D, Lee VM (1993) Altered tau and neurofilament proteins in neuro-degenerative diseases: diagnostic implications for Alzheimer's disease and Lewy body dementias. *Brain Pathol* 3: 45–54

41 Galvin JE, Uryu K, Lee VM, Trojanowski JQ (1999) Axon pathology in Parkinson's disease and Lewy body dementia hippocampus contains alpha-, beta-, and gamma-synuclein. *Proc Natl Acad Sci USA* 96: 13450–13455

42 Ji H, Liu YE, Jia T, Wang M, Liu J, Xiao G, Joseph BK, Rosen C, Shi YE (1997) Identification of a breast cancer-specific gene, BCSG1, by direct differential cDNA sequencing. *Cancer Res* 57: 759–764

43 Nakajo S, Tsukada K, Omata K, Nakamura Y, Nakaya K (1993) A new brain-specific 14-kDa protein is a phosphoprotein. Its complete amino acid sequence and evidence for phosphorylation. *Eur J Biochem* 217: 1057–1063

44 Ninkina NN, Alimova-Kost MV, Paterson JW, Delaney L, Cohen BB, Imreh S, Gnuchev NV, Davies AM, Buchman VL (1998) Organization, expression and polymorphism of the human persyn gene. *Hum Mol Genet* 7: 1417–1424

45 Tobe T, Nakajo S, Tanaka A, Mitoya A, Omata K, Nakaya K, Tomita M, Nakamura Y (1992) Cloning and characterization of the cDNA encoding a novel brain-specific 14-kDa protein. *J Neurochem* 59: 1624–1629

46 Clayton DF, George JM (1998) The synucleins: a family of proteins involved in synaptic function, plasticity, neurodegeneration and disease. *Trends Neurosci* 21: 249–254

47 Hardy J, Gwinn-Hardy K (1998) Genetic classification of primary neurodegenerative disease. *Science* 282: 1075–1079

Neuroscientific Basis of Dementia
C. Tanaka, P.L. McGeer, Y. Ihara (eds)
© 2001 Birkhäuser Verlag Basel/Switzerland

α-Synuclein/NACP and neurodegeneration

Seigo Tanaka, Masanori Takehashi, Naomi Matoh and Kunihiro Ueda

Laboratory of Molecular Clinical Chemistry, Institute for Chemical Research, Kyoto University, Uji, Kyoto, Japan

Introduction

Amyloid deposition in the senile plaque cores is one of the major neuropathological features in the brain with Alzheimer's disease (AD). A non-Aβ component (NAC) of AD amyloid was originally identified in senile plaques as a protein other than Aβ [1]. NAC, consisting of at least 35 amino acids, is supposed to be produced from a precursor protein, termed NACP, by an unkown mechanism. NACP is identical with α-synuclein that belongs to the synuclein family. This family comprises α-, β- and γ-synucleins [2–4], among which only the α-synuclein has a NAC domain. NAC is located in the most hydrophobic portion of the α-synuclein molecule (Fig. 1). The α-synuclein has seven incompletely repeated KTKEGV motifs and no signal peptide sequence or N-linked glycosylation sites.

Recently, two types of mutation, Ala30Pro and Ala53Thr, in the α-synuclein gene were found in some families of Parkinson's disease (PD) [5, 6]. Furthermore, α-synuclein was found to accumulate in Lewy body (LB) [7, 8], which is a hallmark of idiopathic PD. These findings suggest the possible implication of α-synuclein in pathogenesis of both PD and AD.

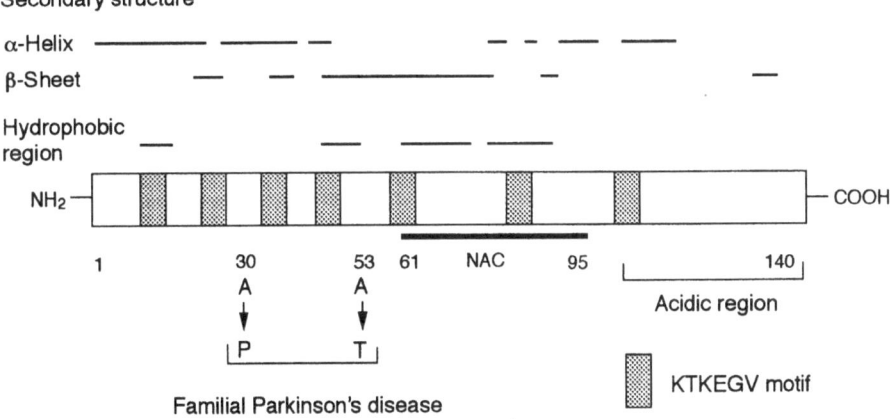

Figure 1. Structure of α-synuclein.

Physiological functions of α-synuclein

α-Synuclein was first isolated by expression screening of a cDNA library with antibodies against purified cholinergic synaptic vesicles from *Torpedo* [2]. The α-synuclein was associated with synaptic vesicles in the rat brain as revealed by immunoelectron microscopy [9]. In order to clarify its physiological functions, we transiently transfected α-synuclein cDNA into PC12 and COS7 cells. We found the α-synuclein to be distributed in the cytosol and neurites of differentiated PC12 cells. A confocal laser microscopic study showed co-localization of α-synuclein with synaptic marker synaptophysin, indicating possible role(s) of α-synuclein in synaptic functions. This view might be supported by the finding that expression of synelfin, a songbird homologue of α-synuclein, was correlated with the plasticity of a developing song control system [10].

Amyloidogenicity of NAC

NAC and Aβ protein are reportedly co-localized in the core of senile plaques. However, immunohistochemical studies showed that anti-NAC antibody stained the central portion more strongly than the peripheral portion of amyloid in senile plaques, whereas anti-Aβ-protein antibody stained the amyloid evenly in AD brain [11]. The ratio of NAC to Aβ in senile plaques was estimated to be less than 1:10 [1]. Structural analysis has shown that NAC has a strong tendency to form a β-sheet structure. We confirmed that fibrils are easily formed from synthetic NAC peptides at 37 °C, and stained with Congo red, give amyloid-suggestive green images under polarized light. We then analyzed the kinetics of amyloid fibril formation by monitoring thioflavine T fluorescence. Formation of $A\beta_{1-40}$ fibrils was facilitated, with no nucleation phase, by the addition of preformed NAC fibrils, suggesting that NAC fibrils could serve as a nucleus for the amyloid formation. Direct interaction between NAC and Aβ was demonstrated *in vitro* [12]. Whether such an interaction occurs *in vivo* remains to be investigated, as does the temporal and topological distribution of the two amyloidogenic proteins in the brain with AD.

In order to analyze the process of NAC production from α-synuclein, we metabolically labeled α-synuclein with [35S]methionine in transfected COS7 cells. The half-life of 35S-labeled α-synuclein in COS7 cells was longer than 24 h. We could detect no proteolytic products either in the cell lysate or culture medium. Although how α-synuclein is processed to NAC remains to be clarified, it is possible that α-synuclein is released from damaged neurites and extracellularly proteolysed to produce NAC.

Neurotoxicity of NAC amyloid

In view of the neurotoxicity of some amyloidogenic peptides, such as Aβ protein and prion protein [13, 14], we investigated the cytotoxicity of NAC fibrils. 3-(4,5-dimethyl-triazol-2-yl)-2,5-diphenyl tetrazolium bromide (MTT) assay showed induction of mitochondrial dysfunction in neuronally differentiated PC12 cells by 100 nM NAC, which produced almost the same cytotoxic effect as Aβ25-35. This finding was confirmed by nuclear staining with Hoechst 33258 and propidium iodide (PI); the former staining all cells and the latter, dead cells. Some nuclei were condensed and others were swollen, indicating that the cytotoxicity was partly apoptotic and partly necrotic. In order to clarify the molecular mechanisms by which NAC exerted the toxic effects, we screened several chemicals for protective effects against NAC toxicity. We found that two antioxidants, propylgallate and N-*t*-butyl-phenyl-nitrone, effectively reduce the NAC cytotoxicity, suggesting a role of reactive oxygen species in inducing the toxicity. Aβ has been reported to produce hydrogen peroxide which mediates the toxicity, and also to enhance the transcriptional activity of NF-κB that is regulated by oxidative stress [15]. The effect of NAC on the NF-κB activity is under investigation.

α-Synuclein and Lewy body

Dementia with LBs (DLB) is considered to be the second commonest form of degenerative dementia in old age after AD [16]. The antibodies against N- and C-terminal portions of α-synuclein stained positively in Lewy bodies (LBs) [17], indicating that full-length α-synuclein is a constituent of LB. Full-length as well as C-terminal truncated forms of α-synuclein were biochemically demonstrated in LBs [8]. Recombinant α-synuclein, like NAC, was shown to form amyloid-like fibrils *in vitro* [18], and the fibril formation was accelerated by mutations found in familial PD (Ala30Pro and Ala53Thr) [19]. Cytochrome c, a component of the mitochondrial electron transport chain, was suggested to contribute to the oxidative stress-induced aggregation of α-synuclein [20]. Since cytochrome c is a mediator of apoptotic signals, the formation of LB might be facilitated by cytochrome c released in apoptotic cells in neurodegenerative diseases.

Conclusion

α-Synuclein is involved in pathogenesis of various neurodegenerative diseases, but the molecular mechanisms might be different (Fig. 2). In AD, NAC is produced from α-synuclein, and then interacts with Aβ protein to form amyloid in senile plaques. In PD and DLB, full-length or partially truncated forms of α-synuclein are a constituent of LB that are present in the degenerating neurons. Recently, oligodendroglial cytoplasmic inclusions in multiple system atrophy were reported to be immunoreactive with anti-α-synuclein antibody [21, 22]. As α-synuclein is a synaptic protein found in neurons, how this protein accumulates in glial cells remains to be examined. However, α-synuclein is suggested to be a common mediator of neurodegenerative diseases.

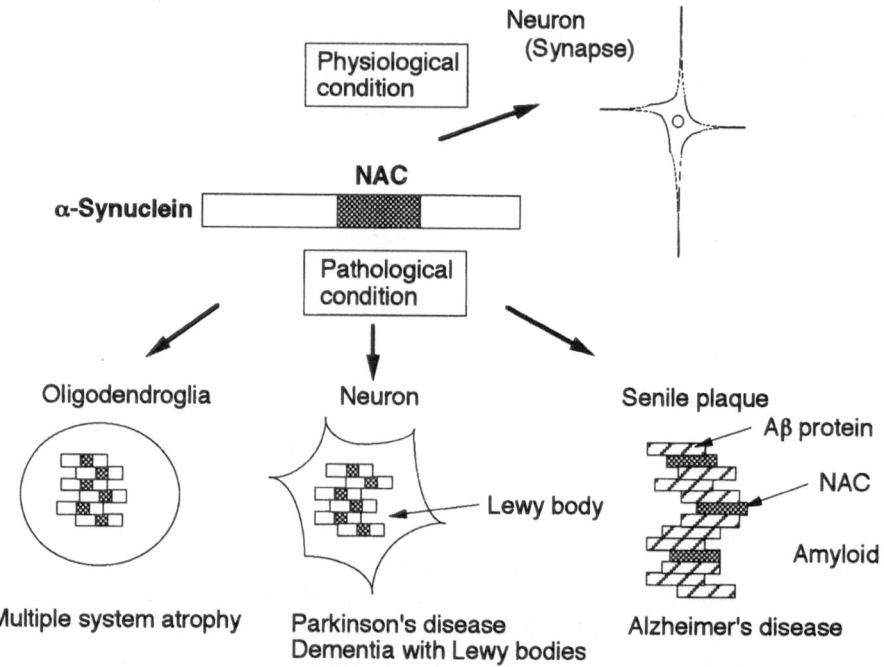

Figure 2. α-Synuclein and neurodegenerative diseases.

References

1 Ueda K, Fukushima H, Masliah E, Xia Y, Iwai A, Yoshimoto M, Otero DAC, Kondo J, Ihara Y, Saitoh T (1993) Molecular cloning of cDNA encoding an unrecognized component of amyloid in Alzheimer disease. *Proc Natl Acad Sci USA* 90: 11282–11286

2 Maroteaux L, Campanelli JT, Scheller RH (1988) Synuclein: A neuron-specific protein localized to the nucleus and presynaptic nerve terminal. *J Neurosci* 8: 2804–2815

3 Nakajo S, Tsukada K, Omata K, Nakamura Y, Nakaya K (1993) A new brain-specific 14-kDa protein is a phosphoprotein: Its complete amino acid sequence and evidence for phosphorylation. *Eur J Biochem* 217: 1057–1063

4 Jia T, Liu YE, Liu J, Shi YE (1999) Stimulation of breast cancer invasion and metastasis by synuclein γ. *Cancer Res* 59: 742–747

5 Polymeropoulos MH, Lavedan C, Leroy E, Ide SE, Dehejia A, Dutra A, Pike B, Root H, Rubenstein J, Boyer R et al (1997) Mutation in the α-synuclein gene identified in families with Parkinson's disease. *Science* 276: 2045–2047

6 Krüger R, Kuhn W, Müller T, Woitalla D, Graeber M, Kösel S, Przuntek H, Epplen JT, Schöls L, Riess O (1998) Ala30Pro mutation in the gene encoding α-synuclein in Parkinson's disease. *Nat Genet* 18: 106–108

7 Trojanowski JQ, Lee VM-Y (1998) Aggregation of neurofilament and α-synuclein proteins in Lewy bodies: Implications for the pathogenesis of Parkinson disease and Lewy body dementia. *Arch Neurol* 55: 151–152

8 Iwatsubo T, Yamaguchi H, Fujimuro M, Yokosawa H, Ihara Y, Trojanowski JQ, Lee VM-Y (1996) Purification and characterization of Lewy bodies from the brains of patients with diffuse Lewy body disease. *Amer J Pathol* 148: 1517–1529

9 Iwai A, Masliah E, Yoshimoto M, Ge N, Flanagen L, De Silva HAR, Kittel A, Saitoh T (1995) The precursor protein of non-Aβ component of Alzheimer's disease amyloid is a presynaptic protein of the central nervous system. *Neuron* 14: 467–475

10 George JM, Jin H, Woods WS, Clayton DF (1995) Characterization of a novel protein regulated during the critical period for song learning in the zebra finch. *Neuron* 15: 361–372

11 Masliah E, Iwai A, Mallory M, Ueda K, Saitoh T (1996) Altered presynaptic protein NACP is associated with plaque formation and neurodegeneration in Alzheimer's disease. *Amer J Pathol* 148: 201–210

12 Yoshimoto M, Iwai A, Kang D, Otero DAC, Xia Y, Saitoh T (1995) NACP, the precursor protein of the non-amyloid β/A4 protein (Aβ) component of Alzheimer disease amyloid, binds Aβ and stimulates Aβ aggregation. *Proc Natl Acad Sci USA* 92: 9141–9145

13 Yankner BA, Dawes LR, Fisher S, Villa-Komaroff L, Oster-Granite ML, Neve RL (1989) Neurotoxicity of a fragment of the amyloid precursor associated with Alzheimer's disease. *Science* 245: 417–420

14 Forloni G, Angeretti N, Chiesa R, Monzani E, Salmona M, Bugiani O, Tagliavini F (1993) Neurotoxicity of a prion protein fragment. *Nature* 362: 543–546

15 Behl C, Davis JB, Lesley R, Schubert D (1994) Hydrogen peroxide mediates amyloid β protein toxicity. *Cell* 77: 817–827

16 McKeith IG, Perry EK, Perry RH (1999) Report of the second dementia with Lewy body international workshop. *Neurology* 53: 902–905

17 Takeda A, Hashimoto M, Mallory M, Sundsmo M, Hansen L, Sisk A, Masliah E (1998) Human NACP/α-synuclein distribution in Lewy body disease. *Lab Invest* 78: 1167–1177

18 Hashimoto M, Hsu LJ, Sisk A, Xia Y, Takeda A, Sundsmo M, Masliah E (1998) Human recombinant NACP/α-synuclein is aggregated and fibrillated *in vitro*: Relevance for Lewy body disease. *Brain Res* 799: 301–306

19 Narhi L, Wood SJ, Steavenson S, Jiang Y, Wu GM, Anafi D, Kaufman SA, Martin F, Sitney K, Denis P et al (1999) Both familial Parkinson's disease mutations accelerate α-synuclein aggregation. *J Biol Chem* 274: 9843–9846

20 Hashimoto M, Hsu LJ, Xia Y, Takeda A, Sisk A, Sundsmo M, Masliah E (1999) Oxidative stress induces amyloid-like aggregate formaiton of NACP/α-synuclein *in vitro*. *NeuroReport* 10: 717–721

21 Arima K, Ueda K, Sunohara N, Arakawa K, Hirai S, Nakamura M, Tonozuka-Uehara H, Kawai M (1998) NACP/α-synuclein immunoreactivity in fibrillary components of neuronal and oligodendroglial cytoplasmic inclusions in the pontine nuclei in multiple system atrophy. *Acta Neuropathol* 96: 439–444

22 Wakabayashi K, Hayashi S, Kakita A, Yamada M, Toyoshima Y, Yoshimoto M, Takahashi H (1998) Accumulation of α-synuclein/NACP is a cytopathological feature common to Lewy body disease and multiple system atrophy. *Acta Neuropathol* 96: 445–452

α-Synuclein fibrillogenesis as target for drug development

Martin Citron, Linda Narhi, Jette Wypych, Jean-Claude Louis and Anja Leona Biere

Amgen, Inc., Thousand Oaks, CA, USA

Introduction

Alzheimer's disease (AD) and Parkinson's disease (PD) are the two most frequent neurode-generative disorders. Current symptomatic treatment of PD is useful for a limited period during disease progression, and the state-of-the-art symptomatic treatment of AD with cholinesterase inhibitors is only marginally effective. Thus, a major unmet medical need exists for both diseases and the ideal therapeutic would halt disease progression and not just alleviate symptoms. Development of a disease-modifying treatment requires some understanding of the disease etiology. For AD the amyloid cascade hypothesis [1] can serve as a framework from which targets for therapeutic intervention can be developed. In contrast, there has been no unifying hypothesis for the pathogenesis of PD, and therapeutic targets still have to be defined. Here we describe basic mechanistic studies which lead us to propose α-synuclein fibrillogenesis as a target for therapeutic intervention for PD and other α-synucleinopathies, e.g., dementia with Lewy bodies.

Why work on α-synuclein?

Despite several decades of intense research, it has remained unclear what causes the loss of the dopaminergic neurons of the substantia nigra that leads to the symptoms of PD. Mutations in two genes have been found to cause familial Parkinsonism. Autosomal recessive PD is caused by deletions and loss-of-function mutations in the Parkin gene. Such mutations have been found in many families from many countries, suggesting that Parkin mutations may be a rather frequent cause of PD. Interestingly, while Parkin mutation cases show the typical PD symptoms and respond nicely to levodopa, they do not show the Lewy body pathology which is a defining criterion for the histological diagnosis of PD in idiopathic PD (for review see [2]). This leads to the semantic question whether PD should be defined by the symptoms and the neuronal loss alone or whether the presence of Lewy bodies should also be required. Whatever the answer may be, it should be noted that idiopathic PD cases always have Lewy bodies [3]. Autosomal dominant PD has been found to be caused by two point mutations in the α-synuclein gene [4, 5]. α-synuclein mutations appear to be a very rare cause of PD, because many investigators have tried to find a mutation in this gene in other families without success. α-synuclein mutant patients show typical PD with good response to levodopa. Lewy bodies were found in the substantia nigra, but some were also detected in the cortex [6]. The finding that mutations in the α-synuclein gene cause familial PD with Lewy bodies has triggered pathology studies focused on the α-synuclein protein, and indeed it now appears that α-synuclein is the major fibrillar component of Lewy bodies [7, 8]. It is therefore tempting to speculate that the formation of Lewy bodies may be causally involved in PD pathogenesis and that the familial α-synuclein mutations somehow enhance formation of Lewy bodies. α-synuclein pathology is not specific for PD, because Lewy bodies are also found in cortical neurons in dementia with Lewy bodies, the second most frequent dementia after AD [9]. In addition, it has been shown recently that α-synuclein is the major component of the filamentous inclusions of multiple system atrophy [10–12]. Thus, a thorough understanding of α-synuclein and its fibrillation could be useful for the treatment of several conditions.

In vitro studies on α-synuclein

The normal biological function of α-synuclein is not well understood. The 140 amino acid protein is highly and widely expressed in the brain. α-synuclein is found predominantly in nerve terminals in close proximity to synaptic vesicles (for review see [13]). The protein lacks any currently defined sequences for intracellular targeting, despite its enrichment in presynaptic terminals. A role in regulation or support of synaptic plasticity seems likely (for review see [14]). Several mechanisms by which α-synuclein mutations cause PD can be envisioned. In the simplest scenario, the mutations could interfere with the unknown function of α-synuclein and thus cause disease. This model would predict that α-synuclein knockout mice should show PD symptoms. In an alternative scenario, the mutant α-synuclein would have gained a toxic function and transgenics but no knockout mice would be expected to exhibit symptoms. Once transgenics and knockouts become available, one will be able to decide which model fits the pathology better. In the absence of *in vivo* models we have chosen to study α-synuclein in *in vitro* systems [15, 16].

We hypothesized that α-synuclein mutations would cause or enhance α-synuclein aggregation and thus lead to the formation of Lewy bodies. Indeed, this most straightforward explanation has been proposed already by Polymeropoulos et al. when they described the first mutation [4]. This hypothesis assumes a gain, rather than a loss of function. For our experiments we used bacterially expressed, purified full-length α-synuclein. We are not aware of posttranslational modifications or proteolytic processing events of this protein that would make it necessary to choose a eukaryotic source or express a truncated form of the protein. Our *E. coli*-expressed material is not a fusion protein and is>99% pure when we do our experiments. We generated 4 forms of α-synuclein: the wild-type protein, the A53T mutant form [4], the A30P mutant form [5] and a fourth construct, A53T/A30P, carrying both mutations. When we compared fresh solutions of all 4 species by CD (circular dichroism) spectroscopy, we found the same natively unfolded structure for all of them, confirming the results of Conway et al. [17]. The FTIR (Fourier transform infrared spectroscopy) spectra of the molecules were also indistinguishable. Initial pilot experiments suggested that α-synuclein aggregation would be very slow and require elevated temperature: therefore, all experiments were run at 37 °C over several days. In our experiments we measured the concentration of material in solution by A280 after precipitation of insoluble material by ultracentrifugation and after a few days all four proteins began to form insoluble aggregates. Using electron microscopy and fluid phase atomic force microscopy we demonstrated that the aggregate consists of fibrillar material and not just amorphous precipitate. The aggregate was

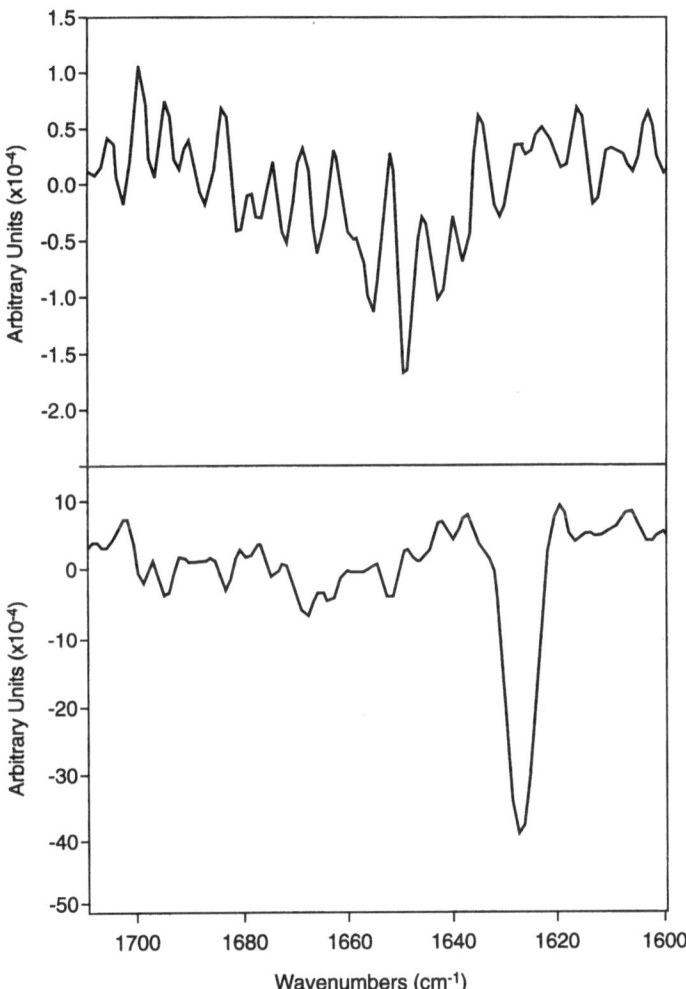

Figure 1. α-synuclein aggregate formation is accompanied by a change in secondary structure from random coil to antiparallel β-sheet. The top panel shows the second derivative FTIR spectrum of the initial solution of wild-type α-synuclein. The spectrum shows primarily random coil (1650 cm^{-1} band). The bottom panel is the second derivative FTIR spectrum of the wild-type α-synuclein aggregate. This spectrum shows the final antiparallel β-sheet structure (1629 cm^{-1} and 1685 cm^{-1}).

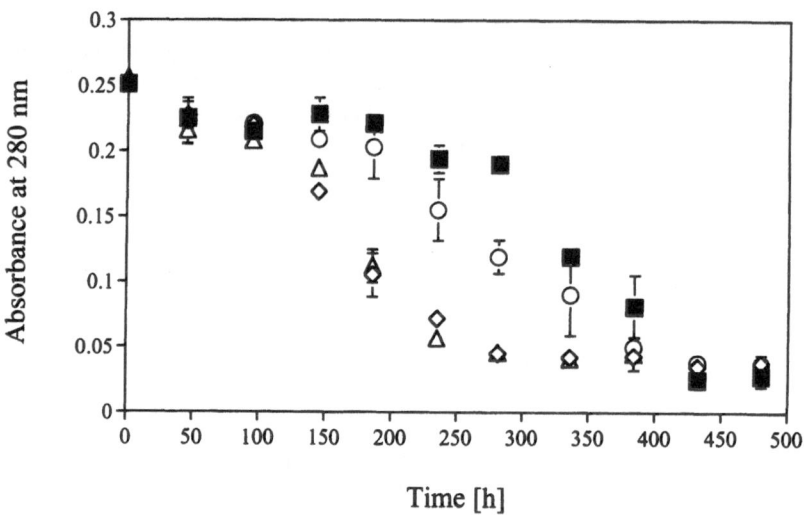

Figure 2. α-synuclein mutations show accelerated aggregation. Aggregate formation of wild-type and mutant α-synuclein solutions (starting concentration 7 mg/ml) was analyzed by measuring the UV absorption at 280 nm of protein in solution after ultracentrifugation. Values are means ± S.E. of three different solutions. Filled square, wild-type; open diamonds, A53T; open circles, A30P; open triangles, A53T/A30P.

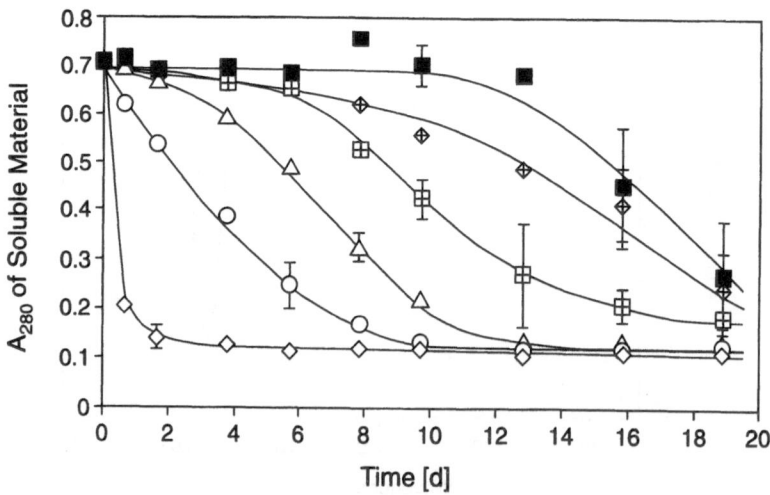

Figure 3. Seeding of α-synuclein aggregation. Aggregate formation of wild-type α-synuclein at 2 mg/ml as monitored by A_{280} of soluble material following ultracentrifugation. A non-seeded control incubation is shown (■) along with incubations containing pre-formed wild-type α-synuclein aggregate as seed. Seed concentrations, expressed as a percentage of the soluble α-synuclein amount, were 0.001% (⬥), 0.01% (⊞), 0.1% (△), 1% (○) and 10% (◇), respectively. Values shown are the average of triplicate incubations ± S.E.M.

further analyzed by FTIR spectroscopy (Fig. 1). This demonstrated that the aggregate formation is accompanied by a structural transition from random coil to antiparallel β-sheet. The structures of the aggregates of all four forms were indistinguishable. While the initial soluble forms and the final aggregates are indistinguishable between wild-type and mutant proteins, we could show a clear difference in the time course of aggregation: all mutant forms aggregate faster than the wild-type protein (Fig. 2). This effect is most striking if one looks at approximate lag times before precipitable material is detected: about 280 h for wild-type protein, about 180 h for A30P and only about 100 h for A53T. The existence of a distinct lag phase immediately suggested that the aggregation may be nucleation-dependent. In such a case addition of exogenous nuclei (seeds) should expedite aggregation, because the long lag phase can be bypassed. Indeed, wild-type α-synuclein could be rapidly aggregated when preformed fibrils were added as seeds (Fig. 3). Interestingly, mutant α-synuclein can cross-seed wild-type protein. This was shown when we added preformed aggregates of A53T α-synuclein to wild-type α-synuclein incubated under our standard conditions and observed a seed concentration dependent acceleration of aggregation (Fig. 4). We conclude that A53T mutant α-synuclein does not only nucleate and consequently aggregate faster than the wild-type form, but in a heterozygous mutant individual, the lag phase of the wild-type protein is reduced to that of the mutant form. We therefore predict that the Lewy bodies of heterozygotes will contain both the wild-type and the mutant protein. In a nucleation-dependent reaction the aggregates grow until a thermodynamic equilibrium between aggregate and monomer is reached. Under these steady-state conditions, the growth equilibrium constant

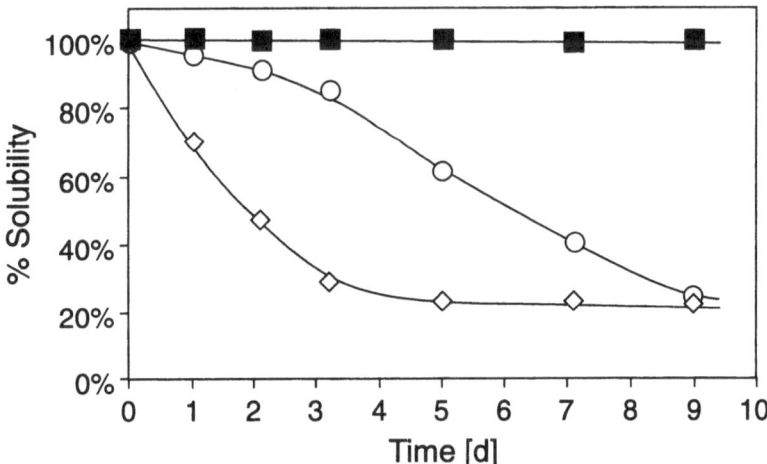

Figure 4. Cross-seeding of wild-type α-synuclein with mutant α-synuclein. Aggregate formation of wild-type α-synuclein at 2 mg/ml as monitored by A_{280} absorption of soluble material following ultracentrifugation. A non-seeded control incubation (■) is shown along with incubations containing mutant α-synuclein aggregate as seed. Seed concentrations, expressed as a percentage of the soluble α-synuclein amount, were 1% (○) and 10% (◇), respectively. Values shown are the average of triplicate incubations.

describes the solubility of the protein which is equivalent to its critical concentration [18]. We determined the critical concentration for wild-type and A53T α-synuclein and found them to be indistinguishable within experimental error. This demonstrates that the accelerated aggregation of the mutant form is a kinetic effect and not due to intrinsically higher fiber stability.

Therapeutic implications

Our *in vitro* data on purified α-synuclein show that all forms of α-synuclein aggregate into fibrils with antiparallel β-sheet structure and that this aggregation is nucleation-dependent. Both α-synuclein mutants aggregate faster than the wild-type form and this may explain why they cause PD. It would therefore be interesting to identify α-synuclein nucleation inhibitors and to try whether such compounds can be developed into useful drugs. It would also be interesting to identify molecules that bind to α-synuclein *in vivo* and protect it from changing into the β-sheet conformation. It has, for example, been demonstrated that α-synuclein can bind to synthetic membranes and that this lipid binding causes a massive increase in α-helicity [19]. Identification of *in vivo* binding partners with similar properties may be helpful in efforts to block α-synuclein fibrillogenesis in Parkinson's disease and dementia with Lewy bodies.

References

1 Hardy J, Allsop D (1991) Amyloid deposition as the central event in the aetiology of Alzheimer's disease. *Trends Pharmacol* 12: 383–388
2 Mizuno Y, Hattori N, Mori H (1999) Genetics of Parkinson's disease. *Biomed Pharmacotherapy* 53: 109–116
3 Adams JH, Duchen LW (1992) *Greenfield's Neuropathology*. Oxford University Press
4 Polymeropoulos MH et al (1997) Mutation in the α-synuclein gene identified in families with Parkinson's disease. *Science* 276: 2045–2047
5 Krüger R et al (1998) Ala30Pro mutation in the gene encoding α-synuclein in Parkinson's disease. *Nat Genet* 18: 106–108
6 Golbe LI et al (1990) A large kindred with autosomal dominant Parkinson's disease. *Ann Neurol* 27: 276–282
7 Spillantini GM et al (1998) α-synuclein in filamentous inclusions of Lewy bodies from Parkinson's disease and dementia with Lewy bodies. *Proc Natl Acad Sci USA* 95: 6469–6473
8 Arai T et al (1999) Argyrophilic glial inclusions in the midbrain of patients with Parkinson's disease and diffuse Lewy body disease are immunopositive for NACP/α-synuclein. *Neurosci Lett* 259: 83–86
9 Dickson DW et al (1987) Diffuse Lewy body disease: neuropathological and biochemical studies of six patients. *Acta Neuropathol* 75: 8–15
10 Arima K et al (1998) NACP/α-synuclein immunoreactivity in fibrillary components of neuronal and oligodendroglial cytoplasmic inclusions in the pontine nuclei in multiple system atrophy. *Acta Neuropathol* 96: 439–444
11 Wakabayashi K et al (1998) Accumulation of α-synuclein/NACP is a cytopathological feature common to Lewy body disease and multiple system atrophy. *Acta Neuropathol* 96: 445–452
12 Spillantini MG et al (1998) Filamentous α-synuclein inclusions link multiple system atrophy with Parkinson's disease and dementia with Lewy bodies. *Neurosci Lett* 251: 205–208
13 Lavedan C (1998) The synuclein family. *Genome Res* 8: 871–880
14 Clayton DF, George JM (1998) The synucleins: a family of proteins involved in synaptic function, plas-

ticity, neurodegeneration and disease. *Trends Neurosci* 6: 249–254

15 Narhi L et al (1999) Both Familial Parkinson's Disease Mutations Accelerate α-Synuclein Aggregation. *J Biol Chem* 274: 9843–9846
16 Wood SJ et al (1999) α-synuclein fibrillogenesis is nucleation-dependent. *J Biol Chem* 274: 19509–19512
17 Conway KA, Harper JD, Lansbury PT (1998) Accelerated *in vitro* fibril formation by a mutant α-synuclein linked to early-onset Parkinson's disease. *Nat Med* 4: 1318–1320
18 Andreu JM, Timasheff SN (1986) *In*: CHW Hirs SN Timasheff (eds): *Methods in Enzymology (Vol. 130)*. Academic Press, 47–59
19 Davidson WS, Jonas A, Clayton DF, George JM (1998) Stabilization of α-synuclein Secondary Structure upon Binding to Synthetic Membranes. *J Biol Chem* 273: 9443–9449

Pathogenesis of dementia – presenilin and amyloid

Neuroscientific Basis of Dementia
C. Tanaka, P.L. McGeer, Y. Ihara (eds)
© 2001 Birkhäuser Verlag Basel/Switzerland

Genetics of early-onset Alzheimer disease

Peter H. St George-Hyslop

Department of Medicine, The Toronto Hospital, University of Toronto, Toronto, Ontario, Canada

Introduction

To date, three different genes have been identified in which selected missense mutations cause early-onset forms of Alzheimer disease (AD). This review examines the biology and genetics of these diseases and describes an emerging and now widely held, but not yet entirely proven, concept that mis-processing of the amyloid precursor protein is an integral part of the pathogenic cascade leading to neurodegeneration in AD.

The amyloid precursor protein

The first gene to be identified in association with inherited susceptibility to AD was the amyloid precursor protein gene (βAPP). The βAPP gene encodes an alternatively spliced transcript which, in its longest isoform, encodes a single transmembrane-spanning polypeptide of 770 amino acids [1–4]. Alternative splicing of exon 7, which encodes a Kurnitz protease-inhibitor protein domain, and exon 8, which encodes a sequence homologous to the ox-2 antigen, results in polypeptides of 695 amino acids (which are expressed predominantly in the brain) and 751 amino acids [5]. There have been considerable advances in the understanding of the processing of the βAPP protein which are reviewed in detail elsewhere [6]. The βAPP precursor protein undergoes a series of endoproteolytic cleavages. One of these, which results from the action of a putative membrane-associated α-secretase, liberates the extracellular N-terminus of βAPP (which was previously identified at protease nexin II) by cleavage within the Aβ peptide domain. It is thus a non-amyloidogenic pathway because this cleavage precludes the formation of Aβ peptide. The other cleavage pathway, which occurs in part in the endosomal-lysosomal compartment, involves the putative β- and γ-secretases which give rise to a series of peptides containing the 40–42 amino acid Aβ peptide. Aβ peptides ending at residue 42 or 43 (long-tailed Aβ) are thought to be more fibrillogenic and more neurotoxic than Aβ ending at residue 40, which is the predominant isoform produced during normal metabolism of βAPP [7–10]. The activity of these enzymes, and especially of the γ-secretase isoform which gives rise to the more fibrillogenic and potentially neurotoxic long-tailed $A\beta_{1-42}$, appears to play a central role in the pathogenesis of AD in both genetic and non-genetic forms (see reviews [6, 11].

The function of βAPP is currently unknown. Knockout of the murine βAPP gene is not very illuminating because it leads only to subtle phenotypes, including minor weight loss, decreased locomotor activity, abnormal forelimb motor activity and minor non-specific degrees of reactive gliosis in the cortex [12]. *In vitro* studies in cultured cells suggest that secreted βAPP (βAPPs) can function as an autocrine factor stimulating cell proliferation and cell adhesion, and supports nerve growth factor (NGF)-induced neurite outgrowth of PC12 cells [13, 14]. Other studies have implied a role for βAPP in signal transduction by association of βAPP with heterotrimeric guanosine triphosphate (GTP)-binding proteins [15].

Direct nucleotide sequencing has led to the discovery of several different missense mutations in exons 16 and 17 of the βAPP gene in families with early-onset AD (Tab. 1). Some of these missense mutations are probably not pathogenic mutations either because they have also been detected in normal elderly relatives or because they are not present in all affected members of these pedigrees. Nevertheless, the missense mutations at codon 670/671 (Swedish mutation) [16], at codon 692 (Flemish mutation) [17], at codon 716 and at codon 717 [18–22] are quite clearly pathogenic. The mutations at codon 670/671 and at codon 692 are rare, having been seen only in single families. Mutations at codon 717 on the other hand have been seen in several unrelated pedigrees (less than 20 pedigrees worldwide) of different ethnic origins. It may be significant, however, that the majority of such mutations at codon 717 have been seen in Anglo-Saxon, Italian and Japanese subjects. The reason for this aggregation is unclear, because within each ethnic group there is no common haplotype of genetic markers within or surrounding the βAPP gene [23]. The absence of a conserved haplotype around the βAPP gene in βAPP$_{717}$ carriers in the same ethnic group argues against a common founder for the βAPP$_{717}$ mutation in each ethnic group. Finally, more recently, a novel βAPP mutation at codon 715 has been described in a French family with early-onset AD [24]. Analysis of the effects of this mutation shows that, paradoxically, this mutation

Table 1. Missense mutations in the βAPP gene2

Codon	Mutation	Phenotype
665	Gln → Asp	Late-onset AD – no segregation
670/671	Lys-Met → Asn-Leu	FAD; increased Aβ production
673	Ala → Thr	No disease phenotype
692	Ala → Gly	FAD + cerebral haemorrhage; increased Aβ
693	Glu → Gly	Late-onset AD, no segregation
	Glu → Gln	HCHWA-D
713	Ala → Val	Schizophrenia – no segregation
715	Val → Met	Early-onset FAD, reduce total Aβ production, particularly Aβ40
	Ala → Thr	AD - no segregation
716	Ile → Val	FAD
717	Val → Ile	FAD; increased long Aβ isoforms
	Val → Phe	FAD
	Val → Gly	FAD

does not cause overproduction of $A\beta$ [24]. However, this mutation does cause both <u>relative</u> overproduction of $A\beta_{42}$ (due to a greater reduction in $A\beta_{40}$ secretion), and overproduction of N-terminally truncated $A\beta$ peptides ending at residue 42 [24].

The exact mechanism by which selected βAPP mutations cause AD is unclear. The most simple explanation is that missense mutations in the βAPP gene result in the overproduction of the $A\beta$ peptide and, in particular, <u>relative</u> overproduction of long-tailed, putatively neurotoxic isoforms ending at residue 42 or 43 (including N-terminally truncated derivatives termed $A\beta_{1-42}$) [8–10, 25–29]. Aggregation of $A\beta$ is increased in the presence of ApoE, heparin sulphate proteoglycan and some heavy metals [30–33]. Multiple molecular mechanisms have been proposed to explain the neurotoxic effects of $A\beta$, including inducing apoptosis both by direct effects on cell membranes and by indirect effects such as potentiating effects of excitatory amino acids, oxidative stress and increases in intracellular calcium and free radicals [34–36]. However, while much attention has been focused upon the possibility that the neurodegenerative effect of βAPP mutations are mediated by overproduction of neurotoxic $A\beta$ peptides, several other putative mechanisms have emerged. For instance, it has been shown that the βAPP717 mutations may induce apoptosis by causing constitutive activation of the heterotrimeric GTP-binding protein G_o [37, 38].

Presenilin 1

After the discovery that βAPP missense mutations were quite rare as a cause of AD, several groups undertook a survey of the remaining non-sex-linked chromosomes other than chromosomes 19 and 21. These studies identified a series of polymorphic genetic markers located on chromosome 14q24.3 (D14S43, D14S71, D14S77 and D14S53) which showed robust evidence of linkage to an early-onset form of familial AD ($z \geq 23.0$ at $p = 0.01$) [39–41]. Subsequent genetic mapping studies narrowed the region containing this third Alzheimer susceptibility locus (AD3) to a region of approximately 10 centiMorgans between the marker D14S271 at the centromeric end, and D14S53 at the telomeric end. The actual disease gene (presenilin 1) was isolated using the positional cloning strategy alluded to earlier [42].

The chromosome 14 AD3 subtype gene, S182 or presenilin 1 (PS1), is highly conserved in evolution, being present in C. elegans [43] and D. melanogaster [44], and appears to encode a polytopic integral membrane protein. Theoretical predictions based upon Kyte-Doolittle hydrophobicity analysis suggest that there are between five and ten membrane-spanning domains, that the N-terminus is acidically charged and that there is a large hydrophilic, acidically charged loop domain between the putative sixth and seventh transmembrane domains [42, 45]. Partial direct experimental support for a polytopic structure has been obtained from studies in transfected cells (see below).

The presenilin 1 gene is transcribed at low levels in many different cell types, both within the central nervous system (CNS) and in non-neurological tissues [42]. In the CNS, PS1 transcripts can be detected by *in situ* hybridization in the neocortex (especially in cortical neurons in layers II and IV), neurons of the CA1-CA3 fields of the hippocampus, granule cell neurons of the dentate gyrus, subiculum, cerebellar Purkinje and granule cells and deep nuclei, as well as lesser amounts in the olfactory bulb, striatum, some brainstem nuclei and thalamus. Despite intense signals on Northern blots of the corpus callosum, there is very little *in situ* hybridization signal detectable in oligodendrocytes in white matter [46].

The genomic structure of the PS1 gene has been elucidated and some of the transcriptional regulatory elements have been defined [47]. As with the βAPP gene, there is evidence for alternate splicing of the PS1 transcript. Thus, there is a variably present four-amino acid VRQS insert which arises from use of an alternate splice donor site at the 3' end of exon 4 [47–49]. In some tissues (especially leukocytes) there is also alternate splicing of exon 9, which encodes a series of hydrophobic residues at the C-terminus of TM6 and the beginning of the TM6-TM7 exposed loop domain [47]. As a result, this splicing event is predicted to significantly alter the functional properties of the TM6-TM7 loop.

Immunoblotting and immunohistochemical studies suggest that the PS1 protein is approximately 50 kDa in size and is predominantly located within intracellular membranes in the endoplasmic reticulum, perinuclear envelope, the Golgi apparatus and some as yet uncharacterized intracytoplasmic vesicles [50, 51]. Studies of the topology of PS1 suggest that the N-terminus and the residues in the TM6-TM7 loop are located in the cytoplasm [50]. The orientation of the C-terminus is not yet resolved [50, 52]. Studies of the PS1 protein in brain tissue, as well as many other peripheral tissues, reveal that only very small amounts of the PS1 holoprotein exist within the cell at any given time [53, 54]. Instead, the holoprotein is actively catabolized, possibly by two different proteolytic mechanisms. One of these mechanisms appears to involve the proteasome [55], while the other involves endoproteolytic cleavage near residue 290 within the TM6-TM7 loop domain. This endoproteolytic cleavage generates a series of N- and C-terminal heterogeneous fragments of approximately 35 kDa and 18 kDa, respectively, in size. It is currently unclear whether the holoprotein, the endoproteolytic fragments, or both, have biological functions. The expression patterns of PS1 protein largely reflect those of the mRNA [56].

The normative function of presenilin 1 has not yet been completely defined. However, several lines of evidence suggest that it plays a direct or indirect role in the proteolytic processing of βAPP and *Notch*. Null mutations in a second presenilin orthologue in *C. elegans* (*sel12*) exert a suppressor effect on abnormalities in vulva progenitor cell fate decisions induced by activated *Notch* mutants [43]. A role for mammalian presenilins in *Notch*-mediated signalling is further supported by the fact that homozygous targeted knock-out of the murine PS1 protein using homologous recombination causes embryonic lethality around day E13 and is associated with severe developmental defects in somite formation and axial skeleton formation and the occurrence of cerebral haemorrhage [57, 58]. Similar phenotypes have been observed in mice with targeted knockouts of the murine *Notch1* gene [59]. These *in vivo* observations have now been supported by studies of *Notch* cleavage in transfected cells using a reporter construct. In the absence of functional PS1 expression, there is failure of the final intramembranous cleavage of *Notch*, which follows ligand binding and liberates the C-terminus of *Notch* necessary for signal transduction [60–62]. The other effect of loss of PS1 function is failure of γ-secretase cleavage of βAPP stubs generated by α- or β-secretase [63]. This results in a profound reduction in the secretion of Aβ and the accumulation of C83 (α-stubs) and C99 (β-stubs) derivatives of α- and β-secretase cleavage, respectively. Both the failure of ligand-dependent cleavage of *Notch* and the γ-secretase cleavage of βAPP would seem to arise from the loss of activity of a novel class of intra-membranous cleavage events similar to the S2 protease-mediated cleavage of sterol responsive element binding protein (SREBP). It is unclear whether this activity conferred by the presenilins is due to a direct enzymatic activity of PS1 [64], or whether PS1 is indirectly involved in the trafficking or activation of the various substrates and components of the "γ-secretase complex". Evidence in favour of a direct enzymatic (aspartyl protease) activity is provided by the observation that mutation of either of two intramembranous aspartate residues (D257 and D385) in PS1 (and PS2) results in failure of endoproteolysis of PS1 and the failure of γ-secretase cleavage of βAPP [64, 65]. These data have been taken to suggest that PS1 might be a novel aspartyl protease. While plausible, there are also some data that indicate that loss of PS1 activity is associated with subtle defects in trafficking of βAPP, APLP and selected other proteins [66, 67].

Other putative roles have included a role in the suppression of apoptosis, suggested by the fact that over-expression of the 3' end of PS2 rescued T-lymphocytes from Fas-ligand-induced apoptosis (see below) [68]. Follow-up studies have to date yielded conflicting evidence as to whether overexpression of full-length wild-type PS1 or wild-type PS2 can cause apoptosis in transfected cells and whether mutations further sensitize these cells to apoptosis [69].

To date, more than 35 different mutations have been discovered in the PS1 gene (Tab. 2). The majority are missense mutations which give rise to the substitution of a single amino acid. These mutations are located predominantly either in highly conserved transmembrane domains, at or near putative membrane interfaces, or in the N-terminal hydrophobic or C-terminal hydrophobic residues of the putative TM6-TM7 loop domain. A single splicing defect mutation has been identified which involves a point mutation in the splice acceptor site at the 5' end of exon 10 [70–72]. Because exon 9 and exon 11 are in-frame, this mutation allows exon 9 to be fused in-frame with exon 11, thereby removing a series of charged residues at the apex of the hydrophilic, acidically-charged, TM6-TM7 loop domain. Interestingly, this mutation removes residues near the endoproteolytic cleavage site at residue 290, and results in the production of higher quantities of uncleaved holoprotein [53]. No deletions, rearrangements or nonsense mutations resulting in truncated proteins, all of which would cause loss-of-function mutations, have yet been found in AD-affected subjects. This raises the question as to whether such mutations might be lethal or might lead to other disease genotypes.

The effect of PS1 mutations is currently being explored. The wide scattering of missense mutations and the absence of null mutations have led to speculation that the effect of the FAD-related mutations is a "gain-of-function" effect [73]. This is partially borne out by preliminary studies of gene-targeted animals which have loss of functional expression of PS1 protein and in which the phenotype is early perinatal mortality without evidence of AD [57]. However, it should be noted that preliminary studies using human PS1 cDNAs in complementation assays of mutant *sel12* mutant in *C. elegans* suggest that the wild-type human PS1, but not mutant human PS1 cDNAs, are able to complement the loss-of-function *sel12* mutants (C. Haass, personal communication). The latter argues that the human PS1 mutants may not be fully functional, but do not preclude a gain-of-function effect as well. The exact

Table 2. Mutations in the presenilin genes

Presenilin I			
Codon	Location	Mutation	Phenotype
79	N-term loop	Ala → Val	FAD, onset 64 years
82	TM1	Val → Leu	FAD, onset 55 years
96	TM1	Val → Phe	FAD, onset 53 years
113–114 (ins)	TM1/TM2 loop	Leu-Thr(ins)-Ile	FAD, onset 35 years
115	TM1/TM2 loop	Tyr → His	FAD, onset 37 years
115	TM1/TM2 loop	Tyr → Cys	FAD, onset 42 years
116	TM1/TM2 loop	Thr → Asn	onset 37 years
117	TM1/TM2 loop	Pro → Leu	FAD, onset 28 yrs
120	TM1/TM2 loop	Glu → Asp	FAD, onset 48 years
120	TM1/TM2 loop	Glu → Lys	FAD, onset 37 years
123	TM1/TM2 loop	Glu → Lys	
135	TM2	Asn → Asp	FAD, onset 36 years
139	TM2	Met → Thr	FAD, onset 49 years
139	TM2	Met → Val	FAD, onset 40 years
139	TM2	Met → Ile	FAD,
139	TM2	Met → Lys	FAD, onset 37 years
143	TM2	Ile → Thr	FAD, onset 35 years
143	TM2	Ile → Phe	FAD, onset 55 years
146	TM2	Met → Leu	FAD, onset 45 years
146	TM2	Met → Val	FAD, onset 38 years
146	TM2	Met → Ile	FAD, onset 40 years
147	TM2	Thr → Ile	FAD, onset 42 years
163	TM3 interface	His → Arg	FAD, onset 50 years
163	TM3 interface	His → Tyr	FAD, onset 47 years
165	TM3	Trp → Cys	FAD, onset 42 years
169	TM3	Ser → Leu	
169	TM3	Ser → Pro	onset 35 years
171	TM3	Leu → Pro	FAD, onset 40 years
173	TM3	leu → Trp	FAD, onset 27 years
177	TM3	Phe → Ser	
178	TM3	Ser → Pro	
206	TM4	Gly → Ser	
209	TM4	Gly → Val	FAD
209	TM4	Gly → Arg	onset 49 years
213	TM4 interface	Ile → Thr	FAD,
213	TM4 interface	Ile → Leu	
219	TM4 interface	Leu → Pro	
222	TM5	Gln → Ala	
231	TM5	Ala → Thr	FAD, onset 52 years
233	TM5	Met → Thr	FAD, onset 35 years
235	TM5	Leu → Pro	FAD, onset 32 years
246	TM6	Ala → Glu	FAD, onset 55 years

Table 2 (Continued)

Presenilin I			
Codon	Location	Mutation	Phenotype
250	TM6	Leu → Ser	onset 53 years
260	TM6	Ala → Val	FAD, onset 40 years
261	TM6	Val → Phe	
262	TM6	Leu → Phe	onset 50 years
263	TM6/TM7 loop	Cys → Arg	FAD, onset 47 years
264	TM6/TM7 loop	Pro → Leu	FAD, onset 45 years
267	TM6/TM7 loop	Pro → Ser	FAD, onset 35 years
269	TM6/TM7 loop	Arg → Gly	
269	TM6/TM7 loop	Arg → His	onset 47 years
278	TM6/TM7 loop	Arg → Thr	onset 37 years
280	TM6/TM7 loop	Glu → Ala	FAD, onset 47 years
280	TM6/TM7 loop	Glu → Gly	FAD, onset 42 years
282	TM6/TM7 loop	Leu → Arg	onset 43 years
285	TM6/TM7 loop	Ala → Val	FAD, onset 50 years
286	TM6/TM7 loop	Leu → Val	FAD, onset 50 years
ƒ291–319	TM6/TM7 loop	short loop	FAD,
318	TM6/TM7 loop	Glu → Gly	onset 35–64 years, polymorphism?
378	TM6/TM7 loop	Gly → Glu	onset 35 years
384	TM6/TM7 loop	Gly → Ala	FAD, onset 35 years
390	TM6/TM7 loop	Ser → Ile	FAD, onset 39 years
392	TM6/TM7 loop	Leu → Val	FAD, onset 25–40 years
410	TM7	Cys → Tyr	FAD, onset 48 years
418	TM7	Leu → Phe	
424	TM7	Leu → Arg	onset 33 years
426	TM7	Ala → Pro	
431	C-term loop	Ala → Glu	
434	C-term loop	Ala → Cys	
436	C-term loop	Pro → Ser	
436	C-term loop	Pro → Gln	
439	C-term loop	Ile → Val	

Presenilin II			
Codon	Location	Mutation	Phenotype
62	N-term loop	Arg → His	Sporadic AD, onset 62 years
141	TM2	Asn → Ile	FAD, onset 50–65 years
239	TM5	Met → Val	FAD, onset variable 45–84 yrs

nature of the putative gain-of-function or loss-of-function effect imparted by PS1 mutations associated with FAD is unclear. It seems likely that one effect is to alter the processing of βAPP by preferentially favouring the production of potentially toxic long-tailed Aβ peptides ending at residue 42 or 43 [74–77]. Thus, fibroblasts from heterozygous carriers of PS1 mutations, various cell lines transfected with βAPP and PS1 cDNAs, as well as the brain from transgenic mice overexpressing mutant PS1 transgenes, all contain or secrete increased quantities of long Aβ peptide isoforms with only a variable but minor increase in short-tailed Aβ peptides [74–77]. Direct measurements of Aβ peptide isoforms in the *post-mortem* brain tissue of patients dying with PS1-linked FAD also show marked increases in the amount of long-tailed Aβ isoforms compared to control brain tissue and brain tissue from subjects with sporadic AD (H. Mori et al., unpublished).

As with mutations in the βAPP gene, there is a preponderance of evidence pointing to a role for over-production of $A\beta_{1-42(43)}$ as the mechanism underlying neurodegeneration in PS1- and PS2-linked FAD. However, as with the βAPP mutations, evidence has emerged that suggests that PS1 and PS2 mutations may modulate cellular sensitivity to apoptosis induced by a variety of factors including staurosporine, Aβ peptide, serum withdrawal, etc. At the current time these data are still evolving and the apparent paradox of a putative "apoptosis promoting" effect for the presenilins and the existence of transgenic mice overexpressing mutant or wild-type presenilin cDNAs but lacking widespread apoptosis remains to be explained.

Presenilin 2

During the cloning of the presenilin 1 gene on chromosome 14, a similar sequence was identified in the public nucleotide sequence databases [78]. Further analysis revealed that this similar nucleotide sequence was derived from a gene on chromosome 1, and encodes a polypeptide whose open reading frame contained 448 amino acids. The sequence of this peptide showed substantial amino acid sequence identity with that of the presenilin 1 protein (overall identity approximately 60%), and would be predicted to have a structural organization very similar to that of PS1 protein. Within the transmembrane (TM) domains, the amino acid sequence identity between this new gene and presenilin 1 was even higher (approximately 90%). However, the pattern of transcription of this novel gene was slightly different from that of presenilin 1, being expressed less homogeneously in the brain and in peripheral tissues, whereas it was maximally expressed in cardiac muscle, skeletal muscle and pancreas. Finally, when the genomic organization of this novel gene was worked out, it was apparent that many of the intron-exon boundaries (especially those relating to the highly conserved transmembrane domains) were identical between this gene and presenilin 1 [79, 80]. Cumulatively, these observations therefore suggested that this novel gene, which became known as presenilin 2, was a homologue of the presenilin 1 gene on chromosome 14.

Mutational analyses have uncovered a very small number of missense mutations in the presenilin 2 gene in families segregating early-onset forms of AD (Tab. 2). The first mutation (Asn141Ile) was detected in a proportion of families of Volga German ancestry [78, 81], in which the FAD locus had been independently mapped by genetic linkage studies to chromosome 1 [81]. The second mutation (Met239Val) was discovered in an Italian pedigree [78]. However, in contrast to the frequency of presenilin 1 mutations, screening of large data sets reveals that presenilin 2 mutations are likely to be rare [80].

Another profound difference between the presenilin 2 mutations and those in the βAPP and PS1 gene is that the phenotype associated with PS2 mutations is much more variable [80, 82]. Thus, the vast majority of heterozygous carriers of missense mutations in the βAPP and PS1 genes develop the illness between the ages of 35 and 65 for PS1 mutations, and between 40 and 65 for βAPP mutations. In contrast, the range of age-of-onset in heterozygous carriers of PS2 mutations is between 40 and 85 years of age, and there is at least one instance of apparent non-penetrance in an asymptomatic octogenarian transmitting the disease to affected offspring [80, 83, 84]. A similar, but less profound variation in age-of-onset within families segregating the βAPP Val717Ile mutation has been ascribed to a modifying effect by the ApoE gene [85–87]. Thus, carriers of the βAPP Val717Ile mutation who have one or more ε4 alleles at ApoE have an earlier onset than do heterozygous carriers of the Val717Ile mutation who have the ε2 allele and no ε4 alleles of ApoE. However, because the effect of ApoE ε4 on the age-at-onset in PS2 mutations is either absent or less profound, modifier loci other than ApoE probably account for much of this variation.

The relationship of the normal function of presenilin 2 to that of presenilin 1 remains an object of investigation. Like PS1, PS2 undergoes endoproteolytic cleavage, and like the PS1-N-terminal fragment (NTF) and PS1-C-terminal fragment (CTF), the NTF and CTF of PS2 are components of high-molecular-weight protein complexes which appear to be independent of the PS1 complexes of similar size [88]. Like PS1, PS2 predominantly resides within the perinuclear envelope, endoplasmic reticulum, Golgi and some yet uncharacterized intracytoplasmic vesicles. PS2 mutations, like PS1 mutations, increase the secretion of long-tailed Aβ peptides [75, 77]. However, despite these similarities, there are substantial differences in tissue-specific patterns of expression of PS1 and PS2 [78, 79]. More importantly, endogenous PS2 does not functionally replace either the βAPP or the *Notch* processing defects in PS1$^{-/-}$ animals. In addition, there are only mild pulmonary defects [89], no significant dorsal axis defects and no significant change in βAPP processing in PS2$^{-/-}$ animals. However, when "PS1$^{-/+}$:PS2$^{-/-}$" mice are created, they display a developmental phenotype as severe as that of "PS1$^{-/-}$:PS2$^{+/+}$" mice [89, 90]. Furthermore, overexpression of PS2 may partially replace the *Notch* defect in PS1$^{-/-}$ mice (D. Westaway et al., unpublished). Cumulatively, these data suggest quite distinct but partially overlapping functions for PS1 and PS2.

References

1 Kang J, Lemaire HG, Unterbeck A, Salbaum JM, Masters CL, Multhap G, Beyreuther K, Muller-Hill B (1987) The precursor of Alzheimer disease amyloid A4 protein resembles a cell surface receptor. *Nature* 325: 733–736
2 Goldgaber D, Lerman MI, McBride OW et al (1987) Characterization and chromosomal localization of a cDNA encoding brain amyloid of Alzheimer's Disease. *Science* 235: 877–880
3 Robakis NK, Lahiri DK, Brown HR, Rubenstein R, Mehta B, Wisniewski H, Goller N (1988) Expression studies of the gene encoding the Alzheimer's Disease and Down Syndrome amyloid peptide *In*: JW Swann (ed.): *Disorders of the Developing Nervous System: changing views on their origins, diagnoses and treatments*. Alan R Liss, New York, 183–193
4 Tanzi RE, Gusella JF, Watkins PC, Bruns GAP, St George-Hyslop PH, Van Keuren ML (1987) Amyloid β-protein gene: cDNA, mRNA distribution and genetic linkage near the Alzheimer locus. *Science* 235: 880–884

5 Kitaguchi N, Takahashi Y, Tokushima Y, Shiojiri S, Ito H (1988) Novel precursor of Alzheimer's Disease amyloid protein shows protease inhibitory activity. *Nature* 331: 530–532
6 Selkoe DJ (1994) Normal and abnormal biology of β-Amyloid Precursor Protein. *Annu Rev Neurosci* 17: 489–517
7 Jarrett JT, Lansbury PT (1993) Seeding "one-dimensional crystallization of amyloid": a pathogenic mechanism in Alzheimer's Disease and scrapie? *Cell* 73: 1055–1058
8 Yankner BA, Duffy LK, Kirschner DA (1990) Neurotrophic and neurotoxic effects of amyloid β protein: reversal by tachykinin neuropeptides. *Science* 250: 279–282
9 Pike CJ, Burdick D, Walencewicz AJ, Glabe CG, Cotman CW (1993) Neurodegeneration induced by beta-amyloid peptides *in vitro*: the role of peptide assembly state. *J Neurosci* 13: 1676–1678
10 Lorenzo A, Yanker BA (1994) β-amyloid neurotoxicity requires fibril formation and is inhibited by Congo red. *Proc Natl Acad Sci USA* 91: 12243–12247
11 Yankner BA (1996) Mechanisms of neuronal degeneration in Alzheimer's disease. *Neuron* 16: 921–932
12 Zheng H, Jiang M, Trumbauer ME, Hopkins R, Sirinathsinghji DJ, Stevens KA, Corner MW, Slunt HH, Sisodia SS, Chen HY et al (1996) Mice deficient for the amyloid precursor protein gene. *Ann N Y Acad Sci* 777: 421–426
13 Saitoh T, Sundsmo M, Roch J-M, Kimura N, Cole G, Schubert D, Oltersdorf T, Schenk DB (1989) Secreted form of amyloid β-protein is involved in the growth regulation of fibroblasts. *Cell* 58: 615–622
14 Milward AE, Papadopoulos R, Fuller SJ, Moir RD, Small D, Beyreuther K, Masters CL (1992) The amyloid protein precursor of Alzheimer disease is a mediator of the effects of NGF on neurite outgrowth. *Neuron* 9: 129–137
15 Nishimoto I, Okamoto T, Matsuura Y, Takahashi S, Murayama Y, Ogata F (1993) Alzheimer amyloid protein precursor complexes with brain GTP-binding protein Go. *Nature* 362: 75–79
16 Mullan MJ, Crawford F, Axelman K, Houlden H, Lilius L, Winblad B, Lannfelt L, Hardy J (1992) A pathogenic mutation for probable Alzheimer's Disease in the APP gene at the N-terminus of β-amyloid. *Nat Genet* 1: 345–347
17 Hendricks M, van Duijn CM, Cras P, Cruts M, van Hul W, Van Harskamp F, Warren A, McInnis M, Antonarakis G, Martin J-J et al (1992) Presenile dementia and cerebral hemorrhage linked to a mutation at codon 692 of the B-amyloid Precursor protein gene. *Nat Genet* 1: 218–221
18 Goate AM, Chartier-Harlin M-C, Mullan M, Brown J, Crawford F, Fidani L, Guiffra L, Haynes A, Hardy JA (1991) Segregation of a missense mutation in the amyloid precursor protein gene with Familial Alzheimer disease. *Nature* 349: 704–706
19 Chartier-Harlin M-C, Crawford F, Holden H, Warren A, Hughes D, Fidani L, Goate A, Rossor M, Hardy J, Mullan M (1991) Early-onset Alzheimer's Disease caused by mutations at codon 717 of the β-amyloid gene. *Nature* 353: 844–846
20 Murrell J, Farlow M, Ghetti B, Benson MD (1991) A mutation in the amyloid precursor protein associated with hereditary Alzheimer's Disease. *Science* 254: 97–99
21 Karlinsky H, Vaula G, Haines JL, Ridgley J, Bergeron C, Mortilla M, Tupler R, Percy M, Robitaille Y, Crapper MacLachlan DR et al (1992) Molecular and prospective phenotypic characterization of a pedigree with Familial Alzheimer disease and a missense mutation in codon 717 of the β-Amyloid Precursor Protein (βAPP) gene. *Neurology* 42: 1445–1453
22 Naruse S, Igarashi S, Kobayashi H, Aoki K, Inuzuki I, Kaneko K, Shimizu T, Iihara K, Kojima T, Miyatake T et al (1991) Missense mutation (Val → Ile) in exon 17 of the amyloid precursor protein gene in Japanese familial Alzheimer disease. *Lancet* 337: 978–979
23 The French Alzheimer's Disease Study Group (1996) No founder effect in three novel Alzheimer's Disease families with APP 717 Val Ile mutation. *J Med Genet* 33: 1–4
24 Ancolio K, Dumanchin C, Barelli H, Warter JM, Brice A, Campion D, Frebourg T (1999) Unusual phenotypic alteration of beta amyloid precursor protein maturation by a new Val715Met betaAPP770 mutation for probable early-onset Alzheimer's Disease. *Proc Natl Acad Sci USA* 96: 4119–4124
25 Citron M, Oltersdorf T, Haass C, McConlogue C, Hung AY, Seubert P, Vigo-Pelfrey C, Lieberburg I, Selkoe DJ (1992) Mutation of the β-amyloid precursor protein in familial Alzheimer's Disease increases β-protein production. *Nature* 360: 672–674
26 Cai XD, Golde TE, Younkin SG (1992) Release of excess amyloid beta protein from a mutant beta protein precursor. *Science* 259: 514–516
27 Susuki N, Cheung TT, Cai X-D, Odaka A, Otvos L, Eckman C, Golde T, Younkin SG (1994) An increased percentage of long amyloid β protein secreted by Familial Amyloid β Protein Precursor (βAPP717) mutants. *Science* 264: 1336–1340

28 Haass C, Hung AY, Selkoe DJ, Teplow DB (1994) Mutations associated with a locus for familial Alzheimer's Disease result in alternative processing of amyloid β-protein precursor. *J Biol Chem* 269: 17741–17748
29 Haas C, Lemere C, Capell A, Citron M, Selkoe D (1995) The Swedish mutation causes early-onset Alzheimer's disease by β-secretase cleavage within the secretory pathway. *Nat Med* 1: 1291–1296
30 Strittmatter WJ, Weisgraber KH, Huang DY, Schemechel D, Saunders AM, Roses AD (1993) Binding of human lipoprotein E to synthetic amyloid beta peptide: isoform-specific effects and implications for late-onset AD. *Proc Natl Acad Sci USA* 90: 8098–8102
31 Strittmatter WJ, Saunders AM, Schmechel D, Goldgaber D, Roses AD et al (1993) Apolipoprotein E: high affinity binding to B/A4 amyloid and increased frequency of type 4 allele in Familial Alzheimers Disease. *Proc Natl Acad Sci USA* 90: 1977–1981
32 Fraser PE, Nguyen J, Chin D, Kirschner DA (1992) Effects of sulphate ions on Alzheimer B/A4 peptide assemblies: implications for amyloid fibril proteoglycan interactions. *J Neurochem* 59: 1531–1540
33 Snow AD, Sekiguchi R, Nochlin D, Fraser P, Kimata K, Arai M, Schreier WA, Morgan DA (1994) An important role of heperin sulfate proteoglycan in a model system for the deposition and persistence of fibrillare Aβ-amyloid in rat brain. *Neuron* 12: 219–234
34 Arispe N, Pollard HB, Rojas E (1993) Giant multilevel cation channels formed by Alzheimer disease amyloid B protein in a bilayer membrane. *Proc Natl Acad Sci USA* 90: 10573–10577
35 Mattson MP, Cheng B, Davis D, Bryant K, Lieberburg I, Rydel R (1992) *Beta*-amyloid peptides destabilize calcium homeostasis and render human cortical neurons vulnerable to excitotoxicity. *J Neurosci* 12: 376–389
36 Mattson MP, Goodman Y (1995) Different amyloidogenic peptides share a similar mechanism of neurotoxicity involving reactive oxygen species and calcium. *Brain Res* 676: 219–224
37 Okamoto T, Takeda S, Murayama Y, Ogata E, Nishimoto I (1995) Ligand-dependent G protein coupling function of amyloid transmembrane precursor. *J Biol Chem* 270: 4205–4208
38 Yamatsuji T, Nishimoto I (1996) G protein-mediated neuronal DNA fragmentation induced by familial Alzheimer's disease associated mutants of APP. *Science* 272: 1349–1352
39 Schellenberg GD, Bird TD, Wijlman EM, Orr HT, Anderson L, Nemens E, White JA, Bonnycastle L (1992) Genetic linkage evidence for a familial Alzheimer's disease locus on chr 14. *Science* 258: 668–670
40 St George-Hyslop P, Haines J, Rogaev E, Mortilla M, Vaula G, Pericak-Vance M, Foncin J-F, Montesi M, Bruni A, Sorbi S et al (1992) Genetic evidence for a novel Familial Alzheimer disease gene on chromosome 14. *Nat Genet* 2: 330–334
41 Van Broeckhoven C, Backhovens H, Cruts M, De Winter G, Bruyland M, Cras P, Martin J-J (1992) Mapping of a gene predisposing to early-onset Alzheimer's Disease to chromosome 14q24.3. *Nat Genet* 2: 335–339
42 Sherrington R, Rogaev E, Liang Y, Rogaeva E, Levesque G, Ikeda M, Chi H, Lin C, Holman K, Tsuda T et al (1995) Cloning of a gene bearing missense mutations in early-onset familial Alzheimer's disease. *Nature* 375: 754–760
43 Levitan D, Greenwald I (1995) Facilitation of lin-12-mediated signalling by sel-12, a *Caenorhabditis elegans* S182 Alzheimer's Disease gene. *Nature* 377: 351–354
44 Boulianne G, Livne-Bar I, Humphreys JM, Rogaev E, St George-Hyslop P (1997) Cloning and mapping of a close homologue of human presenilins in *D. Melanogaster*. *NeuroReport* 8: 1025–1029
45 Slunt HH, Thinkaran G, Lee MK, Sisodia SS (1995) Nucleotide sequence of the chromosome 14 encoded S182 cDNA and revised secondary structure prediction. *Amyloid* 2: 188–190
46 Cribbs DH, Chen LS, Bende SM, LaFerla FM (1996) Widespread neuronal expression of the presenilin 1 early-onset Alzheimer's disease gene in murine brain. *Amer J Pathol* 148: 1797–1806
47 Rogaev EI, Sherrington R, Wu C, Levesque G, Liang Y, Rogaeva EA, Chi H, Ikeda M, Holman K, Lin C et al (1997) Analysis of the 5' sequence, genomic structure and alternative splicing of the presenilin 1 gene associated with early-onset Alzheimer's Disease. *Genomics* 40: 415–424
48 The Alzheimer's Disease Collaborative Group (1995) The structure of the presenilin I gene and the identification of six mutations in early-onset AD pedigrees. *Nat Genet* 11: 219–222
49 Cruts M, Martin J-J, Van Broeckhoven C (1995) Molecular genetic analysis of familial early-onset Alzheimer's Disease linked to chromosome 14q24.3. *Hum Molec Genet* 4: 2363–2371
50 De Strooper B, Beullens M, Contreras B, Craessaerts K, Moechars D, Bollen M, Fraser P, St George-Hyslop P, Van Leuven F (1997) Postranslational modification, subcellular localization and membrane orientation of the Alzheimer's Disease associated Presenilins. *J Biol Chem* 272: 3590–3598
51 Walter J, Capell A, Grunberg J, Pesold B, Schindzielorz A, Prior R, Podlisny MB, Fraser P, St

George-Hyslop P, Selkoe D et al (1996) The Alzheimer's Disease associated Presenilins are differentially phosphorylated proteins located predominantly within the endoplasmic reticulum. *Mol Med* 2: 673–691

52 Thinakaran G, Borchelt DR, Doan A, Slunt HH, Lee MK, Nordstedt C, Seeger M, Gandy SE, Hardy JA, Levey AI et al (1996) Endoproteolytic processing and protein topology of presenilin 1. *Soc Neurosci* 22: 728

53 Thinakaran G, Borchelt DR, Lee MK, Slunt HH, Spitzer L, Kim G, Ratovisky T, Davenport F, Nordstedt C, Seeger M et al (1996) Endoproteolysis of Presenilin 1 and accumulation of processed derivatives *in vivo*. *Neuron* 17: 181–190

54 Podlisny M, Citron M, Amarante P, Sherrington R, Weiming X, Zhang J, Diehl T, Levesque G, Fraser P, Haass C et al (1997) Presenilin proteins undergo heterogeneous endoproteolysis between Thr291 and Ala299 and occur as stable N- and C-terminal fragments in normal brain tissue. *Neurobiol Disease* 3: 325–337

55 Fraser PE, Levesque G, Yu G, Mills L, Thirwell J, Frantseva M, Carlen P, St George-Hyslop P (1998) Presenilin 1 is actively degraded by the 26S proteasome. *Neurobiol Aging*; *in press*

56 Uchihara T, El Hachimi HK, Duyckearts C, Foncin C, Fraser PE, Levesque L, St George-Hyslop P, Hauw JJ (1996) Widespread immunoreactivitry of presenilin in neurons of normal and Alzheimers disease brain: double labeling immunohistochemical study. *Acta Neuropathol* 92: 325–330

57 Wong PC, Zheng H, Chen H, Trumbauer ME, Roskams AJ, Chen HY, Van der Ploeg LH, Price DL, Sisodia SS (1996) Functions of the presenilins: generation and characterization of presenilin-1 null mice. *Soc Neurosci* 22: 728

58 Shen J, Bronson RT, Chen DF, Xia W, Selkoe DS, Tonegawa S (1997) Skeletal and CNS defects in prese-nilin-1 deficient mice. *Cell* 89: 629–639

59 Conlon RA, Reaume AG, Rossant J (1995) *Notch1* is required for the coordinate segmentation of somites. *Development* 121: 1533–1545

60 De Strooper B, Annert W, Cupers P, Saftig P, Craessaerts K, Mumm JS, Schroeter EH, Schrijvers V, Wolfe MS, Ray WJ et al (1999) A presenilin dependent gamma-secretase-like protease mediates release of *Notch* intracellular domain. *Nature* 398: 518–522

61 Ye Y, Lukinova N, Fortini ME (1999) Neurogenic phenotypes and altered Notch processing in *Drosophila* presenilin mutants. *Nature* 398: 525–529

62 Struhl G, Greenwald I (1999) Presenilin is required for activity and nuclear access of *Notch* in *Drosophila*. *Nature* 398: 522–525

63 De Strooper B, Saftig P, Craessaerts K, Vanderstichele H, Guhde G, Annaert W, Von Figura K, Van Leuven F (1998) Deficiency of presenilin 1 inhibits the normal cleavage of amyloid precursor protein. *Nature* 391: 387–390

64 Wolfe MS, Xia W, Ostaszewski BL, Diehl TS, Kimberly WT, Selkoe DS (1999) Two transmembrane aspartates in presenilin 1 required for presenilin endoproteolysis and gamma-secretase activity. *Nature* 398: 513–517

65 Leimer U, Lun K, Romig H, Walter J, Grunberg J, Brand M, Haass C (1999) Zebrafish presenilin promotes aberrant amyloid beta-peptide production and requires a critical aspartate residue for its function in amy-loidogenesis. *Biochemistry* 38: 13602–13609

66 Naruse S, Thinakaran G, Luo JJ, Kusiak JW, Tomita T, Iwatsubo T, Qian X, Ginty DD, Price DL, Borchelt DR et al (1998) Effect of PS1 deficiency on membrane protein trafficking in neurons. *Neuron* 21: 1213–1221

67 Chen F, Yang D-S, Tandon A, Rozmahel R, Yu G, Nishimura M, Kawarai T, Westaway D, Gandy SE, Fraser PE et al (2000) Proteolytic derivatives of Amyloid Precursor Protein accumulate in restricted and unpredicted intracellular compartments in the absence of functional presenilin 1 expression. *J Biol Chem*; *in press*

68 Vito P, Lacana E, D'Adamio L (1996) Interfering with apoptosis: Ca(2+)-binding protein ALG-2 and Alzheimer's disease gene ALG-3. *Science* 271: 521–525

69 Wolozin B, Iwasaki K, Vito P, Ganjei K, Lacana E, Sunderland T, Zhao B, Kusiak JW, D'Adamio L (1996) Participation of presenilin 2 in apoptosis: enhanced basal activity conferred by an Alzheimer mutation. *Science* 274: 1710–1713

70 Sato S, Kamino K, Miki T, Doi A, Li K, St George-Hyslop P, Ogihara T, Sakaki Y (1996) Splicing muta-tion of presenilin 1 gene for early-onset familial Alzheimer's Disease. *Hum Mutat*; *in press*

71 Perez-Tur J, Froelich S, Prihar G, Crook R, Baker M, Duff K, Wragg M, Hardy J, Goate A, Lannfelt L et al (1996) A mutation in Alzheimer's disease destroying a splice acceptor site in the presenilin 1 gene. *Neuroreport* 7: 297–301

72 Kwok JB, Tadder K, Fisher C, Hallup M, Brooks W, Nicholson G, St George-Hyslop P, Fraser PE, Relkin N, Gandy SE (1997) Two novel PS1 mutations in early-onset Alzheimer disease: evidence for association of PS1 mutations with a novel phenotype. *NeuroReport* 8: 1537–1542

73 Van Broeckhoven C (1995) Presenilins and Alzheimer disease. *Nat Genet* 11: 230–232

74 Martins RN, Turner BA, Carroll RT, Sweeney D, Kim KS, Wisniewski HM, Blass JP, Gibson GE, Gandy SE (1995) High levels of amyloid beta-protein from S182 (Glu246) familial Alzheimer's cells. *NeuroReport* 7: 217–220

75 Scheuner D, Eckman L, Jensen M, Sung X, Citron M, Suzuki N, Bird T, Hardy J, Hutton M, Lannfelt L, Selkoe D et al (1996) Secreted amyloid-β protein similar to that in the senile plaques of Alzheimer disease is increased *in vivo* by presenilin 1 and 2 and APP mutations linked to FAD. *Nat Med* 2: 864–870

76 Duff K, Eckman C, Zehr C, Yu X, Prada CH, Pereztur J, Hutton M, Refolo L, Zenk B, Hardy J et al (1996) Increased amyloid beta 42(43) in brains of mice expressing mutant presenilin 1. *Nature* 383: 710–713

77 Citron M, Westaway D, Xia W, Carlson G, Diehl TS, Levesque G, Johnson-Wood K, Lee M, Seubert P, Davis A et al (1997) Mutant presenilins of Alzheimer's Disease increase production of 42 residue amyloid β-protein in both transfected cells and transgenic mice. *Nat Med* 3: 67–72

78 Rogaev EI, Sherrington R, Rogaeva EA, Levesque G, Ikeda M, Liang Y, Chi H, Lin C, Holman K, Tsuda T et al (1995) Familial Alzheimer's disease in kindreds with missense mutations in a novel gene on chromosome 1 related to the Alzheimer's Disease type 3 gene. *Nature* 376: 775–778

79 Levy-Lahad E, Poorkaj P, Wang K, Fu YH, Oshima J, Mulligan J, Schellenberg GD (1996) Genomic structure and expression of STM2, the chromosome 1 Familial Alzheimer disease gene. *Genomics* 34: 198–204

80 Sherrington R, Froelich S, Sorbi S, Campion D, Chi H, Rogaeva EA, Levesque G, Rogaev EI, Lin C, Liang Y et al (1996) Alzheimer's disease associated with mutations in presenilin-2 are rare and variably penetrant. *Hum Molec Genet* 5: 985–988

81 Levy-Lahad E, Wijsman EM, Nemens E et al (1995) A familial Alzheimer's Disease locus on chromosome 1. *Science* 269: 970–973

82 Bird TD, Levy-Lehad E, Poorkaj J, Nochlin D, Sumi SM, Nemens EJ, Wijsman E, Schellenberg GD (1997) Wide range in age of onset for chromosome 1 related familial AD. *Ann Neurol* 40: 932–936

83 Bird TD (1988) Familial Alzheimer's Disease in American descendents of the Volga Germans: probable genetic founder effect. *Ann Neurol* 23: 25

84 Bird TD, Sumi SM, Nemens EJ, Schellenberg GD, Martin G et al (1989) Phenotypic heterogenity in familial Alzheimer's Disease: a study of 24 kindreds. *Ann Neurol* 25: 12–25

85 St George-Hyslop PH, Tsuda T, Crapper McLachlan DR, Karlinsky H, Pollen D, Lippa C (1994) Alzheimer's Disease and possible gene interaction. *Science* 263: 536–537

86 Nacmias B, Latteraga S, Tulen P, Piacentini S, Bracco L, Amaducci L, Sorbi S (1995) ApoE genotype and familial Alzheimer's Disease: a possible influence on age-of-onset in APP717Val → Ile mutated families. *Neurosci Lett* 183: 1–3

87 Sorbi S, Nacmias B, Forleo P, Amaducci L (1995) Epistatic effect of APP717 mutation and apolipoprotein E genotype in familial Alzheimer disease. *Ann Neurol* 38: 124–128

88 Yu G, Chen F, Levesque G, Nishimura M, Zhang D-M, Levesque L, Rogaeva E, Xu D, Liang Y, Duthie M et al (1998) The presenilin 1 protein is a component of a high molecular weight intracellular complex that contains b-catenin. *J Biol Chem* 273: 16470–16475

89 Herreman A, Hartmann D, Annaert W, Saftig P, Craessaerts K, Serneels L, Umans L, Schrijvers V, Checler F, Vanderstichele H et al (1999) Presenilin 2 deficiecy causes a mild pulmonary phenotype and no chnages in amyloid precursor protein processing but enhances the embryonic lethal phenotype of presenilin 1 deficiency. *Proc Natl Acad Sci USA* 96: 11872–11877

90 Donoviel D, Hadjantonalis AK, Ikeda M, Zheng H, St George-Hyslop P, Bernstein A (1999) Mice lacking both presenilin genes exhibit early embryonic patterning defects. *Gene Develop* 13: 2801–2810

Neuroscientific Basis of Dementia
C. Tanaka, P.L. McGeer, Y. Ihara (eds)
© 2001 Birkhäuser Verlag Basel/Switzerland

Lessons from presenilin domain analysis: endoproteolytic processing and enhanced Aβ42 production mediated by FAD-linked variants

Gopal Thinakaran[1], Carlos A. Saura[1], Taisuke Tomita[2], Toshiyuki Honda[3] and Takeshi Iwatsubo[2]

[1] Department of Neurobiology, Pharmacology and Physiology, University of Chicago, Chicago, IL, USA
[2] Department of Neuropathology and Neuroscience, Graduate School of Pharmaceutical Sciences, University of Tokyo, Tokyo, Japan
[3] RIKEN Brain Science Institute, Saitama, Japan

Introduction

Autosomal dominant mutations in genes encoding presenilin 1 (PS1) and presenilin 2 (PS2) predispose individuals to familial early-onset Alzheimer's disease (FAD). FAD is pathologically characterized by the cerebral deposition of 40–42 amino acid β-amyloid (Aβ) peptides, which are generated by the proteolytic processing of amyloid precursor protein (APP). Two lines of evidence demonstrate a critical role for presenilins (PS) in Aβ production: first, cells established from $PS1^{-/-}$ mice do not secrete Aβ [1, 2]; and second, FAD-linked mutations in PS increase the production of highly fibrillogenic Aβ42 [3–6]. PS are polytopic membrane proteins that undergo regulated endoproteolysis to generate saturable levels of stable NH_2- (NTF) and COOH-terminal fragments (CTF), which are the preponderant PS-related species that accumulate in vivo [7, 8]. Although the precise steps involved in the maturation of synthetic PS polypeptides are not clearly defined, properly folded PS polypeptides become stabilized, undergo endoproteolysis and form high-molecular weight complexes. The vast majority of overexpressed full-length PS and transgene-derived polypeptides that correspond to the NTF fail to form stable complexes and are rapidly degraded [9–11]. Whereas endoproteolytic cleavage is not required for biological activity of PS polypeptides [7, 12], several lines of evidence indicate that the NTF and CTF together form the functional PS unit in vivo (reviewed in [6]).

Little is known about the proteolytic cleavage of PS, the mechanism that regulates PS fragment accumulation and the domains of PS that are essential for the pathogenic function of mutant PS. Based on our earlier studies, we speculated that a small fraction of newly synthesized presenilin polypeptides establish proper intramolecular associations during folding and proceed to interact with other components to form a high-molecular-weight complex [8, 11]. One of our interests is to identify the molecular domains of PS that mediate such intra- and intermolecular interactions, and thus regulate the metabolism and pathological activities of PS. To this end, we have initiated a series of studies to examine the role played by the transmembrane (TM) domains and the non–conserved hydrophilic "Loop" (HL) domain of PS.

Deletions of TM domains impair PS metabolism

To assess the importance of the multiple TM domains of PS, we constructed a series of experimental TM deletion mutants (ΔTM) of PS1 (Fig. 1A) and examined their metabolism by generating stable N2a cell lines. Expression of mutant polypeptides was monitored by Western blotting using PS1 antisera, PS1$_{NT}$ and αPS1Loop [7, 13]. We assessed whether the experimental PS1 polypeptides are endoproteolytically processed, and whether their expression interferes with the accumulation of endogenous murine PS1-derived fragments. We

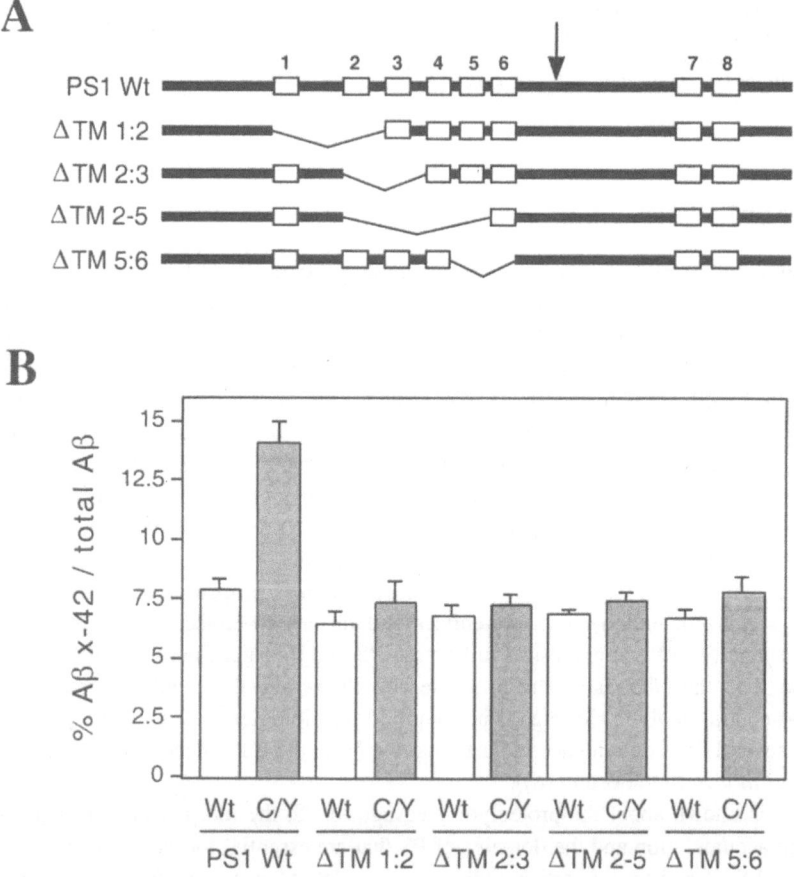

Figure 1. Influence of PS1 TM deletions on Aβ42 production in COS cells. (A) Schematic representation of PS1ΔTM polypeptides. The site of regulated endoproteolysis (*arrow*) and the location of TM domains (*boxes*) are indicated. (B) COS cells were transfected with cDNA encoding full-length or ΔTM polypeptides without or with FAD-linked C410Y substitution, along with cDNA encoding APPswe. Conditioned medium was collected 48 after transfection. The amount of secreted Aβ x-40 and Aβ x-42 was quantified from using two-site ELISAs and Aβ x-42/total Aβ ratio (mean ± S.E. of at least three experiments performed in triplicate) was calculated. PS1 Wt *versus* PS1 C410Y $P < 0.0001$.

found that none of the ΔTM polypeptides were endoproteolytically processed, as shown by the lack of the appearance of a corresponding shorter ΔTM-NTF and human CTF (data not shown). With the exception of ΔTM1:2, expression of the experimental ΔTM polypeptides failed to compromise the accumulation of endogenous PS derivatives. We noticed that the accumulation of endogenous mouse PS1-derived NTF or CTF was diminished by high levels of expression of ΔTM1:2 polypeptides (data not shown). Nevertheless, our results show that the TM domain integrity is essential for the proper endoproteolysis of PS1. Because the subcellular compartments involved in the endoproteolysis of PS1 are not clearly defined, we cannot unambiguously rule out defective intracellular trafficking of the ΔTM polypeptides.

ΔTM polypeptides harboring FAD-linked missense mutations fail to elevate Aβ42 levels

We and others reported previously that FAD-linked PS variants elevate the levels of Aβ42 [3–5]. To assess whether ΔTM polypeptides can increase the levels of Aβ42, we introduced the FAD-linked missense mutation C410Y into each of the ΔTM constructs. Subsequently, african green monkey kidney COS cells were co-transfected with cDNA encoding full-length PS1, ΔTM, or the corresponding C410Y variants and the cDNA that encodes the "Swedish" APP variant (APPswe). The levels of secreted Aβx-40 and x-42 peptides in the conditioned media were quantified by two-site enzyme-linked immunosorbent assay (ELISA) assay as described [5, 14]. As expected from previous studies [4], there was a 2-fold increase in Aβ42/total Aβ ratio in conditioned media collected from cells transfected with PS1 C410Y variant (PS1 Wt was 7.90 ± 0.4 *versus* PS1 C410Y, 14.01 ± 0.99; $P < 0.0001$) (Fig. 1B). However, the ratio of secreted Aβ42 to total Aβ was not influenced by the expression of ΔTM1:2, ΔTM2:3, ΔTM2-5, or ΔTM5:6 deletions without or with the FAD-linked C410Y missense substitution (Fig. 1B). Therefore, the integrity of multiple TM domains is important for the pathogenic properties of FAD-linked PS variants with regard to enhanced production of Aβ42.

Metabolism of PS1 lacking the HL domain

Next, we focused on the role of the cytoplasmic loop (Cys 263–Leu 381) between TM 6 and TM 7, termed the "loop" domain. The NH₂-terminal one-third of the loop domain is rich in hydrophobic residues, highly conserved between PS1 homologues, and harbors several FAD-linked mutations; the site of regulated endoproteolytic cleavage (Met 292) is located within the hydrophobic stretch [15, 16]. The remaining two-thirds of the loop domain is hydrophilic, and only weakly conserved between PS1 homologues. Nevertheless, several findings have suggested that the HL domain may be important for the biological functions of PS. For example, PS1 and PS2 are cleaved by caspases within this non-conserved region [17–20]. In addition, serine residues within the HL domain of PS1 and PS2 are modified by phosphorylation [21–23]; it has been shown that phosphorylation of PS2 within the HL domain can regulate caspase cleavage, and modulate antiapoptotic properties of the PS2 CTF [24, 25]. Finally, several laboratories have independently identified interactions between several members of the armadillo family of proteins and PS within a region that overlaps with the HL domain (reviewed in [26]). To address the importance of the HL domain for the biology of PS, we chose to characterize the biology of PS1 and PS2 polypeptides that lack this region.

To investigate whether the non-conserved HL region of PS plays an essential role in PS1 metabolism, we constructed a cDNA that encodes a human PS1 polypeptide without the HL

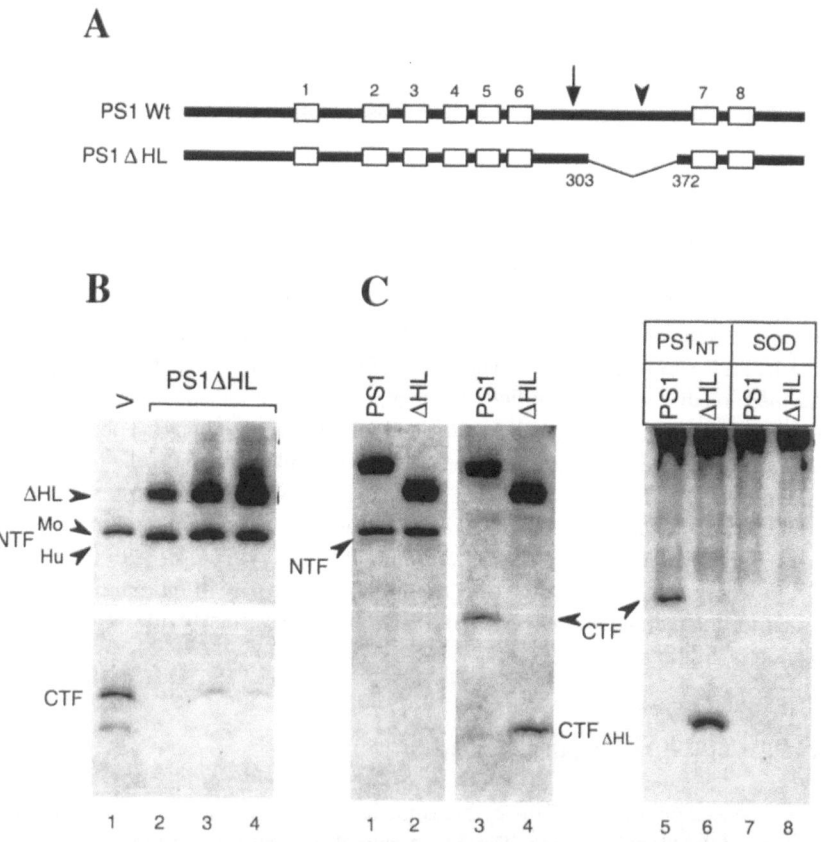

Figure 2. Deletion of the HL domain does not affect PS1 metabolism. (A) Schematic representation of PS1ΔHL. The deletion encompasses Glu304-Gly371. The sites of regulated endoproteolysis (*arrow*) and caspase cleavage (*arrowhead*), and the TM domains (*boxes*) are indicated. (B) Analysis of PS1ΔHL expression in stable N2a cell lines. Detergent lysates prepared from cells transfected with empty vector (V) and PS1ΔHL clones were analyzed by Western blotting with PS1$_{NT}$ (top) or αPS1Loop (bottom). Mouse (Mo) and human (Hu) NTF and "full-length" PS1ΔHL (ΔHL) are indicated. (C) Coimmunoprecipitation analysis of PS1ΔHL derivatives. Detergent lysates prepared from stable cells expressing PS1 or PS1ΔHL were analyzed by immunoblotting with PS1$_{NT}$ (lanes 1 and 2), or COOH-terminal PS1 antibody, PS-C3 (lanes 3 and 4). Lysates prepared under non-denaturing conditions were used for coimmunoprecipitation using PS1$_{NT}$ (lanes 5 and 6) or control SOD antibodies (lanes 7 and 8), and analyzed by immunoblotting with PS-C3. PS1 CTF and CTF$_{ΔHL}$ are indicated.

region (Glu 304–Gly 371), termed PS1ΔHL (Fig. 2A). The site of regulated endoproteolytic cleavage of PS1 (Met 292) [15, 16] is present in PS1ΔHL, but the site of caspase cleavage (Asp 345) has been deleted. Endoproteolysis of PS1ΔHL was examined by generating
independent stable N2a cell lines expressing PS1ΔHL. As shown in Figure 2B, endoproteolysis of PS1ΔHL generated ~30 kDa PS1ΔHL-derived NTF (lanes 2–4), which exhibited
slightly accelerated migration on gels compared to endogenous mouse PS1-derived NTF
(lane 1) and co-migrated with full-length human PS1-derived NTF (Fig. 2C, compare lanes
1 and 2). Thus, the 30 kDa fragment is a derivative of the endoproteolytic processing of
human PS1ΔHL. To determine if the expression of PS1ΔHL also resulted in the diminished
accumulation of endogenous murine PS1 derivatives, we performed Western blot analysis
using αPS1Loop. Because the epitopes recognized by αPS1Loop lie within the HL domain
[7], this reagent does not react with PS1ΔHL or the CTF derived from PS1ΔHL (CTF$_{ΔHL}$).
In each of the stable PS1ΔHL clones, αPS1Loop only detected a weak signal for the mouse
PS1-derived CTFs as compared with cells transfected with empty vector (Fig. 2B, bottom).
Furthermore, PS-C3, an antibody raised against the COOH-terminus of PS1, reacted with
~9 kDa CTF$_{ΔHL}$ derived from PS1ΔHL (Fig. 2B), confirming that the diminution in murine
PS1 CTF was accompanied by the accumulation of CTF$_{ΔHL}$. Finally, we show that PS1$_{NT}$
antiserum could coimmunoprecipitate NTF and CTF$_{ΔHL}$ from detergent extracts of stable
N2a cells expressing PS1ΔHL (Fig. 2C, lanes 5 and 6). These results demonstrate that the
deletion of the HL domain of PS1 does not affect the regulated endoproteolytic processing
of PS1, saturable accumulation of PS1 fragments, or the association of NTF and CTF.

The HL domain is dispensable for FAD-linked PS-mediated elevation of Aβ42 production

We next sought to assess whether the HL domain of PS1 is essential for the FAD mutation-
mediated elevation of Aβ42 production [3, 4]. For these studies, we generated cDNAs that
encode PS1ΔHL polypeptides harboring the FAD-linked missense mutation M146L,
H163R, or C410Y. To measure Aβ production, we cotransfected Wt or mutant PS1ΔHL
cDNAs into COS cells along with a cDNA that encodes APPswe. In parallel, we transfected
COS cells with cDNAs that encode full-length Wt and mutant PS1. As expected from previous studies [4], ELISA quantification revealed a 2-fold increase in Aβ42/total Aβ ratio in
conditioned media collected from cells transfected with mutant full-length PS1 (PS1 Wt was
8.04 ± 0.38 *versus* PS1 mutants, 17.28 ± 0.73; $P < 0.0001$) (Fig. 3). However, Aβ42 ratio
was not significantly different between cells transfected with Wt PS1 and PS1ΔHL (PS1 was
8.04 ± 0.38 *versus* PS1ΔHL, 9.03 ± 0.35; $P = 0.959$), indicating that deletion of the HL
domain, by itself, did not influence the production of Aβ42 (Fig. 3). Notably, Aβ42 ratio was
elevated 1.7-fold by the expression PS1ΔHL harboring FAD-linked mutations M146L,
H163R and C410Y (PS1ΔHL Wt was 9.03 ± 0.35 *versus* PS1ΔHL mutants, 15.7 ± 0.91;
$P = 0.0026$).

Having demonstrated that the deletion of the HL domain of PS1 did not affect Aβ production, we asked whether the same is true for PS2. For these studies, we generated cDNAs

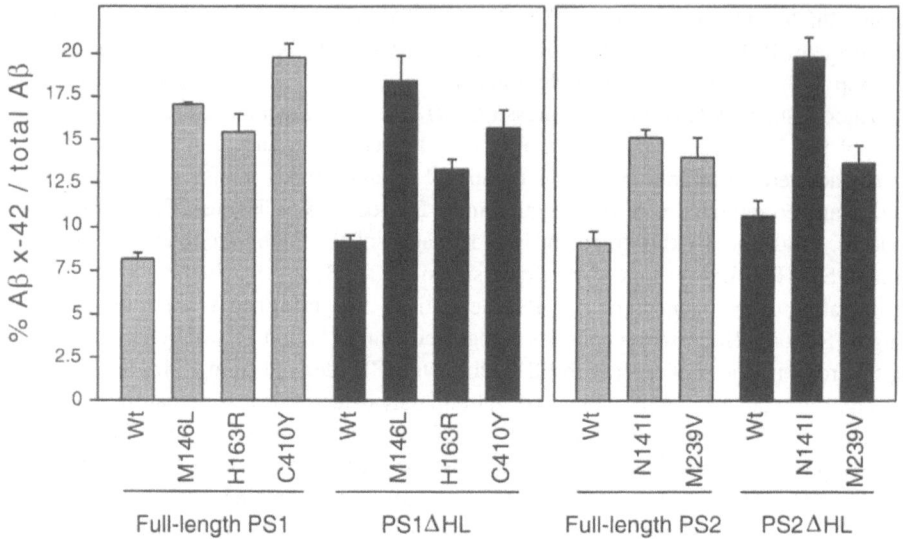

Figure 3. Influence of HL deletion on Aβ production in COS cells. COS cells were transfected with PS1, PS1ΔHL, PS2, or PS1ΔHL cDNA (Wt or FAD mutants as indicated) along with cDNA encoding APPswe and conditioned medium was collected 48 after transfection. The amount of secreted Aβ x-40 and Aβ x-42 was quantified from using two-site ELISAs and Aβ x-42/total Aβ ratio (mean ± S.E. of three transfections) was calculated. PS1 Wt *versus* PS1 mutants $P < 0.0001$; PS1ΔHL *versus* PS1ΔHL mutants $P = .0026$; PS2 Wt *versus* PS2 mutants $P = .0055$; PS2ΔHL *versus* PS2ΔHL mutants $P = 0.0021$.

that encode Wt and mutant PS2 polypeptides lacking the HL domain (PS2ΔHL). As expected, PS2ΔHL polypeptides were also endoproteolytically processed into stable fragments (data not shown). Consistent with the previous studies [5], expression of mutant PS2 significantly altered Aβ42 ratio in transfected COS cells (PS2 Wt was 8.99 ± 0.59 *versus* PS2 mutants, 14.46 ± 0.58; $P = 0.0055$). Similar to our PS1ΔHL findings, deletion of the PS2 HL domain did not significantly influence Aβ42 ratio (PS2 was 8.99 ± 0.59 *versus* PS2ΔHL, 10.57 ± 0.84; $P = 0.813$). Furthermore, expression of PS2ΔHL polypeptides harboring FAD-linked N141I or M239V mutations significantly altered Aβ42 ratio (PS2ΔHL Wt was 10.57 ± 0.84 *versus* PS2ΔHL mutants, 16.57 ± 1.16; $P = 0.0021$). Collectively, these results document that the deletion of the HL domain does not affect FAD-linked mutant PS1- or PS2-mediated production of Aβ42 peptides.

Conclusion

Several lines of evidence indicate that the NTF and CTF together form the functional PS unit *in vivo* (reviewed in [6]). Little is known about the proteolytic cleavage of PS or the mechanism that regulates PS fragment accumulation. Our present study provides three important insights regarding the structure and function of PS. First, we have demonstrated that the integrity of N-terminal TM domains are essential for PS metabolism. When two or more TM domains are deleted, the resulting non-cleaved PS polypeptides do not interfere with the accumulation of endogenous PS fragments. Furthermore, TM-deleted polypeptides harboring a FAD-linked mutation are not pathogenically active in increasing Aβ42 production. Second, we provide evidence that the HL domain of PS1 or PS2 does not provide any structural information critical for the proper folding and maturation of PS polypeptides. Deletion of the HL domain did not affect endoproteolysis, saturable accumulation of PS-derived fragments or the association between NTF and CTF. Finally, we demonstrate that the deletion of PS HL domain, which includes the caspase cleavage site, has little influence on the levels of Aβ production or on the increased production of Aβ42 by PS1ΔHL polypeptides harboring FAD-linked mutations.

Our findings are noteworthy considering the number of proteins that have recently been shown to bind to the "loop" domain of PS1 or PS2. Significantly, members of the armadillo family of proteins, including β-catenin, δ-catenin, p0071 and neuronal-specific plakophilin, bind to PS1 or PS2 through a region within the TM 6 and TM 7 loop that overlaps with the HL domain (reviewed in [26]). Other PS1 CTF-binding proteins reported thus far include Bcl-XL, filamin family of actin-binding proteins and SEL10, a Cdc4p-related protein (reviewed in [26]). It is known that PS NTF and CTF form heteromeric high-molecular-weight complexes *in vivo*, while excess full-length PS polypeptides fail to assemble into stable complexes and are subject to rapid degradation [8–11]. Based on our analysis of PSΔHL metabolism, we predict that HL-interacting proteins may not play a role in the formation of high-molecular-weight PS complexes. Unlike PS polypeptides with substitutions of aspartate residues in TM 6 and TM 7 [27], overexpression of PSΔHL did not affect γ-secretase processing of APP. Furthermore, our analyses of Aβ secreted from cells expressing PS1ΔHL or PS2ΔHL harboring FAD-linked mutations strongly argues against a role for caspase cleavage within the HL domain in mediating the effects of FAD mutations on Aβ42 generation. A similar conclusion regarding the connection between caspase cleavage and Aβ42 production was reached in an earlier study [28]. Although our studies provide clear evidence that under normal culture conditions caspase cleavage of PS does not contribute to enhanced production of Aβ42 by mutant PS, we cannot rule out a role for caspase cleavage of PS in influencing Aβ42 production under pathogenic conditions *in vivo*. We are currently generating transgenic mice that express PSΔHL to characterize the *in vivo* role of PS1 HL and assess whether expression of HL-deleted PS polypeptides can rescue the developmental defects of *PS1*$^{-/-}$ mice.

Acknowledgements
This work was supported by the U.S. Public Health Service, National Institutes of Health Grant 1PO1 AG14248, Alzheimer's Association (G.T.), the Adler Foundation (G.T. and C.A.S.), the Brain Research Foundation (G.T.), an award to the University of Chicago's Division of Biological Sciences under the Research Resources Program for Medical Schools of the Howard Hughes Medical Institute, and grants–in-aid from the Ministry of Health and Welfare, the Ministry of Education, Science, Culture and Sports, Japan, CREST of Japan Science and Technology Corporation (T.I)

References

1 De Strooper B, Saftig P, Craessaerts K, Vanderstichele H, Guhde G, Annaert W, Von Figura K, Van Leuven F (1998) Deficiency of presenilin-1 inhibits the normal cleavage of amyloid precursor protein. *Nature* 391: 387–390

2 Naruse S, Thinakaran G, Luo J-J, Kusiak JW, Tomiata T, Iwatsubo T, Qian X, Ginty DD, Price DL, Borchelt DR et al (1998) Effects of PS1 deficiency on membrane protein trafficking in neurons. *Neuron* 21: 1213–1221

3 Scheuner D, Eckman C, Jensen M, Song X, Citron M, Suzuki N, Bird TD, Hardy J, Hutton M, Kukull W et al (1996) Secreted amyloid beta-protein similar to that in the senile plaques of Alzheimer's disease is increased *in vivo* by the presenilin 1 and 2 and APP mutations linked to familial Alzheimer's disease. *Nat Med* 2: 864–870

4 Borchelt DR, Thinakaran G, Eckman CB, Lee MK, Davenport F, Ratovitsky T, Prada CM, Kim G, Seekins S, Yager D et al (1996) Familial Alzheimer's disease-linked presenilin 1 variants elevate Abeta1-42/1-40 ratio *in vitro* and *in vivo*. *Neuron* 17: 1005–1013

5 Tomita T, Maruyama K, Saido TC, Kume H, Shinozaki K, Tokuhiro S, Capell A, Walter J, Grunberg J, Haass C et al (1997) The presenilin 2 mutation (N141I) linked to familial Alzheimer disease (Volga German families) increases the secretion of amyloid beta protein ending at the 42nd (or 43rd) residue. *Proc Natl Acad Sci USA* 94: 2025–2030

6 Sisodia SS, Kim SH, Thinakaran G (1999) Function and Dysfunction of the Presenilins. *Amer J Hum Genet* 65: 7–12

7 Thinakaran G, Borchelt DR, Lee MK, Slunt HH, Spitzer L, Kim G, Ratovitsky T, Davenport F, Nordstedt C, Seeger M et al (1996) Endoproteolysis of presenilin 1 and accumulation of processed derivatives *in vivo*. *Neuron* 17: 181–190

8 Thinakaran G, Harris CL, Ratovitski T, Davenport F, Slunt HH, Price DL, Borchelt DR, Sisodia SS (1997) Evidence that levels of presenilins (PS1 and PS2) are coordinately regulated by competition for limiting cellular factors. *J Biol Chem* 272: 28415–28422

9 Ratovitski T, Slunt HH, Thinakaran G, Price DL, Sisodia SS, Borchelt DR (1997) Endoproteolytic processing and stabilization of wild-type and mutant presenilin. *J Biol Chem* 272: 24536–24541

10 Steiner H, Capell A, Pesold B, Citron M, Kloetzel PM, Selkoe DJ, Romig H, Mendla K, Haass C (1998) Expression of Alzheimer's disease-associated presenilin-1 is controlled by proteolytic degradation and complex formation. *J Biol Chem* 273: 32322–32331

11 Saura CA, Tomita T, Davenport F, Harris CL, Iwatsubo T, Thinakaran G (1999) Evidence that intramolecular associations between presenilin domains are obligatory for endoproteolytic processing. *J Biol Chem* 274: 13818–13823

12 Steiner H, Romig H, Grim MG, Philipp U, Pesold B, Citron M, Baumeister R, Haass C (1999) The biological and pathological function of the presenilin-1 Deltaexon 9 mutation is independent of its defect to undergo proteolytic processing. *J Biol Chem* 274: 7615–7618

13 Thinakaran G, Regard JB, Bouton CML, Harris CL, Price DL, Borchelt DR, Sisodia SS (1998) Stable association of presenilin derivatives and absence of presenilin interactions with APP. *Neurobiol Disease* 4: 438–453

14 Suzuki N, Cheung TT, Cai XD, Odaka A, Otvos L Jr, Eckman C, Golde TE, Younkin SG (1994) An increased percentage of long amyloid beta protein secreted by familial amyloid beta protein precursor (beta APP717) mutants. *Science* 264: 1336–1340

15 Podlisny MB, Citron M, Amarante P, Sherrington R, Xia W, Zhang J, Diehl T, Levesque G, Fraser P, Haass C et al (1997) Presenilin proteins undergo heterogeneous endoproteolysis between Thr291 and Ala299 and occur as stable N- and C-terminal fragments in normal and Alzheimer brain tissue. *Neurobiol Disease* 3: 325–337

16 Steiner H, Romig H, Pesold B, Philipp U, Baader M, Citron M, Loetscher H, Jacobsen H, Haass C (1999) Amyloidogenic function of the Alzheimer's disease-associated presenilin 1 in the absence of endoproteolysis. *Biochemistry* 38: 14600–14605

17 Kim TW, Pettingell WH, Jung YK, Kovacs DM, Tanzi RE (1997) Alternative cleavage of Alzheimer-associated presenilins during apoptosis by a caspase-3 family protease. *Science* 277: 373–376

18 Vito P, Ghayur T, D'Adamio L (1997) Generation of anti-apoptotic presenilin-2 polypeptides by alternative transcription, proteolysis, and caspase-3 cleavage. *J Biol Chem* 272: 28315–28320

19 Loetscher H, Deuschle U, Brockhaus M, Reinhardt D, Nelboeck P, Mous J, Grunberg J, Haass C, Jacobsen

H (1997) Presenilins are processed by caspase-type proteases. *J Biol Chem* 272: 20655–20659

20 Grunberg J, Walter J, Loetscher H, Deuschle U, Jacobsen H, Haass C (1998) Alzheimer's disease associated presenilin-1 holoprotein and its 18–20 kDa C-terminal fragment are death substrates for proteases of the caspase family. *Biochemistry* 37: 2263–2270

21 Seeger M, Nordstedt C, Petanceska S, Kovacs DM, Gouras GK, Hahne S, Fraser P, Levesque L, Czernik AJ, George-Hyslop PS et al (1997) Evidence for phosphorylation and oligomeric assembly of presenilin 1. *Proc Natl Acad Sci USA* 94: 5090–5094

22 Walter J, Grunberg J, Capell A, Pesold B, Schindzielorz A, Citron M, Mendla K, George-Hyslop PS, Multhaup G, Selkoe DJ et al (1997) Proteolytic processing of the Alzheimer disease-associated presenilin-1 generates an *in vivo* substrate for protein kinase C. *Proc Natl Acad Sci USA* 94: 5349–5354

23 Walter J, Grunberg J, Schindzielorz A, Haass C (1998) Proteolytic fragments of the Alzheimer's disease associated presenilins-1 and -2 are phosphorylated *in vivo* by distinct cellular mechanisms. *Biochemistry* 37: 5961–5967

24 Walter J, Schindzielorz A, Grunberg J, Haass C (1999) Phosphorylation of presenilin-2 regulates its cleavage by caspases and retards progression of apoptosis. *Proc Natl Acad Sci USA* 96: 1391–1396

25 Vito P, Lacana E, D'Adamio L (1996) Interfering with apoptosis: Ca(2+)-binding protein ALG-2 and Alzheimer's disease gene ALG-3. *Science* 271: 521–525

26 Thinakaran G (1999) The role of presenilins in Alzheimer's disease. *J Clin Invest* 104: 1321–1327

27 Wolfe MS, Xia W, Ostaszewski BL, Diehl TS, Kimberly WT, Selkoe DJ (1999) Two transmembrane aspartates in presenilin-1 required for presenilin endoproteolysis and gamma-secretase activity. *Nature* 398: 513–517

28 Brockhaus M, Grunberg J, Rohrig S, Loetscher H, Wittenburg N, Baumeister R, Jacobsen H, Haass C (1998) Caspase-mediated cleavage is not required for the activity of presenilins in amyloidogenesis and NOTCH signaling. *Neuroreport* 9: 1481–1486

Neuroscientific Basis of Dementia
C. Tanaka, P.L. McGeer, Y. Ihara (eds)
© 2001 Birkhäuser Verlag Basel/Switzerland

Amyloid and presenilins in the pathobiology of Alzheimer's disease

Takeshi Iwatsubo and Taisuke Tomita

Department of Neuropathology and Neuroscience, Graduate School of Pharmaceutical Sciences, University of Tokyo, Tokyo, Japan

Alzheimer's disease (AD) is characterized pathologically by a massive deposition of amyloid β peptides (Aβ), which are proteolytically produced from β-amyloid precursor proteins (βAPP) through two sequential cleavages by β-secretase (recently identified as BACE) and γ-secretase. Two major forms of Aβ with distinct carboxyl (C) termini ending at the 40th and 42nd residues (Aβ40 and Aβ42, respectively) have been identified; these are differentially cleaved by γ-secretase(s) [1]. Aβ42 aggregates much faster than Aβ40 *in vitro* [2] and Aβ42 is the initially and predominantly deposited Aβ species in the brains of patients with AD and Down's syndrome [3, 4]. Moreover, missense mutations in βAPP genes, a rare cause of familial AD (FAD), lead to increased production of Aβ42, strongly implicating Aβ42 in the pathogenesis of AD [1].

Presenilin (PS) 1 and PS2 genes were identified as the major causative genes for early-onset FAD [5, 6], which encode homologous polytopic membrane proteins spanning the membrane 8 times [7]. Although a major proportion of nascent PS is rapidly degraded, a small fraction of PS is stabilized and undergoes endoproteolysis resulting in a heterodimeric complex of N- and C-terminal derivatives (NTF and CTF, respectively, [8]) with an unusually long half-life. Overexpression of exogenous PS results in the replacement of endogenous PS fragments, suggesting that stabilization of PS is a saturable process competing for a limiting cellular factor [8].

The finding that ablation of PS1 in mice dramatically decreased γ-cleavage of βAPP indicated that PS1 physiologically serves as a co-activator of γ-secretase [9]. Moreover, data from studies in *Caenorhabditis elegans* [10], PS1 knock-out mice [11] and *Drosophila melanogaster* [12] suggest that PS1 facilitates *Notch* signaling by activating a γ-secretase-like protease to release Notch intracellular domain.

More than 50 missense mutations in PS1, and two in PS2, have been identified in FAD pedigrees [13]. Accumulating data suggest that PS mutations cause AD by promoting the secretion of Aβ42 [14–17], although the mechanism whereby mt PS leads to the increased production of Aβ42 remains unknown.

We previously showed that C-terminally truncated PS2 harboring the N141I FAD mutation (mt PS2) corresponding to endoproteolytic NTF lost the capacity to increase secretion of Aβ42 in N2a stable cells [18]. These data prompted us to postulate that a subdomain in the PS C terminus mediates Aβ42 overproduction and to undertake molecular dissection studies to identify this subdomain. To define the minimal PS2 C-terminal region required for the overproduction of Aβ42, we constructed cDNAs encoding mt PS2 (full-length, 448 residues) with small C-terminal deletions ending at residues 411, 441 or 445. The percentage of Aβ42 (%Aβ42) as a fraction of total Aβ secreted by stably transfected N2a cells with these cDNAs was ~10%, and this was similar to the %Aβ42 secreted by cells expressing full-length (FL), wild-type (wt) PS2, whereas the %Aβ42 secreted from cells expressing FL mt PS2 was constantly elevated to ~35–55% (Fig. 1A).

We next replaced each of the three C-terminal residues of PS2 with Ala and examined their effects on Aβ42 secretion. mt PS2/L446A or /Y447A increased the %Aβ42 to comparable levels seen in N2a cells with FL mt PS2 (~45–55%), whereas %Aβ42 from cells expressing mtPS2/I448A was ~25%, which was intermediate between the levels detected in the N2a cells with FL wt and mt PS2 (Fig. 1B). We then focused on the role of residue I448 by replacing it with amino acids having different properties. mt PS2/I448V and /I448F enhanced secretion of Aβ42 at comparable levels to those in cells expressing FL mt PS2. In sharp contrast, %Aβ42 secreted from cells expressing mtPS2/I448R was ~10%, which was similar to those in cells with wt PS2 (Fig. 1C).

We next examined the effects of the addition of amino acids to the C terminus of mt PS2 using six His residues (CHis) or a repetition of six C-terminal residues of PS2 (CDup; Ser-His-Gln-Leu-Tyr-Ile). When stably transfected into N2a cells, neither mt PS2/CHis nor mt PS2/CDup retained the capacity to increase secretion of Aβ42, and the %Aβ42 was ~10%, which was similar to cells with wt PS2 (Fig. 1D).

To gain insights into the role of the N terminus of PS2 in Aβ42 production, we constructed cDNAs encoding two types of N-terminally truncated PS2, i.e., dAS lacking the N-terminal 20 residues encompassing the acidic-stretch, and dN lacking the entire N-terminal cytoplasmic domain corresponding to residues 1-75. When stably expressed in N2a cells, mt PS2/dAS as well as /dN both increased the %Aβ42 at similar levels to that of cells with FL mt PS2 (Fig. 1E).

It has not been definitively proven whether nascent holoproteins or stable complexes of endoproteolytic fragments (NTF and CTF) of PS that are capable of replacing endogenous PS are the biologically active forms. To investigate the relationship between replacement by the C- or N-terminally modified forms of PS2 and their pathological overproduction of Aβ42, we examined the levels of endogenous PS1 in N2a stable cells expressing C- or N-terminally modified PS2. The amounts of ~30 kDa NTF and ~23 kDa CTF of endogenous mouse PS1 were decreased in cells expressing FL wt or mt PS2, as well as in cells expressing PS2/L446A or PS2/dN, whereas they were maintained at similar levels to those in mock-transfectants in cells expressing wt or mt PS2/445stop, PS2/I448R or PS2/CHis (Fig. 2). These results clearly showed that the pathologically active forms of mt PS2 that promote overproduction of Aβ42 are the stable complex forms that can replace endogenous PS, whereas replacement does not occur with C-terminally modified PS2 lacking the Aβ42 promoting effects.

How can we interprete the pathomechanism of Aβ42 overproduction caused by mt PS with or without C-terminal modifications? In cultured cells transfected with mt PS, the exogenous PS replaces the endogenous PS and occupies the cellular limiting factor that is essential for the stabilization and function of PS. Therefore, mt PS that gained the abnormal function to promote Aβ42 generation becomes functionally predominant, and the secretion

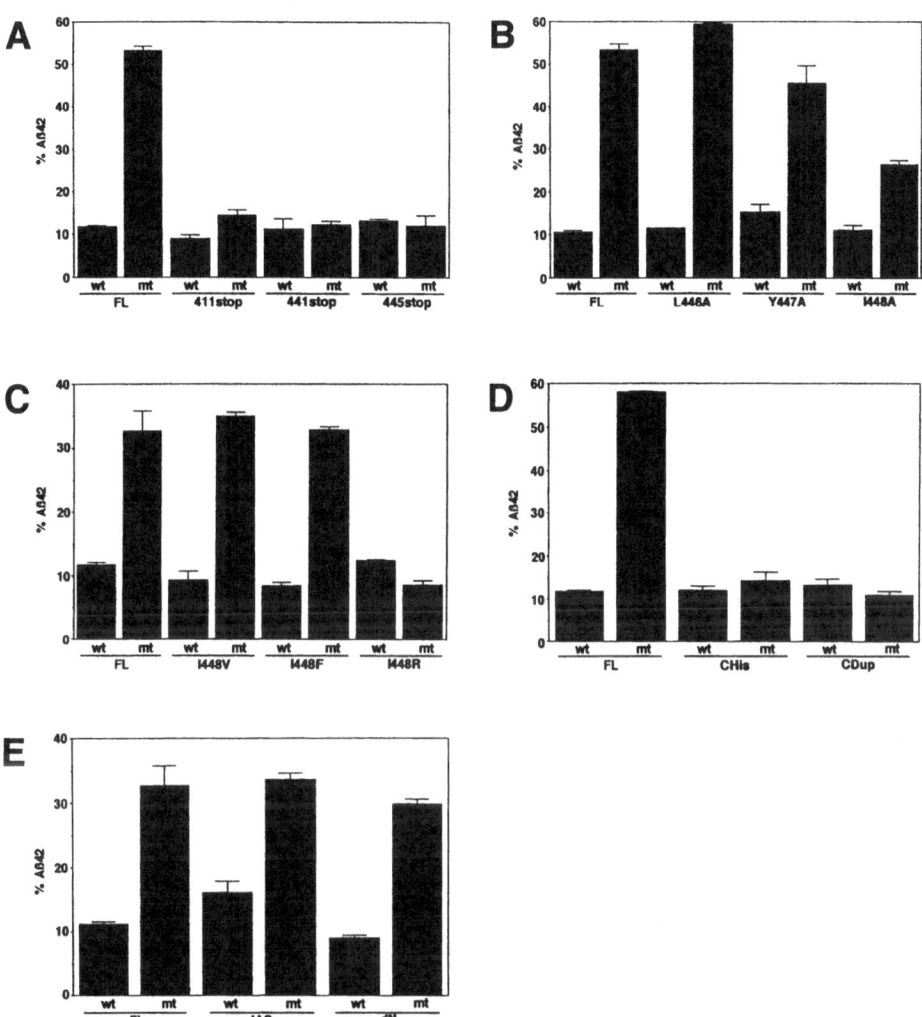

Figure 1. Percentages of Aβ42 secreted from cells expressing C- or N-terminally truncated PS2. Percentages of Aβx-42 as a fraction of total Aβ (= Aβx-40 + Aβx-42) (%Aβ42) secreted from N2a cells stably transfected with full-length (FL) or C-terminally truncated (411stop, 441 stop and 445 stop) (A), with Ala substitution at Leu446 (L446A), Tyr447 (Y447A) or Ile448 (I448A) (B) or substitution at Ile448 to Val (I448V), Phe (I448F) or Arg (I448R) (C), with additional amino acids at the C terminus (CHis: six His residues; CDup: duplication of the C-terminal six amino acid residues of PS2) (D), or with N-terminally truncated (E; dAS: PS2 lacking the N-terminal acidic stretch corresponding to residues 1–20. dN: PS2 lacking the entire N-terminal cytoplasmic domain corresponding to residues 1-75) PS2 genes with (mt) or without (wt) N141I FAD mutation quantified by two-site ELISAs. Mean values ± SE in four independent experiments are shown. Transfected PS2 cDNAs are indicated below the columns.

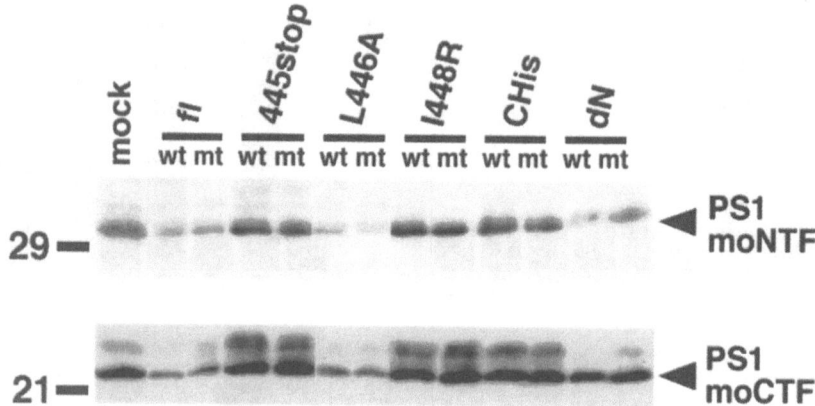

Figure 2. Replacement of endogenous PS1 by PS2 derivatives overexpressed in N2a cells. Western blot analysis of the levels of NTF (PS1 moNTF, upper panel) and CTF (PS1 moCTF, lower panel) of endogenous mouse PS1 in N2a cells stably transfected with C- or N-terminally modified PS2. Cell lysates (20 μg protein) from N2a cells transfected with each cDNA were fractionated by SDS-PAGE and analyzed by immunoblotting with anti-PS1N (upper panel) or anti-G1L3 (lower panel) antibodies. The names of the transfected cDNA constructs are indicated at the top of each lane. mock: cells transfected with an empty vector alone. Molecular mass standards are shown in kilodaltons.

of Aβ42 is upregulated. In contrast, C-terminally modified PS lacks the capacity to associate with the limiting factor and so they are not stabilized as functional proteins. How does the C terminus of PS govern the functional complex formation of PS? One possibility is that the carboxy-tip serves as a binding site for the limiting factor protein; however, it is still possible that this hydrophobic portion plays some role in the proper folding of PS proteins that is essential for its stabilization and function.

In conclusion, the C terminus of PS plays a critical role in the formation of the stabilized complexes of PS which are the physiologically and pathologically active forms of PS, and these forms of PS in turn lead to increased production of Aβ42 in FAD. Further systematic mutagenesis studies should be an effective approach to resolve the complex functions of PS.

Acknowledgments
The authors thank Rie Takikawa, Akihiko Koyama, Yuichi Morohashi, Nobumasa Takasugi, Takaomi C. Saido and Kei Maruyama for their invaluable contribution to this study. A full description of the data presented here has been published as ref 19.

References

1 Suzuki N, Cheung TT, Cai X-D, Odaka A, Otvos L Jr, Eckman C, Golde TE, Younkin SG (1994) An increased percentage of long amyloid β-protein is secreted by familial amyloid β-protein precursor (βAPP717) mutants. *Science* 264: 1336–1340
2 Jarrett JT, Lansbury PT Jr, (1993) Seeding "one-dimensional crystallization" of amyloid: a pathogenic mechanism in Alzheimer's disease and scrapie? *Cell* 73: 1055–1058

3 Iwatsubo T, Odaka A, Suzuki N, Mizusawa H, Nukina N, Ihara Y (1994) Visualization of Aβ42(43) and Aβ40 in senile plaques with end-specific Aβ monoclonals: evidence that an initially deposited species is Aβ42(43). *Neuron* 13,45–53

4 Iwatsubo T, Mann DMA, Odaka A, Suzuki N, Ihara Y (1995) Amyloid β protein (Aβ) deposition: Aβ42(43) precedes Aβ40 in Down syndrome. *Ann Neurol* 37: 294–299

5 Sherrington R, Rogaev EI, Liang Y, Rogaeva EA, Levesque G, Ikeda M, Chi H, Lin C, Li G, Holman K et al (1995) Cloning of a gene bearing missense mutations in early-onset familial Alzheimer's disease. *Nature* 375: 754–760

6 Levy-Lahad E, Wasco W, Poorkaj P, Romano DM, Oshima J, Pettingell WH, Yu C-E Jondro PD, Schmidt SD, Wang K et al (1995) Candidate gene for the chromosome 1 familial Alzheimer's disease locus. *Science* 269: 973–977

7 Doan A, Thinakaran G, Borchelt DR, Slunt HH, Ratovitsky T, Podlisny M, Selkoe DJ, Seegar M, Gandy SE, Price DL et al (1996) Protein topology of presenilin 1. *Neuron* 17: 1023–1030

8 Thinakaran G, Borchelt DR, Lee MK, Slunt HH, Spitzer L, Kim G, Ratovitsky T, Davenport F, Nordstedt C, Seeger M et al (1996) Endoproteolysis of presenilin 1 and accumulation of processed derivatives *in vivo*. *Neuron* 17: 181–190

9 De Strooper B, Saftig P, Craessaerts K, Vanderstichele H, Guhde G, Annaert W, Von Figura K, Van Leuven F (1998) Deficiency of presenilin-1 inhibits the normal cleavage of amyloid precursor protein. *Nature* 391: 387–390

10 Levitan D, Greenwald I (1995) Facilitation of lin-12-mediated signalling by sel-12, a *Caenorhabditis elegans* S182 Alzheimer's disease gene. *Nature* 377: 351–354

11 De Strooper B, Annaert W, Cupers P, Saftig P, Craessaerts K, Mumm JS, Schroeter EH, Schrijvers V, Wolfe MS, Ray WJ et al (1999) A presenilin-1-dependent γ-secretase-like protease mediates release of Notch intracellular domain. *Nature* 398: 518–522

12 Struhl G, Greenwald I (1999) Presenilin is required for activity and nuclear access of Notch in *Drosophila*. *Nature* 398: 522–525

13 Hardy J (1997) The Alzheimer family of diseases: many etiologies, one pathogenesis? *Proc Natl Acad Sci USA* 94: 2095–2097

14 Duff K, Eckman C, Zehr C, Yu X, Prada C-M, Perez-tur J, Hutton M, Buee L, Harigaya Y, Yager D et al (1996) Increased amyloid-β42(43) in brains of mice expressing mutant presenilin 1. *Nature* 383: 710–713

15 Borchelt DR, Thinakaran G, Eckman CB, Lee MK, Davenport F, Ratovitsky T, Prada CM, Kim G, Seekins S, Yager D et al (1996) Familial Alzheimer's disease-linked presenilin 1 variants elevate Aβ1-42/1-40 ratio *in vitro* and *in vivo*. *Neuron* 17: 1005–1013

16 Citron M, Westaway D, Xia W, Carlson G, Diehl T, Levesque G, Johnson-Wood K, Lee M, Seubert P, Davis A et al (1997) Mutant presenilins of Alzheimer's disease increase production of 42-residue amyloid β-protein in both transfected cells and transgenic mice. *Nat Med* 3: 67–72

17 Tomita T, Maruyama K, Saido TC, Kume H, Shinozaki K, Tokuhiro S, Capell A, Walter J, Grünberg J, Haass C et al (1997) The presenilin 2 mutation (N141I) linked to familial Alzheimer disease (Volga German families) increases the secretion of amyloid β protein ending at the 42nd (or 43rd) residue. *Proc Natl Acad Sci USA* 94: 2025–2030

18 Tomita T, Tokuhiro S, Hashimoto T, Aiba K, Saido TC, Maruyama K, Iwatsubo T (1998) Molecular dissection of domains in mutant presenilin 2 that mediate overproduction of amyloidogenic forms of amyloid β peptides. Inability of truncated forms of PS2 with familial Alzheimer's disease mutation to increase secretion of Aβ42. *J Biol Chem* 273: 21153–21160

19 Tomita T, Takikawa R, Koyama A, Morohashi Y, Takasugi N, Saido TC, Maruyama K, Iwatsubo T (1999) C terminus of presenilin is required for overproduction of amyloidogenic Aβ42 through stabilization and endoproteolysis of presenilin. *J Neurosci* 19: 10627–10634

Neuroscientific Basis of Dementia
C. Tanaka, P.L. McGeer, Y. Ihara (eds)
© 2001 Birkhäuser Verlag Basel/Switzerland

Role of presenilin in APP processing and Aβ production

Weiming Xia and Dennis J. Selkoe

Department of Neurology and Program in Neuroscience, Harvard Medical School and Center for Neurologic Diseases, Brigham and Women's Hospital, Boston, MA, USA

Introduction

Mutations in presenilin (PS) 1 [1] and 2 [2–4] genes are the major causes of early-onset Alzheimer's disease (AD). More than 60 missense mutations in PS1 (for review, see [5–7]) and three in PS2 [2–4, 8] have been identified. PS1 and PS2 are 467 and 448 amino acid polypeptides with ~60% homology. An 8 transmembrane (TM) domain model for PS1 suggests that the N-terminus, TM 6→7 loop, and C-terminus all oriented toward the cytoplasm [9]. The holoproteins can be detected in transfected cells at ~44 kDa (PS1) and ~50 kDa (PS2) [10–12]. The half-life of both proteins is brief (30–60 min), and they undergo phosphorylation but no glycosylation or sulfation [13, 14]. Very low levels of holoprotein are detected in untransfected cells and tissues; the major forms existing in cells are the endoproteolytic N-terminal fragments (NTF) and C-terminal fragments (CTF) [11, 15]. PS1 is primarily cleaved at or close to amino acid 292; a point mutation at residue 292 completely abolished the endoproteolysis of PS1 [16]. This confirms the direct radiosequencing of the CTF of PS1 [15]. A secondary cleavage site is between residues 298 and 299, another site discovered by radiosequencing of the CTF [15]. The protease (presenilinase) responsible for the initial proteolytic cleavage around residue 292 is not identified; recent studies suggest that the PS1 itself might be the presenilinase [17]. Levels of PS1 and PS2 NTFs and CTFs are tightly regulated, because overexpression of holoprotein in transfected cells or transgenic mice essentially does not change the total levels of fragments [11]. PS holoproteins can also undergo caspase-mediated cleavage at alternative (slightly C-terminal) sites, a process that is believed to relate to apoptosis [18]. The NTF and CTF form high-molecular-weight complexes that may associate with β- and other catenins and appear to represent the biologically active form of the presenilins [19–21].

Abbreviations
AB, amyloid β-protein; PS, presenilin; AD, Alzheimer's disease; FAD, familial AD; APP, β-amyloid precursor protein; ELISA, enzyme-linked immunosorbent assay; ER, endoplasmic reticulum; CHO, Chinese hamster ovary; IP, immunoprecipitation; KO, knockout; NTF, N-terminal fragment; CTF, C-terminal fragment; TM, transmembrane; WT, wild-type; CM, conditioned media.

FAD-causing mutations in presenilin specifically increase $A\beta_{42}$ production and oligomerization

All familial AD (FAD) mutations in PS1 and PS2 studied to date selectively enhance the production of amyloid β protein ($A\beta_{42}$) in transfected cells and the brains of transgenic mice [12, 22–25], in the brain tissue of humans carrying PS mutations [26] and in the plasma of such carriers [27]. The elevation of $A\beta_{42}$ appears to be a direct result of the expression of mutant PS genes that does not require any other features of the AD state, as it occurs acutely even in non-neural, non-human cells in culture [12]. These conclusions are exemplified in Figure 1. Here, multiple stable cell lines overexpressing β-amyloid precursor protein (APP) and wild-type (WT) or FAD-mutant PS1 or PS2 were used to study the effect of mutant PS on $A\beta$ production in Chinese hamster ovary (CHO) cells [12]. When levels of $A\beta_{40}$ and $A\beta_{42}$ in conditioned media (CM) were immunoprecipitated (IP) with antibodies that are highly specific for each of these species or measured by enzyme-linked immunoassay orbent assay (ELISA), only a specific increase of $A\beta_{42}$ (not $A\beta_{40}$) in cell lines carrying FAD-mutant PS1 or PS2 were observed (Fig. 1). The increased $A\beta_{42}$ levels were comparable to that observed in the CM from CHO cells overexpressing FAD-mutant APP_{V717F}. Consistent with a previous report [28], Sodium dodecyl sulfate (SDS)-stable oligomers of $A\beta_{42}$ in the CM of the APP_{V717F} cell line were detected at higher levels than in CHO cells expressing WT APP. These oligomers also rose in CM from cells transfected with PS1 and PS2 bearing FAD

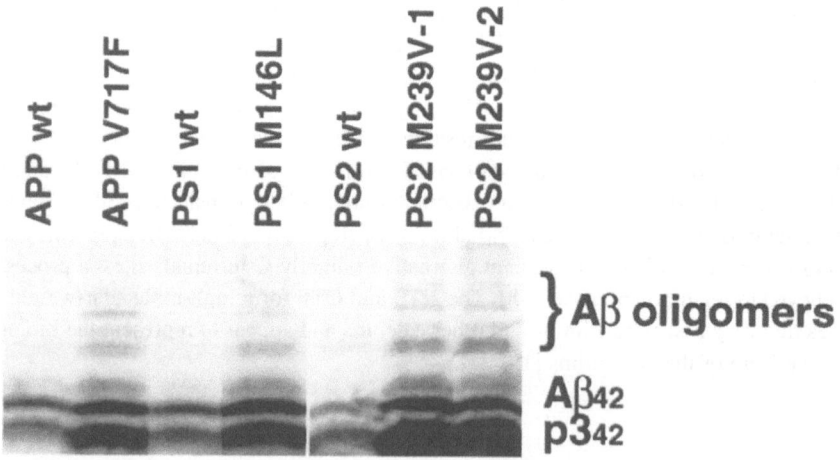

Figure 1. Increased $A\beta_{42}$ production and oligomerization in CM of FAD-linked mutant APP, PS1 or PS2 cells vs WT cells. Cells were metabolically labeled with ^{35}S-methonine, and CM was precipitated with the antibody specific for $A\beta$ peptides ending at residue 42. Mutant PS1 or PS2 cell lines show significant $A\beta_{42}$ increases, accompanied by the appearance of low molecular weight $A\beta_{42}$ oligomers, which can also be seen in the CM of the APP_{V717F} mutant line.

mutations (Fig. 1). Therefore, FAD mutations in PS1 and PS2 increase the production of Aβ$_{42}$, which apparently promotes the self-assembly of Aβ$_{42}$ into low-molecular-weight oligomeric species (principally dimers and trimers). It is known that the presence of two additional hydrophobic residues at the C-terminus of Aβ$_{42}$ increase its fibrillogenic potential *in vitro* and lead to its seeding of the further Aβ$_{40}$ aggregation [29]. Our observation of enhanced levels of these SDS-stable oligomers in the CM of cells overexpressing FAD-mutant PS1 or PS2 suggests that patients carrying FAD mutations have an accelerated Aβ aggregation and deposition in brains.

Deletion of PS1 inhibits γ-secretase activity and blocks Aβ production

De Strooper et al. reported that PS1 is required for normal Aβ generation from APP C-terminal fragments [30]. The cleavage of APP by γ-secretase in neurons cultured from PS1 knockout (KO) mice is markedly inhibited; the C-terminal fragments of APP (C83 and C99) accumulate dramatically, and Aβ production drops to ~25% of WT levels [30]. The specific accumulation of APP C99/C83 has also been observed in brains and fibroblasts from PS1 KO mice [31]. A modest increase in C99/C83 occurred in heterozygous (+/–) brains, indicating that partial reduction of PS1 expression is sufficient to alter APP C99/C83 catabolism [31]. To examine any alteration of C99/C83 distribution in these cells, subcellular fractionation of PS1 WT (+/+) or KO (–/–) fibroblasts was performed (Fig. 2). Previous characteri-

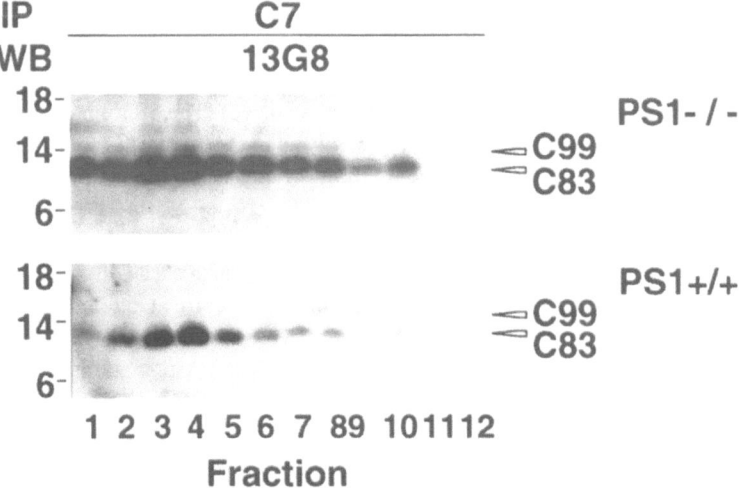

Figure 2. Lack of alteration in subcellular distribution of C99/C83 in PS1 KO vs WT fibroblasts. Microsomal vesicles prepared from PS1 KO or WT fibroblasts were subjected to subcellular fractionation. C99/C83 were visualized by immunoprecipitation of individual fractions with APP polyclonal antibody C7 followed by Western blotting with APP monoclonal antibody 13G8.

zation of these gradient fractions indicate that fractions 1–4 are enriched in endoplasmic reticulum (ER) vesicles (Calnexin-rich fractions), and fractions 5–8 are enriched in Golgi/TGN vesicles (fractions rich in β-1,4-galactosyltransferase activity) [31]. As expected, a marked accumulation of C99/C83 was observed in vesicles of the KO fibroblasts (Fig. 2). Importantly, the subcellular distribution of C83 and C99 after gradient fractionation of total microsomes was indistinguishable in the presence (+/+) *versus* the complete absence (–/–) of PS1 in mouse fibroblasts (Fig. 2); only an increase in amounts of these derivatives occurred in the PS1 –/– fractions, consistent with the findings in PS1 KO mouse brains. These results confirm that PS1 is required for proper γ-secretase activity and Aβ generation, but do not provide evidence for an alteration in the trafficking/subcellular distribution of C83 and C99, the substrates of the γ-secretase reaction.

Specific requirement of two conserved aspartates in PS1 for endoproteolytic and γ-secretase activities

Our recent findings indicate that two unusual aspartate residues in TM domains 6 and 7 (D257 and D385), which are conserved in all known PS homologues, are critical for endoproteolytic and γ-secretase activities [17]. When we mutated each of the two Asp residues to Ala, these "asp-mutant" PS1 holoproteins did not undergo endoproteolysis to form NTFs and CTFs. It is generally believed that PS1 NTF/CTF levels are tightly regulated by competition for limiting cellular factors [11]. In CHO cells stably expressing human WT PS1, endogenous hamster NTFs and CTFs were partially or fully replaced by the human NTFs and CTFs. The extent of replacement was directly dependent on the level of WT human PS1 expressed. However, in our cells expressing asp-mutant human PS1, no endoproteolysis of PS1 to human NTF and CTF was detected, and the formation of endogenous PS1 NTFs and CTFs was partially inhibited (i.e., replacement occurred). Therefore, mutation of either the TM6 or the TM7 Asp prevents PS1 endoproteolysis in the TM6/TM7 loop. Asp-mutant PS1 holoproteins appear to compete for the limiting cellular factors and suppress the endoproteolysis of endogenous PS1 holoprotein.

The mutations in either aspartate residue also specifically blocked γ-secretase activity [17]. Expression of D257A or D385A PS1 consistently increased C99 and C83 levels compared to those seen with WT PS1. The extent of increase correlated roughly with the expression level of Asp→Ala holoprotein. Consistent with the results of deleting PS1 entirely (above), the accumulation of C99 and C83 caused by the Asp→Ala mutations was due to interference with γ-secretase activity and not to increased α- and β-secretase cleavage of holoAPP, as the levels of *N*- and *N*+*O*-glycosylated APP holoproteins, α-APP$_s$ and β-APP$_s$ were not significantly changed in stable cell lines expressing Asp→Ala mutant PS1 [17]. Accordingly, all of the cell lines stably expressing Asp→Ala mutant PS1 secreted much less total Aβ and Aβ$_{42}$ than did WT-PS1 expressing cell lines, as quantified by a sensitive sandwich ELISA. Reduction of Aβ secretion in the asp-mutant PS1-containing cell lines was also confirmed by [^{35}S]-Met labeling of cells followed by IP of CM with Aβ antibody. Consistent with the increase of C99/C83, the extent of Aβ lowering among the different stable mutant lines correlated roughly with the degree of expression of the mutant holoprotein. When the asp-mutant PS1 is expressed at very high levels in particular stable clones, a marked inhibition of γ-secretase activity results in a reduction of Aβ close to the level of that obtained by knocking out the entire PS1 gene. The residual Aβ production in asp-mutant cells is presumed to be due to PS2 and residual amounts of endogenous PS1 NTFs and CTFs. It has recently been shown that mutating the PS2 TM aspartate D366 also results in accumulation of C99/C83 and reduced Aβ generation [32, 33]. Apparently, PS2 can contribute to Aβ generation in a manner qualitatively similar to PS1 but quantitatively less significant. The presence of low levels of endogenous PS1 and PS2 in asp-mutant cells (and low endogenous PS2 levels in PS1 KO cells) allows for the production of a limited amount of Aβ despite the lack of a functional PS1 molecule. Therefore, our Asp mutations act as dominant negatives with respect to endogenous PS1. Such effects have not been observed with any other FAD–linked missense mutations.

To distinguish the role of two aspartate residues in endoproteolytic and γ-secretase activities, the D385A mutation was introduced into PS1 containing the exon 9 deletion (PS1ΔE9). Cells overexpressing PS1ΔE9/D385A were found to undergo a similar accumulation of C99/C83 [17]. Therefore, mutation of a TM Asp abolished γ-secretase activity, even in a PS1 variant that does not require endoproteolysis to function in APP processing. We conclude that the reduced γ-secretase activity was not a sole consequence of eliminating endoproteolysis of the mutant PS1 holoproteins, strongly suggesting that Asp residues are independently critical for both endoproteolysis and γ-secretase cleavage.

PS1 may directly participate in γ-secretase cleavage of APP

Findings of a selective increase of Aβ$_{42}$ in FAD-mutant PS cells and a marked decrease of Aβ in asp-mutant PS cells suggest that PS1 may be directly involved in the γ-secretase cleavage complex. An interaction of some APP and PS molecules in intact cells has been reported [34–36], and complex formation could be detected at endogenous protein levels in nontransfected cells [35]. Using distinct PS and APP antibodies, small amounts of APP can be co-immunoprecipitated with PS1. The co-precipitating APP was principally the *N*-glycosylated form (Fig. 3). The cytoplasmic domain of APP was dispensable for the PS1 interaction, as APP-PS complex can still be observed in cells expressing APP with a deletion of almost all of the cytoplasmic domain. Complex formation was not significantly altered by FAD mutations in APP; APP carrying the KM670/671NL (Swedish) or V717F missense mutations still precipitate with PS1 [35]. In addition, in COS cells transiently transfected with cDNAs encoding C100 and presenilins, Pradier et al. detected the complex formation of C100 and PS1 or PS2 [36]. Interaction of PS1 with APP and its C-terminal derivatives suggest that PS1 may directly participate in the γ-secretase cleavage complex.

Additional evidence from studies on PS1-dependent Aβ generation *in vitro* further supports a direct involvement of PS1 in γ-secretase cleavage of APP. Microsomes were prepared from the CHO cells stably expressing WT or Asp→Ala mutant PS1 and added to a cell-free transcription/translation system in which C99 is expressed from DNA. When incubation at

IP: pre   X81   1G5

WB: 8E5

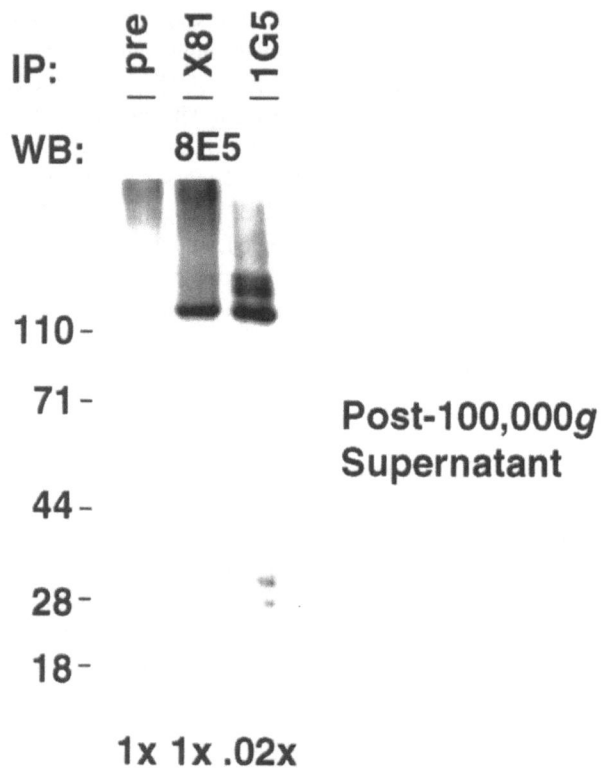

110 -

71 -   **Post-100,000*g***
              Supernatant

44 -

28 -

18 -

1x 1x .02x

Figure 3. Complex formation between PS1 and APP. Cells were lysed in 1% NP-40 and cell lysates were co-immunoprecipitated with PS1 antibody X81, APP antibody 1G5, or preimmune serum (pre). APP-PS1 complexes were detected (but not with preimmune serum) in the post-100,000 *g* supernatant of the cell lysates. Importantly, the amount of *N*-APP that is co-immunoprecipitated with PS1 is a small fraction of the total APP (1/50th the amount of lysates used for precipitation by APP antibody 1G5).

37°C was performed at neutral pH, C99 was well expressed, but no Aβ was detected. However, under mildly acidic conditions, a small amount of new Aβ appeared when microsomes purified from WT PS1-expressing cell lines were present, but no new Aβ was generated with microsomes from asp-mutant cells [17]. These results in isolated, washed microsomes mirror the inhibition of Aβ production observed in whole Asp→Ala mutant cells and in intact neurons of PS1 KO mice and suggest that PS1 plays a direct role in γ-secretase cleavage rather than influencing the trafficking of the component proteins of this reaction.

Discussion

Involvement of PS1 in APP processing and Aβ generation appears to be one of the roles PS1 plays in cellular metabolism. Information on PS function has been obtained from genetic studies and biochemical analysis using cultured mammalian cell lines, *C. elegans*, *Drosophila*, and PS1 KO mice. The PS1 homologous gene in *C. elegans*, *sel-12*, has been cloned and the gene product was found to facilitate the function of lin-12, a member of the Notch receptor family that participates in cell-cell recognition during development [37]. A lethal *sel-12* mutation in *C. elegans* can be rescued by expression of WT human PS1 [38, 39], indicating that WT PS1 has a similar function as SEL-12. It is possible that the primary function of the presenilins in development is to facilitate Notch signaling [38, 39]. Notch undergoes proteolytic processing within its TM domain in response to binding extracellular ligand; this cleavage event is required for signal transduction [40]. Similar to the finding of C99/C83 accumulation, a constitutively cleaved derivative of Notch, mNotch1ΔE, does not undergo further processing in cells cultured from PS1 –/– mice [41]. Furthermore, Notch proteolysis is inhibited by γ-secretase inhibitors based on the APP cleavage site [41] and by mutation of the critical aspartate residues in PS1 or PS2 [42]. Therefore, Notch appears to undergo a γ-secretase-like intramembranous processing event that requires the same TM aspartate residues required for APP.

In summary, presenilins are likely to participate directly in γ-secretase proteolytic complexes, possibly as the γ-secretase itself or else as a unique "diaspartyl" co-factor for γ-secretase that requires mildly acidic pH to operate. The PS1-dependent cleavage may occur with a variety of substrates, including APP, Notch and certain other single TM proteins. Because the γ-secretase cleavage site of APP and Notch are in or very close to the TM domains, understanding the role of PS1 in APP processing will shed light on the novel molecular mechanism of the intramembranous cleavage of TM proteins.

Acknowledgments
This work was supported by grants from the NIH (AG15379 and AG06173 to D.J.S.), the Alzheimer's Association (to W. X.), and the Foundation for Neurologic Diseases.

References

1 Sherrington R, Rogaev EI, Liang Y, Rogaeva EA, Levesque G, Ikeda M, Chi H, Lin C, Li G, Holman K et al (1995) Cloning of a novel gene bearing missense mutations in early-onset familial Alzheimer disease. *Nature* 375: 754–760
2 Levy-Lahad E, Wijsman EM, Nemens E, Anderson L, Goddard AB, Weber JL, Bird TD, Schellenberg GD (1995) A familial Alzheimer's disease locus on chromosome 1. *Science* 269: 970–973
3 Levy-Lahad E, Wasco W, Poorkaj P, Romano DM, Oshima J, Pettingell H, Yu C, Jondro PD, Schmidt SD, Wang K et al (1995) Candidate gene for the chromosome 1 familial Alzheimer's disease locus. *Science* 269: 973–977
4 Rogaev EI, Sherrington R, Rogaeva EA, Levesque G, Ikeda M, Liang Y, Chi H, Lin C, Holamn K, Tsuda T et al (1995) Familial Alzheimer's disease in kindreds with missense mutations in a gene on chromosome 1 related to the Alzheimer's disease type 3 gene. *Nature* 376: 775–778
5 Kovacs DM, Tanzi RE (1998) Monogenic determinants of familial Alzheimer's disease: presenilin-1 mutations. *Cell Mol Life Sci* 54: 902–909

6 Hardy J (1997) The Alzheimer family of diseases: many etiologies, one pathogenesis? *Proc Natl Acad Sci USA* 94: 2095–7

7 Cruts M, Van Broeckhoven C (1998) Presenilin mutations in Alzheimer's disease. *Hum Mutat* 11: 183–190

8 Cruts M, van Duijn CM, Backhovens H, Van den Broeck M, Wehnert A, Serneels S, Sherrington R, Hutton M, Hardy J, St George-Hyslop PH, Hofman A et al (1998) Estimation of the genetic contribution of presenilin-1 and -2 mutations in a population-based study of presenile Alzheimer disease. *Hum Mol Genet* 7: 43–51

9 Li X, Greenwald I (1998) Additional evidence for an eight-transmembrane-domain topology for *Caenorhabditis elegans* and human presenilins. *Proc Natl Acad Sci USA* 95: 7109–7114

10 Kovacs DM, Fausett HJ, Page KJ, Kim T-W, Moir RD, Merriam DE, Hollister RD, Hallmark OG, Mancini R, Felsenstein KM et al (1996) Alzheimer-associated presenilins 1 and 2: Neuronal expression in brain and localization to intracellular membranes in mammalian cells. *Nat Med* 2: 224–229

11 Thinakaran G, Borchelt DR, Lee MK, Slunt HH, Spitzer L, Kim G, Rotovitsky T, Davenport F, Nordstedt C, Seeger M et al (1996) Endoproteolysis of presenilin 1 and accumulation of processed derivatives *in vivo*. *Neuron* 17: 181–190

12 Xia W, Zhang J, Kholodenko D, Citron M, Podlisny MB, Teplow DB, Haass C, Seubert P, Koo EH, Selkoe DJ (1997) Enhanced production and oligomerization of the 42-residue amyloid β-protein by Chinese hamster ovary cells stably expressing mutant presenilins. *J Biol Chem* 272: 7977–7982

13 Walter J, Capell A, Grunberg J, Pesold B, Schindzielorz A, Prior R, Podlisny M, Fraser P, St George Hyslop P, Selkoe D et al (1996) The Alzheimer's disease associated presenilins are differentially phosphorylated proteins located predominantly within the endoplasmic reticulum. *Mol Med* 2: 673–691

14 De Strooper B, Beullens M, Contreras B, Levesque L, Craessaerts K, Cordell B, Moechars D, Bollen M, Fraser P, St George-Hyslop P et al (1997) Phosphorylation, subcellular localization, and membrane orientation of the Alzheimer's disease associated presenilins. *J Biol Chem* 272: 3590–3598

15 Podlisny MB, Citron M, Amarante P, Sherrington R, Xia W, Zhang J, Diehl T, Levesque G, Fraser P, Haass C et al (1997) Presenilin proteins undergo heterogeneous endoproteolysis between Thr291 and Ala299 and occur as stable N- and C-terminal fragments in normal and Alzheimer brain tissue. *Neurobiol Disease* 3: 325–337

16 Steiner H, Romig H, Pesold B, Philipp U, Baader M, Citron M, Loetscher H, Jacobsen H, Haass C (1999) Amyloidogenic function of the Alzheimer's disease-associated presenilin 1 in the absence of endoproteolysis. *Biochemistry* 38: 14600–14605

17 Wolfe MS, Xia W, Ostaszewski BL, Diehl TS, Kimberly WT, Selkoe DJ (1999) Two transmembrane aspartates in presenilin-1 required for presenilin endoproteolysis and γ-secretase activity. *Nature* 398: 513–517

18 Kim TW, Pettingell WH, Jung YK, Kovacs DM, Tanzi RE (1997) Alternative cleavage of Alzheimer-associated presenilins during apoptosis by a caspase-3 family protease. *Science* 277: 373–376

19 Zhou J, Liyanage U, Medina M, Ho C, Simmons AD, Lovett M, Kosik KS (1997) Presenilin 1 interaction in the brain with a novel member of the Armadillo family. *Neuroreport* 8: 2085–2090

20 Yu G, Chen F, Levesque G, Nishimura M, Zhang DM, Levesque L, Rogaeva E, Xu D, Liang Y, Duthie M et al (1998) The presenilin 1 protein is a component of a high molecular weight intracellular complex that contains β-catenin. *J Biol Chem* 273: 16470–16475

21 Capell A, Grunberg J, Pesold B, Diehlmann A, Citron M, Nixon R, Beyreuther K, Selkoe DJ, Haass C (1998) The proteolytic fragments of the Alzheimer's disease-associated presenilin-1 form heterodimers and occur as a 100–150-kDa molecular mass complex. *J Biol Chem* 273: 3205–3211

22 Borchelt D, Thinakaran G, Eckman C, Lee M, Davenport F, Ratovitsky T, Prada C, Kim G, Seekins S, Yager D et al (1996) Familial Alzheimer's disease-linked presenilin 1 variants elevate Ab1-42/1-40 ratio *in vitro* and *in vivo*. *Neuron* 17: 1005–1013

23 Duff K, Eckman C, Zehr C, Yu X, Prada C, Perez-tur J, Hutton M, Buee L, Harigaya Y, Yager D et al (1996) Increased amyloid-b42(43) in brains of mice expressing mutant presenilin 1. *Nature* 383: 710–713

24 Citron M, Westaway D, Xia W, Carlson G, Diehl T, Levesque G, Johnson-Wood K, Lee M, Seubert P, Davis A et al (1997) Mutant presenilins of Alzheimer's disease increase production of 42-residue amyloid β-protein in both transfected cells and transgenic mice. *Nat Med* 3: 67–72

25 Tomita T, Maruyama K, Saido TC, Kume H, Shinozaki K, Tokuhiro S, Capell A, Walter J, Grunberg J, Haass C et al (1997) The presenilin 2 mutation (N141I) linked to familial Alzheimer diseae (Volga German families) increases the secretion of amyloid β protein ending at the 42nd (or 43rd) residue. *Proc Natl Acad Sci USA* 94: 2025–2030

26 Lemere CA, Lopera F, Kosik KS, Lendon CL, Ossa J, Saido TC, Yamaguchi H, Ruiz A, Martinez A, Madrigal L et al (1996) The E280A presenilin 1 Alzheimer mutation produces increased Aβ42 deposition

and severe cerebellar pathology. *Nat Med* 2: 1146–1148

27 Scheuner D, Eckman C, Jensen M, Song X, Citron M, Suzuki N, Bird TD, Hardy J, Hutton M, Kukull W et al (1996) Secreted amyloid β-protein similar to that in the senile plaques of Alzheimer's disease is increased *in vivo* by the presenilin 1 and 2 and APP mutations linked to familial Alzheimer's disease. *Nat Med* 2: 864–870

28 Podlisny MB, Ostaszewski BL, Squazzo SL, Koo EH, Rydell RE, Teplow DB, Selkoe DJ (1995) Aggregation of secreted amyloid β-protein into SDS-stable oligomers in cell culture. *J Biol Chem* 270: 9564–9570

29 Jarrett JT, Berger EP, Lansbury Jr PT (1993) The carboxy terminus of the beta amyloid protein is critical for the seeding of amyloid formation: Implications for the pathogenesis of Alzheimer's disease. *Biochemistry* 32: 4693–4697

30 De Strooper B, Saftig P, Craessaerts K, Vanderstichele H, Gundula G, Annaert W, Von Figura K, Van Leuven F (1998) Deficiency of presenilin-1 inhibits the normal cleavage of amyloid precursor protein. *Nature* 391: 387–390

31 Xia W, Zhang J, Ostaszewski B, Kimberly W, Seubert P, Koo E, Selkoe D (1998) Presenilin 1 Regulates the Processing of APP C-terminal Fragments and the Generation of Amyloid β-Protein in ER and Golgi. *Biochemistry* 37: 16465–16471

32 Steiner H, Duff K, Capell A, Romig H, Grim M, Lincoln S, Hardy J, Yu X, Picciano M, Fechteler K (1999) A loss of function mutation of presenilin-2 interferes with amyloid-peptide production and notch signaling. *J Biol Chem* 274: 28669–28673

33 Kimberly W, Xia W, Rahmati T, Wolfe M, Selkoe D (2000) The transmembrane aspartates in presenilin 1 and 2 are obligatory for γ-secretase activity and amyloid β-protein generation. *J Biol Chem* 275: 3173–3178

34 Weidemann A, Paliga K, Durrwang U, Czech C, Evin G, Masters CL, Beyreuther K (1997) Formation of stable complexes between two Alzheimer's disease gene products: presenilin-2 and β-amyloid precursor protein. *Nat Med* 3: 328–332

35 Xia W, Zhang J, Perez R, Koo EH, Selkoe DJ (1997) Interaction between amyloid precursor protein and presenilins in mammalian cells: Implications for the pathogenesis of Alzheimer's disease. *Proc Natl Acad Sci USA* 94: 8208–8213

36 Pradier L, Carpentier N, Delalonde L, Clavel N, Bock M, Buee L, Mercken L, Tocque B, Czech C (1998) Mapping the APP/presenilin (PS) binding domains: the hydrophilic N-terminus of PS2 is sufficient for interaction with APP and can displace APP/PS1 interaction. *Neurobiol Disease* 6: 43–55

37 Levitan D, Greenwald I (1995) Facilitation of *lin-12*-mediated signalling by *sel-12*, a *Caenorhabditis elegans* S182 Alzheimer's disease gene. *Nature* 377: 351–354

38 Levitan D, Doyle TG, Brousseau D, Lee MK, Thinakaran G, Slunt HH, Sisodia SS, Greenwald I (1996) Assessment of normal and mutant human presenilin function in *Caenorhabditis elegans*. *Proc Natl Acad Sci USA* 93: 14940–14944

39 Baumeister R, Leimer U, Zweckbronner I, Jakubek C, Grunberg J, Haass C (1997) Human presenilin-1, but not familial Alzheimer's disease (FAD) mutants, facilitate *Caenorhabditis elegans* notch signalling independently of proteolytic processing *Genes and Function* 1: 149–159

40 Schroeter EH, Kisslinger JA, Kopan R (1998) Notch-1 signalling requires ligand-induced proteolytic release of intracellular domain. *Nature* 393: 382–386

41 De Strooper B, Annaert W, Cupers P, Saftig P, Craessaerts K, Mumm JS, Schroeter EH, Schrijvers V, Wolfe MS, Ray WJ (1999) A presenilin-1-dependent γ-secretase-like protease mediates release of Notch intracellular domain. *Nature* 398: 518–522

42 Ray WJ, Yao M, Mumm J, Schroeter E, Saftig P, Wolfe M, Selkoe D, Kopan R, Goate AM (1999b) Cell surface presenilin-1 participates in the g-secretase-like cleavage of Notch. *J Biol Chem* 274: 36801–36807

Neuroscientific Basis of Dementia
C. Tanaka, P.L. McGeer, Y. Ihara (eds)
© 2001 Birkhäuser Verlag Basel/Switzerland

Impairment of response to ER stress in presenilin 1 mutant

Takashi Kudo[1], Kazunori Imaizumi[2], Taiichi Katayama[2], Naoya Sato[2], Yuka Nakano[3], Yuka Jinno[1], Yuko Segawa[1], Junji Takeda[3], Masaya Tohyama[2] and Masatoshi Takeda[1]

[1] Department of Clinical Neuroscience and Psychiatry, [2] Department of Anatomy and Neuroscience, [3] Department of Social and Environmental Medicine, Osaka University, Graduate School of Medicine, Osaka, Japan

Introduction

Alzheimer's disease (AD) is the most common neurodegenerative disorder resulting in dementia in the elderly, and is characterized by amyloid deposition, neurofibrillary tangles and neuronal loss with dystrophic neurites in wide areas of the cerebral cortex. Some AD cases show familial AD (FAD), indicating genetic factors involving in the pathogenesis of AD. Missense mutations have been reported in the amyloid precursor protein (APP), the pre-senilin-1 (PS1) and the presenilin-2 (PS2) gene on chromosome 21,14 and 1, respectively. At present, more than 50 different kinds of mutations have been reported in the PS1 gene, and PS1 mutations are believed to be the most frequent mutations in FAD.

However, the mechanism by which mutations in presenilins predispose individuals to FAD has not yet been determined. FAD-linked PS1 variants alter proteolytic processing of APP [1, 2], and mutations in PS1 increase cellular susceptibility to apoptosis induced by various insults, including withdrawal of trophic factors and exposure to Aβ [3, 4]. The mechanism by which PS1 mutations promote cell death is not known, but cell-culture studies have revealed perturbed calcium homeostasis and increased production of free radicals in affected cells. These apoptotic stimuli provoke the accumulation of unfolded proteins in the endoplasmic reticulum (ER) a type of 'ER stress'. As PS1 is an integral membrane protein and is localized mainly to the ER [5], it is possible that PS1 mutations could have a significant role in vulnerability to ER stress. Normal cells respond to ER stress by increasing transcription of genes encoding ER-resident chaperones such as GRP78/1Bip, GRP94 and protein disuiphide isomerase (PDI) to facilitate protein folding. This induction system is termed the unfolded-protein response (UPR). One of the mediators of the UPR is a sensor of ER stresses, an ER-resident transmembrane kinase, IRE1. IRE1 senses the perturbed environment in the ER, and leads to downstream signaling by a process that is thought to depend on oligomerization and autophosphorylation of its kinase domain [6–8]. Here we study the mechanisms by which cells expressing PS1 mutants are vulnerable to ER stresses. Our data show that a PS1 mutation alters UPR signaling and indicate that this alteration may contribute to the pathogenic mechanism in the brains of FAD patients.

Susceptibility of PS1 mutants to ER stress

We stably transfected SK-N-SH neuroblastoma cells with complementary DNA constructs encoding wild-type PS1, PS1 with an alanine-to-glutamate mutation (A246E), or PS1 with a deletion of exon 9 (ΔE9). The A246E and ΔE9 mutations are FAD-linked mutations. We then added tunicamycin, which induces ER stress by preventing protein glycosylation, to the medium. Tunicamycin induced increased cell death in cells expressing mutant PS1 as compared with cells expressing the wild-type protein. The PS1 mutations also increased susceptibility to other ER stresses, such as a calcium ionophore (A23187), which depletes intracellular calcium stores. The effect of PSI mutations on the expression of GRP78/1Bip. To determine the molecular mechanisms responsible for vulnerability to ER stress in cells with PS1 mutations, we studied the levels of messenger RNA encoding GRP78, a molecular chaperone, after ER stress was induced by addition of tunicamycin. Basal levels of GRP78 mRNA were not altered in SK-N-SH cell lines stably expressing each construct. In contrast, when cells were stimulated with 3 mg/ml tunicamycin for 6 h, induction of GRP78 mRNA expression was significantly inhibited in cells expressing mutant PS1 (the A246E or the ΔE9 variant) compared with those transfected with wild-type PS1 or empty vector (Fig. 1A). To confirm that FAD-linked mutants generally cause inhibition of GRP78 mRNA induction, we

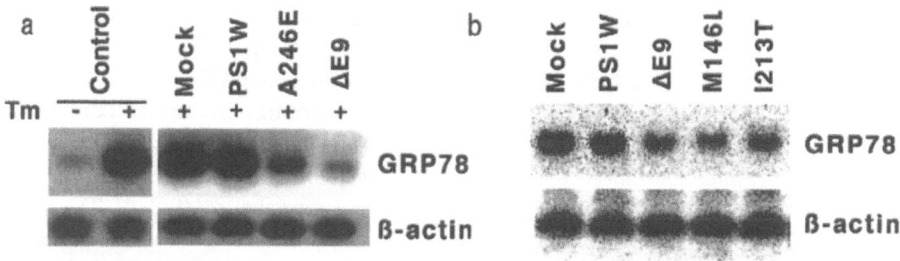

Figure 1. Psi mutations alter expression of GRP78 mRNA. (a) SK-N-SH cells were stably transfected with the indicated PS1 cDNA constructs. (b) HEK293T cells were transiently transfected with the indicated expression vectors. Cells were exposed to 3 mg/ml (SK-N-SH cells) or 5 mg/ml (293 T cells) Tm for 6 h. Total RNA was isolated from each cell line, and subjected to northern blotting with probes for GRP78 mRNA (upper panels) or β-actin mRNA (lower panels).

transiently transfected other FAD-linked PS1 mutants, such as M146V and I213T, into HEK293T cells. We used 293 T cells as they show a transfection efficiency of nearly 100%. These mutations also suppressed the induction of GRP78 mRNA (Fig. 1b). The expression of β-actin mRNA as an internal control was not changed in cells transfected with each PS1 cDNA or empty vector, indicating that most PS1 mutations specifically impair the induction of expression of GRP78 mRNA in response to ER stress.

PS1 interacts with IRE1 on the ER membrane

Our further investigation of the function of PS1 mutations in the regulation of UPR signaling was the analysis of the effects of PS1 mutants on the function of IRE1, the most upstream component in the UPR signaling pathway. To determine whether PS1 and IRE1 interact physically, we performed immunoprecipitation followed by western blotting analyses of SK-N-SH cells stably expressing IRE1-Flag. We prepared immunoprecipitates using an anti-Flag-epitope monoclonal antibody, and immunoblotted the precipitates with an antibody directed towards the PS1 amino terminus to detect endogenous-PS1. Western blotting with the anti-PS1 antibody showed that fiill-length PS1 co-immunoprecipitated with IRE1-Flag. However, the N-terminal fragment of PS1 did not immunoprecipitate with IRE1. As the reverse experiment, we prepared immunoprecipitates using the anti-PS1 antibody and used the anti-Flag antibody for immunoblotting. The IRE1 co-immunoprecipitated with endogenous PS1. We also carried out an immunoprecipitation analysis using SK-N-SH cells stably expressing wild-type PS1 or the A246E or ΔE9 mutants. Endogenous IRE1 co-immunoprecipitated with each PS1 construct to an equivalent extent. Bcl-x L, which is a transmembrane protein of the ER and mitochondria, did not co-immunoprecipitate with IRE1. These results indicate that full-length wild-type and mutant PS1 may specifically interact with IRE1 on the membrane of the ER. Previous subcellular fractionation analyses have shown that full-length PS1 is distributed in the ER fraction, whereas the PS1 fragments were localized predominantly to the Golgi fraction [9]. As IRE1 is an ER-resident protein, it is reasonable that only the full-length PS1 protein (not the PS1 fragments, which were predominantly localized to the Golgi fraction) can interact with IRE1 in manimalian cells.

Attenuation of UPR signaling by PS1 mutants

Having shown that PS1 binds to IRE1 and that mutations in PS1 affect the UPR signaling pathway, we wanted to determine whether PS1 mutants alter the function of IRE1 as a stress sensor in the ER. IRE1 leads to downstream signaling by a process that depends on oligomerization and autophosphorylation of its kinase domain [6–8]. Therefore, we studied the effects of mutations in PS1 on autophosphorylation of IRE1. We transfected HEK293T cells with IRE1-Flag and either wild-type or mutant PS1, metabolically labeled these cells by addition of ^{32}P to the medium for 2 h, and then subjected the cells to immunoprecipitation with the anti-Flag antibody. In cells expressing wild-type PS1, IRE1 showed robust autophosphorylation equivalent to that observed in cells transfected with empty vector (Fig. 2a). In contrast, the levels of autophosphorylated IRE1 were slightly diminished in cells expressing the A246E mutant, and were markedly diminished (by ~70%) in cells expressing ΔE9 (Fig. 2a). These effects were consistent with the efficiency of inhibition of GRP78 mRNA induction and the vulaerability to ER stress induced by the different PS1 forms. In cells transfected with PS1 mutants, phosphorylation of Tau or glycogen synthase kinase-3 β (GSK-3 β) was not changed compared with that in mocktransfected cells or those transfected with wild-type PS1 (Fig. 2b). These results indicate that attenuated levels of IRE1 phos-

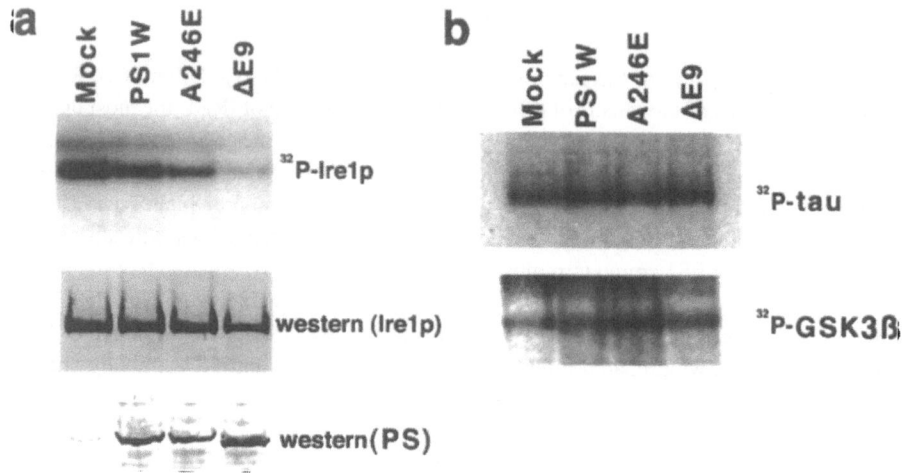

Figure 2. Altered phosphorylation of IRE1 and UPR signaling by PS1 mutants. (a) 293 T cells cultured in six-well plates were co-transfected with 0.2 μg IRE1-flag and 0.5 μg of an empty vector, wild-type PS1 (PS1W), A246E, or ΔE9. The next day, cells were labelled with ^{32}P for 2 h. Upper panel, IRE1-flag proteins were recovered by immunoprecipitation with anti-flag antibody and subjected to SDS-PAGE and autoradiography. Lower panel, western blotting confirmed that equivalent amounts of IRE1-flagand each PS1 protein were expressed in each cell line. (b) phosphorylation of tau and GSK-3 β in cells co-transfected with tau or GSK-3 β and each PS1 construct. Phosphorylated tau and GSK-3 β were recovered by anti-tau and anti-GSK-3 β immunoprecipitation, respectively. Mutant PS1 did not affect phosphorylation of tau or GSK-3 β in contrast to IRE1.

phorylation levels may be related to the effects of PS1 mutants on downregulation of UPR signaling and cellular susceptibility to ER stress.

Effects of exogenous GRP78 expression on sensitivity to ER stress

To confirm that vulnerability to ER stress in cells expressing PS1 mutants is based on atten-uated induction of GRP78 mRNA by inhibition of IRE1 function, we infected 5K-N-SH cells bearing mutant PS1 with recombinant GRP78 using Semliki forest virus (SFV-GRP78 fusion). We then studied responses to ER stress. Sensitivity to ER stress caused by treatment with tunicamycin or A23187 in neuroblastoma lines expressing PS1 mutants was reversed by infection with recombinant GRP78. These results indicated that expression of GRP78 protects against neuronal death caused by ER stress, and that the reduction in GRP78 gene expression causes vulnerability to ER stress in cells expressing PS1 mutants.

Establishment of PS1 'knock-in' mice to confirm our hypothesis

The studies shown above were based on cells having overexpression of PS1. To test our hypothesis dfrectly and to exclude the possibility that the effects were caused by overexpression of PS1 mutants, we established PS1 mutant 'knock-in' mice, which express the PS1 (I213T) protein at normal physiological levels. Many different transgenic mice expressing FAD mutants have been generated to elucidate the pathophysiology of AD. The mutant transgenic mice all show an increased Aβ 42(43)/Aβ 40 ratio; however, no neuronal degeneration has been clearly demonstrated. Transgenic mice carry multiple copies of transgenes in addition to the endogenous gene; this is clearly different from the genetic status of FAD patients, whose disease is triggered in an autosomal dominant fashion.

Considering the limitations of the transgenic strategy, we generated 'knock-in' mice of a PS1 missense mutation identified in Japanese FAD patients [10]. The methods to generate PS1 'knock-in' mice were fully described previously [11]. To introduce the I213T missense mutation into the mouse PS1 gene locus, a targeting vector for homologous recombination was constructed. A silent mutation was also introduced for creating a new MboI site into exon 7 to facilitate the detection of mutation. A neomycin resistance gene flanked by two loxP sites was inserted into intron 6 for selection. Homologous recombination was confirmed by Southern blot analysis. Three homologous recombinants were obtained and germline-competent chimeric miec were produced from one clone. To remove the neomycin-resistant gene from the targeted allele, the chimeric mice were mated with CAG-cre mice, which express Cre recombinase in the whole body, including germ cells. The mice with the mutated PS1 allele were born at the expected frequency and appeared to be healthy. Northern blot analysis revealed that the levels of PS1 mRNA were equal in the wild-type (wild) and the mutant PS1 homozygous (homo) mice. The level of mRNA from the mutant PS1 allele measured by RT-PCR was equal to that from the wild-type allele in mutant PS1 heterozygous (hetero) mice, which were revealed by MboI digestion. These results demonstrate that the PS1 silent and missense mutations were successfully introduced into the mouse PS1 locus without affecting the efficiency of transcription.

Mice of each genotype were sacrificed at the age of 16–20 weeks and the amounts of Aβ peptides were assessed by sandwich enzyme-linked immunosorbent assays (ELISAs) specific for Aβ 40 and Aβ 42(43), as described previously [12]. The levels of brain Aβ 40 were about equal in all three genotypes (Fig. 3a). In contrast, the level of Aβ 42 was significantly increased in mutant PS1 mice (Fig. 3a). Because the efficiency of transcription from the mutant PS1 allele was similar to that from the wild PS1 allele, the accumulation of gene products of mutant PS1 in homo mice was considered to be double that in hetero mice. These results indicate that the PS1 'knock-in' mice are animal models that closely mimic AD pathophysiology.

To confirm the vulnerability of mutant PS1 to ER stress in its physiological expression level, we prepared primary cultured neurons of foetal 'knock-in' mouse at 17.5 days of gestation. LDH assay showed that tunicamycin induced increased cell death in homo cells as compared with cells of another genotype expressing the wild-type protein (data not shown). The induction of GRP78 mRNA was slightly decreased in neurons from hetero mice, and

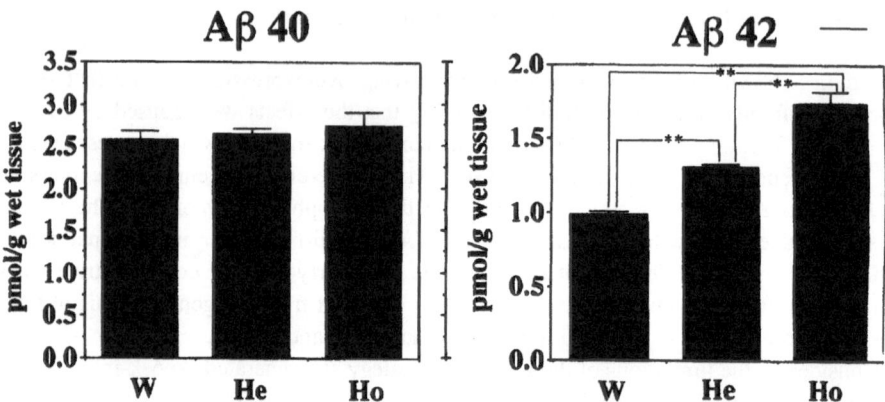

Figure 3. Elevation of brain A42(43) in PS1 'knock-in' mice. Brain homogenate were prepared from wild-type (W) (n = 8), heterozygous (He) (n = 8), and Thomozygous (Ho) (n = 6) PS1 'knock-in' mice at the age of 16–20 weeks. The amounts of Aβ 42(43) and Aβ 40 were measured independently. **p < O.Ol. Error bars indicate SEMs.

was significantly decreased in those from homo mice. It indicates that the reduction of the expression of GRP78 mRNA was dependent on the expression level of mutant PSI.

Conclusion

Our results indicate a new mechanism by which PS1 mutations may affect the sensing of ER stress. Experimental manipulation of IRE1 phosphorylation or GRP78 expression might allow the development of therapeutic strategies for FAD.

References

1 Borchelt DR et al (1996) Familial Alzheimer's disease-linked presenilin 1 variants elevate Aβ 1-42/1-40 ratio *in vitro* and *in vivo*. *Neuron* 17: 1005–1013
2 Duff K et al (1996) Increased amyloid-β 42(43) in brains of mice expressing mutant presenilin 1. *Nature* 383: 710–713
3 Guo Q et al (1997) Alzheimer's presenilin mutation sensitizes neural cells to apoptosis induced by trophic factor withdrawal and amyloid β-peptide: involvement of calcium and oxyradicals. *J Neurosci* 17, 42124222
4 Guo Q et al (1999) Increased vulnerability of hippocampal neurons to excitotoxic necrosis in presenilin-1 mutant knock-in mice. *Nat Med* 5: 101–106
5 Kovacs D et al (1996) Alzheimer-associated presenilin 1 and 2: neuronal expression in brain and localization to intracellular membranes in mammalian cells. *Nat Med* 2: 224–229
6 Sidrauski C et al (1998) The unfolded protein response: an intracellular signaling-pathway with many suprising features. *Trends Cell Biol* 8: 245–249
7 Tirasophon W et al (1998) A stress response pathway from the endoplasmic reticulum to the nucleus requires a novel bifunctional protein kinase/endoribonuclease (Irelp) in mammalian cells. *Gene Develop*

12: 1812–1824
8   Wang X-Z et al (1998) Cloning of mammalian Irel reveals diversity in the ER stress responses. *EMBO J*
    17: 5708–5717
9   ZThang J et al (1998) Subcellular distribution and turnover of presenilins in transfected cells. *J Biol Chem*
    273, 12436–12442
10 Kamino K et al (1996) Three different mutations of presenilin 1 gene in early-onset Alzheimer's disease
    families. *Neurosci Lett* 208,195–198
11 Nakano Y et al (1999) Accumulation of murine amyloid beta42 with a gene-dosage dependent manner in
    PS1 'knock-in' mice. *Eur J Neurosci* 11: 2577–2581
12 Asami Odaka A et al (1995) Long amyloid beta-protein secreted from wild-type human neuroblastoma
    IMR-32 cells. *Biochemistry* 34: 10272–10278

Neuroscientific Basis of Dementia
C. Tanaka, P.L. McGeer, Y. Ihara (eds)
© 2001 Birkhäuser Verlag Basel/Switzerland

Mechanism of neuron death in Alzheimer's disease

Akihiko Takashima, Ohoshi Murayama, Toshiyuki Honda, Xiaoyan Sun and Shinji Sato

Laboratory for Alzheimer's Disease, RIKEN Brain Science Institute, Saitama, Japan

Introduction

Alzheimer's disease (AD) is a progressive neurodegenerative disorder associated with devastating memory loss. Major pathological features are the accumulation of Aβ called senile plaque, aggregation of highly phosphorylated tau in neurons known as neurofibrillary tangles (NFT) and loss of neurons. Loss of neurons may explain the progressive dementia in this disease. From the genetic studies of early-onset familial AD (FAD), the causative genes for FAD have been cloned. One is amyloid protein precursor (APP) on chromosome 21 and the others are presenilin 1 and 2 on chromosome 14 and 1. If the gene is mutated, the gene product develops AD with 100% penetration. Thus, the neuron death through these mutations may appear to be one of the mechanisms responsible for the dementia in AD.

APP mutations to neuron death

APP with FAD mutations [1–3] has been known to produce Aβ much more than wild-type [4, 5], suggesting that Aβ may induce neuron death and the formation of NFT, which is called the amyloid hypothesis [6]. Based on this hypothesis, the Aβ-inducing signal cascade leads to neuron death and NFT formation. From the analysis of NFT, tau is highly phosphorylated in AD brain. Thus, Aβ may activate tau kinase and phosphorylate tau to the paired helical filament (PHF)-state. We have identified a tau kinase which fulfils the above condition. GSK-3β is a kinase which can phosphorylate tau *in vivo* and *in vitro* to the PHF-state [7].

Neuron death through GSK-3β in response to Aβ exposure

Aβ exposure to rat hippocampal primary culture shows the enhancement of tau phosphorylation in proportion to GSK-3β activity [8, 9]. Figure 1 shows the time course of tau phosphorylation and GSK-3β activity in hippocampal primary culture. The level of GSK-3β activity increased approximately to 2 times that of control after Aβ exposure for 6 h. The phosphorylation of tau in hippocampal culture is also enhanced approximately by 2.3 times that of control level by Aβ exposure. The change of GSK-3β level activity correlates most

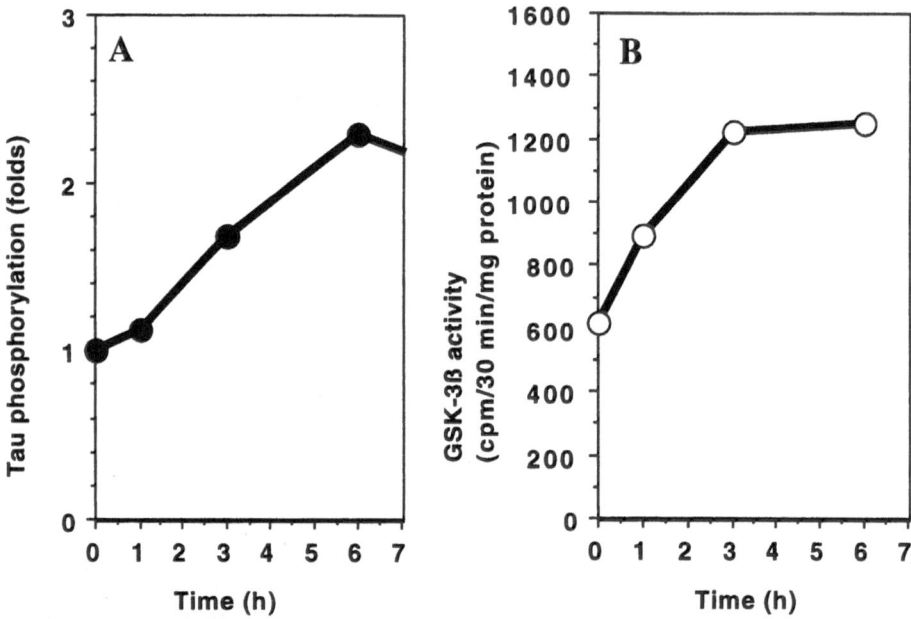

Figure 1. Enhancement of tau phosphorylation and GSK3-β activity in Aβ25-35 treated hippocampal culture. Primary hippocampal culture was exposed 20 μM Aβ25-35 for the indicated period, and phosphorylation of Ser396 of tau (A) and activity of GSK-3β (B) were analyzed in each lysate.

closely with the observed changes in tau phosphorylation. The correlation coefficient of GSK-3β activity and tau phosphorylation is more than 0.9, suggesting that an increase in GSK-3β activity may directly induce the enhancement of tau phosphorylation in the Aβ-treated hippocampal culture. To confirm the effect of GSK-3β on tau phosphorylation in the Aβ-treated hippocampal culture, we pretreated the culture with GSK-3β antisense oligonucleotide (Fig. 2). In GSK-3β sense oligonucleotide-treated culture, Aβ exposure showed an increase of both GSK-3β and tau phosphorylation. The GSK-3β antisense oligonucleotide treatment prevents both the increase of GSK-3β and the enhancement of tau phosphorylation in response to Aβ. These results indicate that GSK-3β plays a role in the phosphorylation of tau in response to Aβ exposure.

The substrates of GSK-3β other than tau have been reported [10]. One is pyruvate dehydrogenase (PDH). PDH is an important enzyme to maintain cellular energy metabolism using glycolysis, which converts pyruvate to acetyl-coA. PDH is inhibited by phosphorylation through GSK-3β. The inhibition of PDH induces apoptosis by intracellular ATP depletion. Based on these results, we propose two pathways for neuron death through the activation of GSK-3β by Aβ exposure (Fig. 3). Aβ directly affects neurons and activates GSK-3β. The activated GSK-3β phosphorylates both tau and PDH. On the one hand, the inactivation of PDH leads to activation of apoptosis. On the other hand, the fully phosphorylated tau by

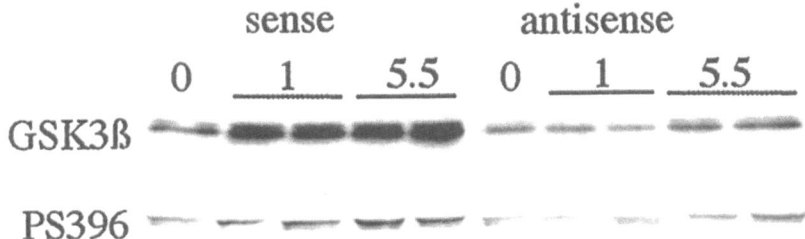

Figure 2. Effects of GSK-3β antisense oligonucleotide on Aβ(25-35)-induced phosphorylation of tau. Hippocampal cells were pretreated with oligonucleotide corresponding to the cloned rat GSK-3β sense or anti-sense sequence at 2.5 μM, and then they were treated with Aβ (25-35)(20 μM) for indicated periods.

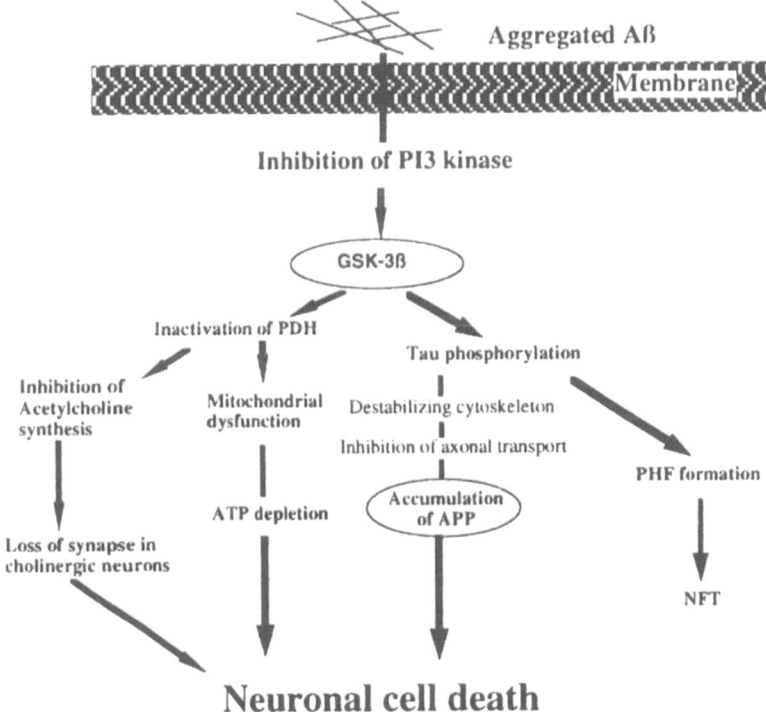

Figure 3. Mechanism of Aβ-induced neuron death through GSK-3β. Aβ affects neurons and activates GSK-3β. Activation of GSK-3β results in tau phosphorylation, leading to the production of PHF and NFT. On the other hand, the phosphorylated tau loses the ability to bind to microtubules, which leads to destabilization of the cytoskeleton and inhibits axonal transport, resulting in accumulation of APP in cells. Pyruvate dehydrogenase (PDH) is the substrate for GSK-3β other than tau. Phosphorylated PDH inhibited its activity. Inactivated PDH cannot convert pyruvate to acetyl-Co A, resulting in the dysfunction of mitochondria, which activates an apop-totic cascade through ATP depletion. The inactivation of PDH results in the reduction of Ach synthesis in cholin-ergic neurons, which leads to loss of cholinergic synapse and neuronal death by trophic factor deprivation.

GSK-3β loses an ability to bind microtubules, leading to the accumulation of tau in cytoplasm and the instability of microtubules. The accumulation of tau may facilitate NFT formation. The instability of microtubule deteriorates the protein trafficking, resulting in neuron death [11].

PS1 and GSK-3β

Recent studies reveal that presenilins with FAD mutation affect APP processing and enhance Aβ42 production [12–15]. However, the increase in Aβ42 by each mutant PS1 does not correlate with the onset-age of AD [16], suggesting that factors other than Aβ may also be involved in development of AD through PS1 mutation. We found that PS1 bound to GSK3β, tau and β-catenin [17, 18]. PS1 mutants increased an affinity for GSK-3β but not for tau and β-catenin. Soluble β-catenin was regulated by GSK-3β activity. Transient expression of PS1 reduced the amount of soluble β-catenin, which was restored by LiCl treatment, indicating that PS1 regulates the amount of soluble β-catenin through GSK-3β. These results suggest that PS1 may act as a scaffold protein for GSK-3β and regulate the efficiency of phosphorylation of tau and β-catenin (Fig. 4). It is well known that PS1 is related to γ-secretase activity [19, 20]. PS1 might facilitate its regulating effect on Aβ through GSK-3β. LiCl, an inhibitor for GSK-3β, reduced Aβ40 and Aβ42 production in the transfected cells, suggesting that GSK-3β may regulate APP processing. Thus, GSK-3β may be an essential factor for developing AD through PS1 mutation.

Neuron death through tau mutation

Recently, the genetic studies of familial frontal temporal dementia parkinsonism (FTDP) have shown that tau is a causative gene in familial FTDP. The patients with tau-bearing mutation exhibit the formation of NFT and neuron loss in their brains [21, 22]. The similarity of these pathological features in tauopathy to AD has raised the question whether the same mechanism for neuronal death through tau mutation occurs in AD. From this aspect, we produced mutant tau cDNA using polymerase chain reaction (PCR)-based site-directed mutagenesis to analyze neuronal death through tau mutations. Each mutant and wild tau cDNA was transfected into COS-7 cells, which did not show any morphological changes. Although it has been reported that mutant tau reduced the ability of microtubule binding *in vitro*, the level of tau in cytoplasm and microtubules did not show significant differences between wild-type and mutant forms. The study of phosphorylation of mutant tau by endogenous kinase in COS-7 cells did not reveal any significant change as compared to wild-type tau, except R406L mutant tau. R406L mutation of tau lost the immunoreactivity against anti-phospho Ser 404 of the tau antibody. However, it is not clear whether R406L tau cannot be phosphorylated at Ser404 by kinase or whether the antibody cannot recognize that site due to conformational change by the existence of a mutation at the proximal site for antibody recognition. As far as an analysis of mobility shift on western blot is concerned, there were

PS1 may act as a molecular tether, connecting GSK-3b with important substrates

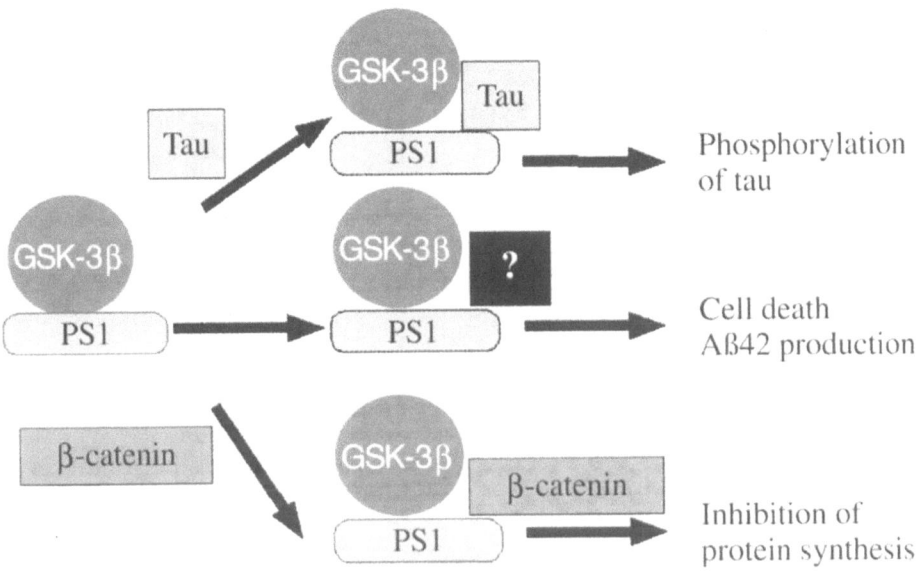

Figure 4. Role of PS1/GSK-3β association on development of AD.

no significant differences between wild and all mutant tau. Phosphorylation of tau under co-transfection with GSK-3β or Cdk-5 activator also showed no differences between wild and all mutant tau. These results suggest that tau mutation may not affect GSK-3β activity. Mutant tau itself can affect neuron death with no relation to GSK-3β activation. To further explore the pathological role of mutant tau, we established stable cell lines expressing each mutant tau using rat pheochromocytoma (PC)12 cells, which are known to differentiate to sympathetic neuron-like cells in response to nerve growth factor (NGF). All cell lines start to extend neurites within 24 h in response to 50 ng/ml NGF, and make a network. In morphological observations, the rate of neurite extension looked the same in all mutant cell lines. These results suggest that tau mutation may not affect neurite extension. In other words, the overexpression of tau mutation did not affect the stabilization of microtubules. The cause of neuron death by tau mutation may not be due to destabilization of microtubules, but to the other functions of tau. In fact, PC12 cells expressing mutant tau were vulnerable to H_2O_2 exposure as compared to control and the cells expressing 3repeat tau. We further examined the level of scavengers for reactive oxygen species (ROS). The catalase, which converts H_2O_2 to H_2O and O_2, showed a lower level in mutant tau-expressing cells, suggesting that mutant tau-expressing cells may lose the ability of protection from ROS. Thus, the accumu-

Role of tau in cell death

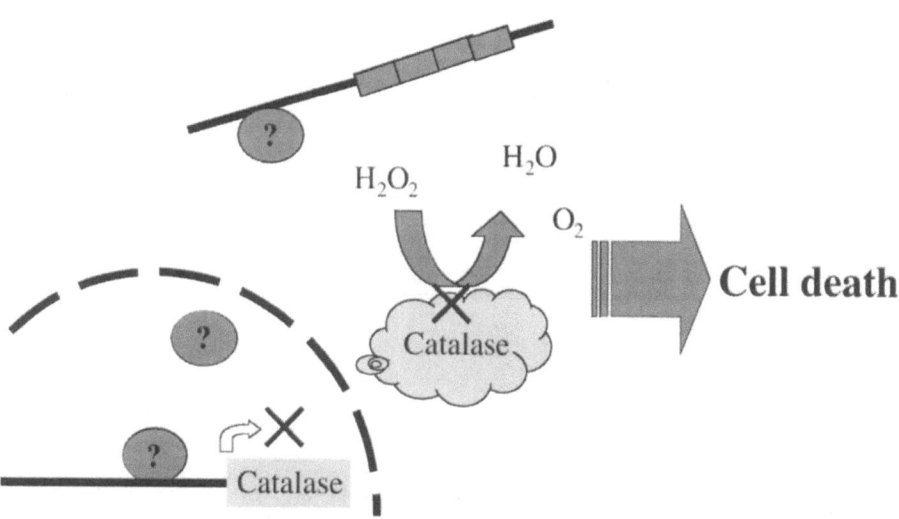

Figure 5. Mechanism of cell death through cytoplasmic tau accumulation.

lation of tau in cytoplasm, caused by phosphorylation or mutation, reduces catalase level, leading to upregulation of ROS metabolism and results in tau aggregation and neuron death (Fig. 5). The precise mechanism is under investigation.

References

1 Goate A, Chartier-Harlin MC, Mullan M, Brown J, Crawford F, Fidani L, Giuffra L, Haynes A, Irving N, James L et al (1991) Segregation of a missense mutation in the amyloid precursor protein gene with familial Alzheimer's disease. *Nature* 349: 704–706

2 Mullan M, Crawford F, Axelman K, Houlden H, Lilius L, Winblad B, Lannfelt L (1992) A pathogenic mutation for probable Alzheimer's disease in the APP gene at the N-terminus of β-amyloid. *Nat Genet* 1: 345–347

3 Murrell J, Farlow M, Ghetti B, Benson MD (1991) A mutation in the amyloid precursor protein associated with hereditary Alzheimer's disease. *Science* 254: 97–99

4 Cai XD, Golde TE, Younkin SG (1993) Release of excess amyloid beta protein from a mutant amyloid β protein precursor. *Science* 259: 514–516

5 Citron M, Oltersdorf T, Haass C, McConlogue L, Hung AY, Seubert P, Vigo-Pelfrey C, Lieberburg I, Selkoe DJ (1992) Mutation of the β-amyloid precursor protein in familial Alzheimer's disease increases β-protein production. *Nature* 360: 672–674

6 Hardy JA, Higgins GA (1992) Alzheimer's disease: the amyloid cascade hypothesis. *Science* 256: 184–185

7 Ishiguro K, Shiratsuchi A, Sato S, Omori A, Arioka M, Kobayashi S, Uchida T, Imahori K (1993)

Glycogen synthase kinase 3 β is identical to tau protein kinase I generating several epitopes of paired helical filaments. *FEBS Lett* 325: 167–172

8   Takashima A, Honda T, Yasutake K, Michel G, Murayama O, Murayama M, Ishiguro K, Yamaguchi H (1998) Activation of tau protein kinase I/glycogen synthase kinase-3 β by amyloid β peptide (25–35) enhances phosphorylation of tau in hippocampal neurons. *Neurosci Res* 31: 317–323

9   Takashima A, Noguchi K, Hoshino T, Imahori K (1993) Tau protein kinase I is essential for amyloid β-protein-induced neurotoxicity. *Proc Natl Acad Sci USA* 90: 7789–7793

10 Hoshi M, Takashima A, Noguchi K, Murayama M, Sato M, Kondo S, Saitoh Y, Ishiguro K, Hoshino T, Imahori K (1996) Regulation of mitochondrial pyruvate dehydrogenase activity by tau protein kinase I/glycogen synthase kinase 3 β in brain. *Proc Natl Acad Sci USA* 93: 2719–2723

11 Takashima A, Yamaguchi H, Noguchi K, Michel G, Ishiguro K, Sato K, Hoshino T, Hoshi M, Imahori K (1995) Amyloid beta peptide induces cytoplasmic accumulation of amyloid protein precursor via tau protein kinase I/glycogen synthase kinase-3 β in rat hippocampal neurons. *Neurosci Lett* 198: 83–86

12 Borchelt DR, Thinakaran G, Eckman CB, Lee MK, Davenprt F, Ratovitsky T, Prada CM, Kim G, Seekins S, Yager D et al (1996) Familial Alzheimer's disease-linked presenilin 1 variants elevate Abeta1-42/1-40 ratio *in vitro* and *in vivo*. *Neuron* 17: 1005–13

13 Duff K, Eckman C, Zehr C, Yu X, Prada CM, Perez-tur J, Hutton M, Buee L, Harigaya Y, Yager D et al (1996) Increased amyloid-β 42(43) in brains of mice expressing mutant presenilin 1. *Nature* 383: 710–3

14 Lemere CA, Lopera F, Kosik KS, Lendon CL, Ossa J, Saido TC, Yamaguchi H, Ruiz A, Martinez A, Madrigal L et al (1996) The E280A presenilin 1 Alzheimer mutation produces increased A beta 42 deposition and severe cerebellar pathology. *Nat Med* 2: 1146–50

15 Suzuki N, Cheung TT, Cai XD, Odaka A, Otvos LJ, Eckman C, Golde TE, Younkin SG (1994) An increased percentage of long amyloid β protein secreted by familial amyloid β protein precursor (β APP717) mutants. *Science* 264: 1336–1340

16 Murayama O, Tomita T, Nihonmatsu N, Murayama M, Sun X, Honda T, Iwatsubo T, Takashima A (1999) Enhancement of amyloid beta 42 secretion by 28 different presenilin 1 mutations of familial Alzheimer's disease. *Neurosci Lett* 265: 61–63

17 Murayama M, Tanaka S, Palatino J, Murayama O, Honda T, Nihonmatsu N, Wolozin B, Takashima A (1998) Direct association of presenilin-1 with β-catenin. *FEBS Lett* 433: 73–77

18 Takashima A, Murayama M, Murayama O, Kohno T, Honda T, Yasutake K, Nihonmatsu N, Mercken M, Yamaguchi H, Sugihara S et al (1998) Presenilin 1 associates with glycogen synthase kinase-3 β and its substrate tau. *Proc Natl Acad Sci USA* 95: 9637–9641

19 De Strooper B, Saftig P, Craessaerts K, Vanderstichele H, Guhde G, Annaert W, VF, K, Van Leuven F (1998) Deficiency of presenilin-1 inhibits the normal cleavage of amyloid precursor protein. *Nature* 391: 387–390

20 Wolfe MS, Xia W, Ostaszewski BL, Diehl TS, Kimberly WT, Selkoe DJ (1999) Two transmembrane aspartates in presenilin-1 required for presenilin endoproteolysis and gamma-secretase activity. *Nature* 398: 513–517

21 Hutton M, Lendon CL, Rizzu P, Baker M, Froelich S, Houlden H, Pickering-Brown S, Chakraverty S, Isaacs A, Grover A et al (1998) Association of missense and 5'-splice-site mutations in tau with the inherited dementia FTDP-17. *Nature* 393: 702–705

22 Spillantini MG, Murrell JR, Goedert M, Farlow MR, Klug A, Ghetti B (1998) Mutation in the tau gene in familial multiple system tauopathy with presenile dementia. *Proc Natl Acad Sci USA* 95: 7737–7741

Neuroscientific Basis of Dementia
C. Tanaka, P.L. McGeer, Y. Ihara (eds)
© 2001 Birkhäuser Verlag Basel/Switzerland

Notch3 gene in CADASIL syndrome: mutation frequencies in Japanese and its expression and processing

Keikichi Takahashi, Satoshi Kotorii, De-Hua Chui, Keiro Shirotani and Takeshi Tabira

Department of Demyelinating Disease and Aging, National Institute of Neuroscience, National Center of Neurology and Psychiatry, Tokyo, Japan

CADASIL syndrome

Cerebral autosomal dominant arteriopathy with subcortical infarcts and leukoencephalopathy (CADASIL) is an inherited cerebrovascular disease characterized by recurrent subcortical ischemic strokes starting in mid-adulthood, often leading to pseudobulbar palsy and dementia [1–4]. Other clinical findings include mood disturbance and recurrent attacks of severe migraine. However, vascular risk factors, including arterial hypertension, are usually absent. Magnetic resonance imaging (MRI) of the brain of affected persons shows numerous small subcortical infarcts on T1-weighted images and diffuse white matter changes on T2-weighted images. The pathological basis of the disease is a non-atherosclerotic, non-amyloid microangiopathy affecting the deep vessels perforating the white matter and basal ganglia. A basophilic, periodic acid-Schiff (PAS)-positive material and, in electron microscopy, granular osmiophilic material (GOM) accumulates between degenerated vascular smooth muscle cells (VSMC) in the brain arterial walls. Although CADASIL is a central nervous system disease and no extra-cerebral manifestations have been described so far, the arteriopathy has been also observed in other organs. Recently, the Notch3 gene, located on chromosome 19p13.1, has been identified as the causative gene for CADASIL and 28 different missense mutations have been reported in 56 unrelated Caucasian patients [5–6]. Although one of these mutations was found in a Japanese family [7], there have been no available data on the frequency of Notch3 mutations in non-Caucasian populations.

Mutational analysis of the Notch3 gene in patients with suspected CADASIL syndrome

To investigate the genetic contribution of Notch mutation in familial cases with vascular leukoencephalopathy, we studied 13 affected individuals from 11 unrelated families of Japanese (9 families), Canadians (1 family) and Iranians (1 family). These subjects were selected on the basis of a family history compatible with an autosomal dominant trait, MRI scanning examination showing multiple infarcts, leukoaraiosis and the absence of vascular risk factors. They also displayed one or more of the following clinical criteria: recurrent subcortical ischemic events or strokes, migraine with or without aura, mood disturbance or severe depressive episodes (character change), gait disturbance or pseudobulbar palsy, progressive dementia and relatively early-onset of the disease. All exons of the Notch3 gene except for exon 1 and exon 26 were amplified from genomic DNAs of patients by using intronic primer pairs as previously described [7] and PCR products were determined by a combination of SSCP (single strand conformation polymorphism) and direct sequencing.

In these analyses, we identified 4 missense mutations in exon 3 and exon 4 (Tab. 1). Two mutations (Arg90Cys and Arg133Cys) observed in patients 2, 3, 4 and 5 were as found in Caucasian patients [6], and another two (Arg213Lys and Cys174Phe) in patients 1 and 6 were novel missense mutations. Because three mutations in four families caused addition or loss of a cysteine residue in epidermal growth factor (EGF)-like repeats of Notch3 protein, it is likely that the mutations are involved in the CADASIL pathogenesis. Based on these data, we roughly calculated the mutation frequency of the Notch3 gene to be 35% among autosomal dominant forms of vascular dementia and leukoencephalopathy in the population studied.

Table 1. Mutations in the Notch3 gene

Patient (family)	Exon	Domain	Nucleotide substitution	Amino acid substitution
1 (MU)	4	EGF5	716G→A	Arg213Lys
2 (IR)	3	EGF2	346C→T	Arg90Cys
3 (KT)	4	EGF3	475C→T	Arg133Cys
4 (KMA)	4	EGF3	475C→T	Arg133Cys
5 (KMA)	4	EGF3	475C→T	Arg133Cys
6 (KW)	4	EGF4	599G→T	Cys174Phe

An intriguing question is whether a missense mutation (Arg213Lys) is a rare polymorphism or is related to the CADASIL pathogenesis, because the diagnosis of CADASIL for patient 1 was strongly supported by ultrastructural examination of brain autopsy samples revealing numerous deposits of GOM in meningeal vessels [8]. It is also important to note that most of the cysteine additions (15/18) found in CADASIL patients are created by substitution for arginine residues, although six codons (Try, Phe, Ser, Trp, Gly and Arg) are possibly changed to a cysteine codon by a single nucleotide mutation. Therefore, there is the possibility that the deletion of arginine residues may be involved in the pathogenesis of CADASIL. It cannot be excluded that the high prevalence of arginine mutations in the Notch3 gene merely results from a certain mechanism of somatic mutation, such as the methylation of DNA, because all point mutations in arginine codons occur at a C of CpG dinucleotide, which is highly methylated in mammalian cells. Further studies will be needed to determine the significance of arginine mutations.

We have identified 8 patients (7 families) without CADASIL mutations, although their family history and symptoms were similar to those of individuals with CADASIL syndrome.

Therefore, the additional mutation(s) in the Notch3 gene may occur in such phenocopies [9]. Alternatively, there is the possibility that the causative gene of phenocopies is different from the Notch3 gene. Recent reports suggest a genetic heterogeneity in familial vascular leukoencephalopathy and multi-infarct dementia, and some are not associated with chromosome 19 [10, 11]. As in Alzheimer's disease, there seems to be more than one locus for hereditary cerebral vascular encephalopathy.

The function of the Notch receptor

The Notch gene family encodes large transmembrane receptors that are involved in cell fate decisions during embryonic development in vertebrates and invertebrates [12]. At present, four Notch genes have been cloned in human and mouse, and their protein structures are highly conserved from insects to vertebrates. The Notch3 protein consists of 2321 amino acid residues containing, in the extracellular domain, 34 repeats of an EGF sequence motif and three repeats of a Notch/Lin-12 (N/L) sequence motif, and in the intracellular domain, six ankyrin/Cdc10 repeats and a PEST-containing region [5]. Based on data from *Drosophila*, the EGF-like repeats are thought to function as the ligand-binding region, while the intracellular domain is involved in modulation of signal transduction, resulting in inhibition of differentiation and transduction of inducible signals. However, the normal biological role(s) of Notch3 in adult tissues and the mechanism(s) by which the CADASIL-associated mutations exert their effect remain unknown. Because all missense mutations found in CADASIL patients lead to loss or addition of a cysteine residue in the EGF-like repeats [6], an abnormal disulfide bridge formation may cause improper protein conformation, which affects specificity of ligand binding or signal transduction.

Expression and processing of Notch3

It has been reported that Notch receptors undergo at least two proteolytic cleavages during their maturation and upon activation by ligand binding [13–16]. In mammalian cells, the large precursors of Notch1 and Notch2 are cleaved at the extracellular region by furin-like proteinase in the trans-Golgi network. This process generates two fragments that remain associated and form the functional receptor. Following ligand binding, Notch1 and Notch2 are further cleaved close to or within the transmembrane (TM) domain, releasing the intracellular domain (ICD) from the membrane. The ICD translocates to the nucleus and modifies transcription of target genes through its association with CSL proteins (CBF-1, Su(H), Lag-1]. Recently, it was shown that presenilin 1 and 2, causative genes for early-onset familial Alzheimer's disease, regulate the endoproteolytic cleavage of Notch1 within the TM domain [17]. However, expression and processing of Notch3 have not been studied.

To address the molecular basis of maturation and activation of Notch3, we determined the expression and processing of human Notch3 in COS and 293 cells using various cDNA constructs and four specific antibodies. In COS cells transiently transfected with the HN-FL construct, western blot analyses detected a major band of 250 kDa by all antibodies and a minor band of 80 kDa by C1 and C2 antibodies raised against the ICD of Notch3 (Fig. 1). The 80 kDa band was also observed in cells transfected with HN-ΔN/L, and comigrated with a IC polypeptide expressed in COS cells. This indicates that the 80 kDa polypeptide consists of the proteolytic C-terminal fragment which is cleaved closed to the TM domain. However, other than these two bands, we could identify no band in HN-FL transfected cells with an increased level as compared with control cells transfected with the HN-R. In 293 cells transfected with HN-FL, the 250 kDa polypeptide, but not the 80 kDa protein, was detected (data not shown). The 80 kDa polypeptide was absent in 293 cells transfected with HN-ΔN/L. The expression patterns of the Notch3 in myoblast C2C12 cells were the same as those in 293 cells (data not shown). Thus, the extracellular processing of Notch3 did not occur in COS or 293 cells, and the cleavage close to or within the TM domain was confined in COS cells. Since the extracellular processing of Notch1 is observed in 293 cells [14], there is the possibility that Notch3 lacks the potential recognition site of furin-like protease. A recent report has indicated that the complete abolition of the Notch1 extracellular processing is required for mutations in not only the consensus cleavage site [RX(K/R)R] of furin but also in the two

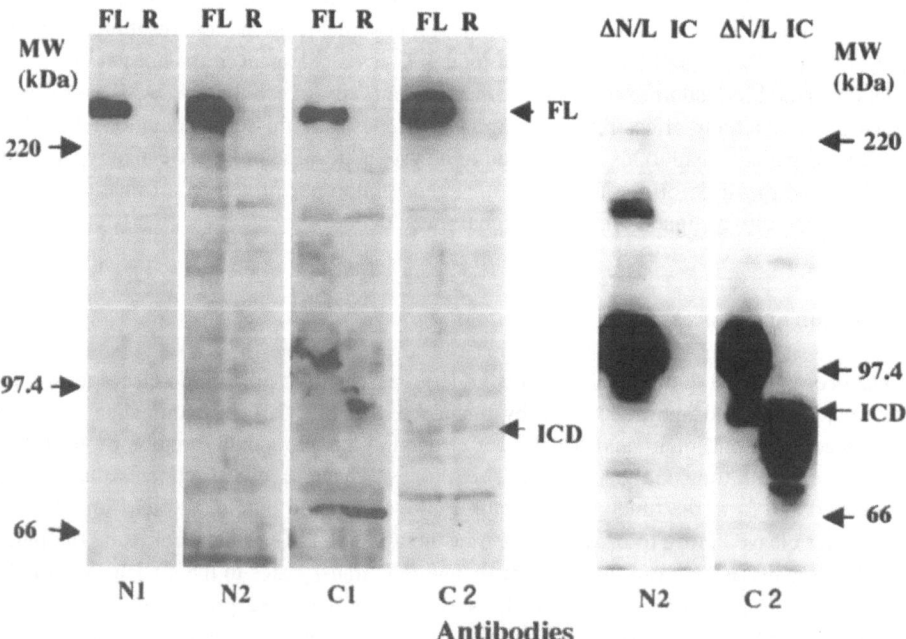

Figure 1. Expression and endoproteolytic cleavage of Notch3 and its derivatives in COS-7 cells. Cells were plated in either 3.5 cm dishes or 10 cm dishes, and transfected with plasmid DNA using a calcium phosphate method. After 48 h, cells were harvested and processed for western blot analysis. Antibodies used are indicated below panels. FL and ICD represent the full-length protein and a proteolytically-cleaved C-terminal fragment with 80 kDa, respectively, and R is the control construct with a reverse orientation of FN3-FL.

secondary sites (KR and RK) [14]. Interestingly, Notch3 possesses the consensus cleavage site at a similar position (RARR, amino acids 1567–1570), while the secondary potential sites cannot be found between the Notch/Lin repeats and the TM domain. Thus, the RARR sequence in Notch3 might not be a recognition site of furin. On the other hand, it is unexpected that the intracellular processing of Notch3 takes place in COS cells but not in 293 cells transfected with the ΔN/L constructs, because the truncated polypeptides of Notch paralogs, which lack the EGF repeats and the N/L repeats, are constitutively cleaved within the TM domain to release the ICD in many cell lines, including 293 cells [15–17]. Therefore, it is likely that the intracellular processing of Notch3 is exerted by the cell-type specific protease(s), which is probably distinct from the processing enzyme for Notch1 and Notch2.

Finally, it is important to note that the full-length Notch3 consistently underwent intracellular processing in COS cells. The same process was observed in cells transfected with CADASIL mutants (Fig. 2), but there seemed to be no difference in the size and amount of the ICD among wild-types and mutants. It is unclear whether this intracellular cleavage of

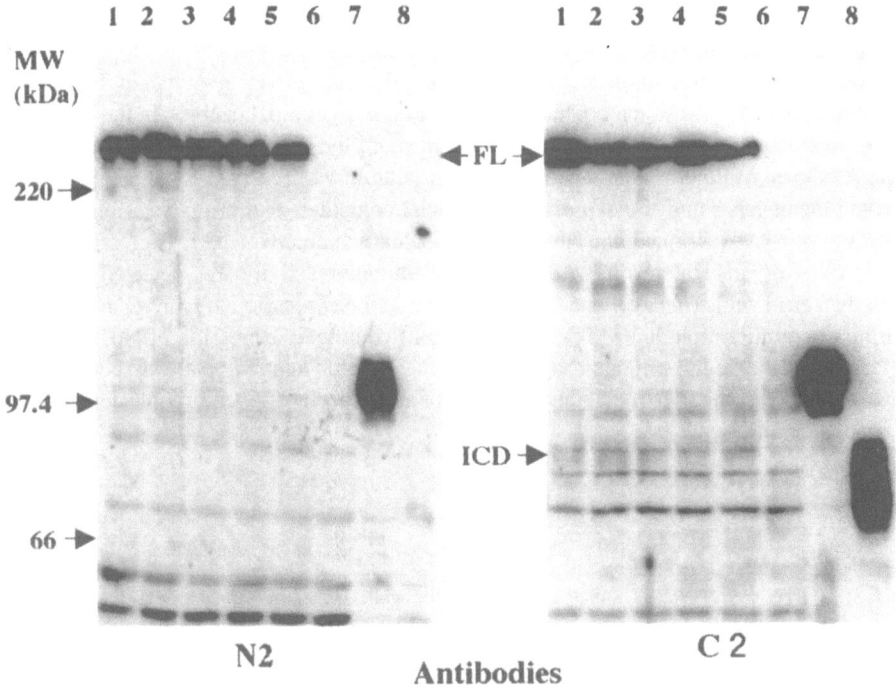

Figure 2. Expression and endoproteolytic processing of Notch3 with CADASIL mutations in COS-7 cells. Notch3 cDNAs with CADASIL mutations (R90C, R133C, C185R and R213K) were prepared using a site-directed mutagenesis procedure. Cells were transfected with cDNA constructs with and without mutations and subjected to western blot analysis. Lane 1, wild-type (HN3-FL), lane 2, R90C, lane 3, R133C, lane 4, C185R, lane 5, R213K, lane 6, HN3-R, lane 7, HN3-ΔN/L, lane 8, HN3-IC.

Notch3 is mediated by ligand interaction, since there are no available data for the properties of Notch3 ligand(s). Further experiments will be needed to investigate whether Notch3 mutations result in constitutively activated receptors by aberrant processing of the ICD.

Notch3 immunoreactivity in the human brain

CADASIL is pathologically characterized by arterial lesions in the brain, although the arteriopathy can be found, to a lesser extent, in other organs [4]. There are the defining changes in the brain arterial walls, including thickening of arterial media, fragmentation and reduplication of the internal elastic lamella and depositions of PAS-positive granules on light microscopy, and the accumulation of GOM between degenerating VSMC on electron microscopy. Therefore, we examined the distribution of Notch3 in normal human brains by immunohistochemical analysis using Notch3 antibodies (N_2, C1 and C2). The immunostaining showed moderate staining of vessel walls in cerebral white matter (Fig. 3). The immunoreactivity was predominantly located either between VSMCs and endothelium or in the tunica media consisting of VSMCs. Since some vessels were not stained with antibodies, the expression of Notch3 might be restricted to specific arteries. This is supported by the finding that Notch3 immunoreactivity was mainly detected in vessels with similar sizes. In contrast to Notch1 and Notch2, which are expressed in pyramidal neurons in the hippocampal formation and cortex, neither cortical neurons nor glial cells were Notch3 immunoreactive. As shown in Figure 4, an identical staining pattern was observed in vessels of the cortex and meninges. Thus, the expression of Notch3 coincides with primary lesions found in the brain of CADASIL patients, although it is unclear whether Notch3 is expressed in arteries of other organs. Interestingly, the strongest immunoreactivity was detected in arteries with thickened walls (Fig. 4 B and C), indicating that the expression of Notch3 may increase during the proliferation of VSMCs. Our finding of continued expression of Notch3 protein in vessels of the adult brain also suggests that the Notch3 signaling pathway could remain involved in determination of cell fate decisions of arterial cells after embryonic development.

Figure 3. (upper panel) Notch3 immunoreactivity in the white matter of normal human brain. Immunohistochemical staining was performed. Specimens (A, B, and C) were stained with Notch3 antibodies N_2, C2, and C1, respectively. Specimen D is a consecutive section of C and stained with antibody C1 preabsorbed with GST-C1 fusion protein. Red asterisks represent vessels stained with antibodies. A white asterisk indicates the same vessel, which was stained with antibody C1 but not the preabsorbed antibody. No immunoreactivity was observed in some vessels (red arrows).

Figure 4. (lower panel) Expression of Notch3 in the cortex (A, B and C) and meninges (D) of human brain. Specimens of A, B, and C were stained with antibodies N_2, C1, and C2, respectively. Specimen D was stained with antibody C2. Red asterisks represent vessels with Notch3 immunoreactivity. Vessels that were not stained are indicated by red arrows.

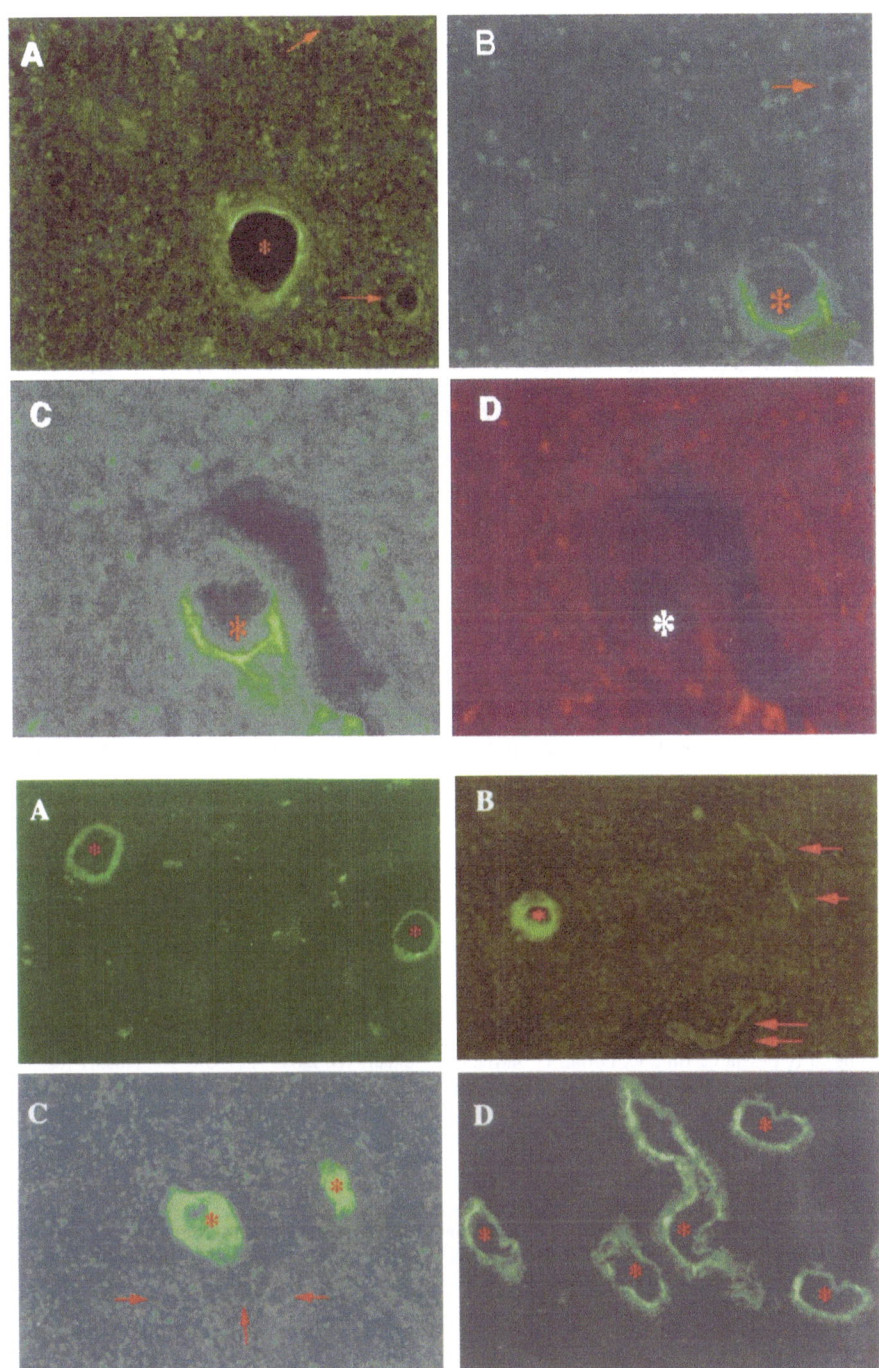

Acknowledgments
We thank the medical doctors who referred cases and provided us with DNA samples. We also thank Ms. K. Kubo for her excellent technical assistance. This study was supported by a grant from the Ministry of Health and Welfare (Aging and Health Science), by a Grant-in-Aid for Scientific Research from the Ministry of Education, Science, Culture, and Sports of Japan, and by a grant (COE) from the Science and Technology Agency of Japan.

References

1 Baudrimont M, Dubas F, Joutel A, Tournier-Lasserve E, Bousser M-G (1993) Autosomal dominant leukoencephalopathy and subcortical ischemic stroke. A clinicopathological study. *Stroke* 24: 122–125
2 Chabriat H, Vahedi K, Iba-Zizen MT, Joutel A, Nibbio A, Nagy TG, Krebs TG, Julien J, Dubois B, Ducrocq X et al (1995) Clinical spectrum of CADASIL: a study of 7 families. *Lancet* 346: 934–939
3 Chabriat H, Tournier-Lasserve E, Vahedi K, Leys D, Joutel A, Nibbio A, Escaillas JP, Iba-Zizen MT, Bracard S, Tehindrazanarivelo A et al (1995) Autosomal dominant migraine with MRI white-matter abnormalities mapping to the CADASIL locus. *Neurology* 45: 1086–1091
4 Lammie GA, Rakshi J, Rossor MN, Harding AE, Scaravilli F (1995) Cerebral autosomal dominant arteriopathy with subcortical infarcts and leukoencephalopathy (CADASIL)-confirmation by cerebral biopsy in 2 cases. *Clin Neurophathol* 14: 201–206
5 Joutel A, Corpechot C, Ducros A, Vahedi K, Chabriat H, Mouton P, Alamowitch S, Domenga V, Cecillion M, Marechal E et al (1996) Notch3 mutations in CADASIL, a hereditary adult-onset condition causing stroke and dementia. *Nature* 383: 707–710
6 Joutel A, Vahedi K, Corpechot C, Troesch A, Chabriat H, Vayssiére C, Cruaud C, Maciazek J, Wessenbach J, Bousser M-G et al (1997) Strong clustering and stereotyped nature of Notch3 mutations in CADASIL patients. *Lancet* 350: 1511–1515
7 Kamimura K, Takahashi K, Uyama E, Tokunaga T, Kotorii S, Uchino M, Tabira T (1999) Identification of Notch3 mutation in a Japanese CADASIL family. *Alz Dis Assoc Disord* 13: 222–225
8 Nishio T, Arima K, Eto K, Ogawa M, Sunohara N (1997) Cerebral autosomal dominant arteriopathy with subcortical infarcts and leukoencephalopathy: report of an autopsied Japanese case. *Clin Neurol (Japan)* 37: 910–916
9 Mellies JK, Baumer T, Muller JA, Tournier-Lasserve E, Chabriat H, Knobloch O, Hackeloer HJ, Goebel HH, Wetzig L, Haller P (1998) SPECT study of a German CADASIL family: a phenotype with migraine and progressive dementia only. *Neurology* 50: 1715–1721
10 St Clair D, Bolt J, Morris S, Doyle D (1995) Hereditary multi-infarct dementia unlinked to chromosome 19q12 in a large Scottish pedigree: evidence of probable locus heterogeneity. *J Med Genet* 32: 57–60
11 Utatsu Y, Takashima H, Michizono K, Kanda N, Endou K, Mitsuyama Y, Fujimoto T, Nagai M, Umehara F, Higuchi I et al (1997) Autosomal dominant early-onset dementia and leukoencephalopathy in a Japanese family: clinical, neuroimaging and genetic studies. *J Neurol* 147: 55–62
12 Artavanis-Tsakonas S, Matsuno K, Fortini ME (1995) Notch signaling. *Science* 268: 225–23224
13 Blaumueller CM, Qi H, Zagouras P, Artavanis-Tsakonas S (1997) Intracellular cleavage of Notch leads to a heterodimeric receptor on the plasma membrane. *Cell* 90: 281–291
14 Logeat F, Bessia C, Brou C, LeBail O, Jarriault S, Seidah NG, Israel A (1998) The Notch1 receptor is cleaved by a furin-like convertase. *Proc Natl Acad Sci USA* 95: 8108–8112
15 Kopan R, Schroeter EH, Weintraub H, Nye JS (1996) Signal transduction by activated mNotch: Importance of proteolytic processing and its regulation by the extracellular domain. *Proc Natl Acad Sci USA* 93: 1683–1688
16 Schroeter EH, Kisslinger JA, Kopan R (1998) Notch-1 signaling requires ligand-induced proteolytic release of intracellular domain. *Nature* 393: 382–386
17 De Strooper B, Annaert W, Cupers P, Saftig P, Craessaerts K, Mumm JS, Schroeter EH, Schrijvers V, Wolfe MS, Ray WJ et al (1999) A presenilin-1-dependent g-secretase-like protease mediates release of Notch intracellular domain. *Nature* 398: 518–522

Neuroscientific Basis of Dementia
C. Tanaka, P.L. McGeer, Y. Ihara (eds)
© 2001 Birkhäuser Verlag Basel/Switzerland

Etiological roles of Aβ and carboxyl terminal peptide fragments of amyloid precursor protein in Alzheimer disease

Yoo-Hun Suh[1], Hye-Sun Kim[1], Cheol Hyoung Park[1], Ji-Heui Seo[1], Jean-Pyo Lee[1], Sung-Jin Jeong[1], Sung-Soo Kim[2], Jun-Ho Lee[1], Se Hoon Choi[1], Keun-A Chang[1], Jong-Cheol Rah[1] and Sung-Su Kim[3]

[1] *Department of Pharmacology, College of Medicine and Neuroscience Research Institute, Medical Research Center, Seoul National University and Biomedical Brain Research Center, National Institute of Health, Seoul, Korea*
[2] *Department of Pharmacology, College of Medicine, Kang Won National University, Seoul, Korea*
[3] *Department of Anatomy, College of Medicine, Chung Ang University, Seoul, Korea*

Introduction

Alzheimer's disease (AD) is a neurodegenerative disorder of the brain characterized by the presence of neuritic amyloid plaques and neurofibrillary tangles in various areas of the brain [1]. Many lines of evidence indicate that Aβ, a 39–43 amino acid, proteolytically derived fragment from an integral membrane protein termed amyloid precursor protein (APP), which is a principal constituent of senile plaques, may play an important role in the pathogenesis of AD [2]. Proteolytic processing of APP by several proteases occurs in cells via several metabolic pathways and can lead to the formation of various fragments.

Many studies have shown that Aβ is toxic to neurons *in vitro* and *in vivo*. However, a relatively high concentration (20 μM) of Aβ is needed to induce toxicity and some studies still failed to demonstrate the toxicity of Aβ *in vivo* [3]. Aβ deposition has been found without accompanying neurodegeneration and neurodegeneration could occur in areas with no Aβ deposition. Furthermore, it has been reported that under certain conditions in culture, Aβ promotes neurite outgrowth instead of exerting toxic action. Thus, Aβ may not be the sole fragment in the neurotoxicity associated with AD. Consequently, the possible effects of other cleaved products of APP need to be explored.

Recently it has been reported that carboxyl (C)-terminal fragments of APP are found in plaque microvessels, media and cytosol of lymphoblastoid cells obtained from patients with early- or late-onset familial AD (FAD) and Down's syndrome. These C-terminal fragments, which contain the complete Aβ sequence, appear to be toxic to neurons in culture, although the mechanism of action is not understood. Furthermore, transgenic mice that over-expressed the C-terminal 100 peptide were reported to show extensive neuronal degeneration in the hippocampal area, with cognitive impairments and impairment of long-term potentiation (LTP). Recently, it has been reported that APP mutations that cause FAD increase the intracellular accumulation of potentially amyloidogenic C-terminal fragments of βAPP in neurons [4]. While the C-terminal fragments of amyloid precursor protein (APP) have been presumed to exert its toxic action through secondary generation of Aβ, the role of the C-terminal peptide itself in the pathogenesis of AD is poorly understood. Previously, we have reported that a recombinant C-terminal 105 amino-acid (CT_{105}) itself caused direct neurotoxicity in PC12 cells and primary cortical neurons [5, 6], induced strong non-selective inward currents in *Xenopus* oocytes [5, 7], planar lipid bilayers [8] and Purkinje cells [9], and blocked the later phase of LTP in rat hippocampus *in vivo* [10]. In this manuscript, we show that CT105 impaired calcium homeostasis by inhibiting microsomal calcium uptake by Mg^{2+}-Ca^{2+} ATPase in the rat brain microsomes, Na^+-Ca^{2+} exchanger activity in SK-N-SH cells but Aβ did not [11, 12]. In addition, we found that intracellular injection of CT_{105} impaired learning and memory and was toxic to animals [13] and that CT_{105} induced MAPKs- and NF-κB-dependent astrocytosis and iNOS induction (unpublished observation).

Neurotoxicity of CT fragments

It has been reported that amyloidogenic CT fragments of βAPP appear to be toxic to neurons in culture [14, 15], transgenic mice and the brains of mice transplanted with CT-transfected cells.

Neurotoxicity of an amyloidogenic CT fragment caused by overexpression of its cDNA in differentiated neurons derived from a PC12 transformant was originally reported by Yankner et al. [15].

Fukuchi et al. [16] reported that overexpression of an Aβ-bearing CT region of βAPP led to degeneration of P19-derived neurons. Higaki et al. [17] demonstrated that in Chinese hamster ovary (CHO) cells overexpressing $βAPP_{695}$, the Aβ-bearing CT fragment accumulates with treatment with MDL28170 (a calpain protease inhibitor) and that the degeneration of the accumulated fragment is inhibited by monensin, suggesting that an early endosomal compartment is involved in the processing of the CT fragment.

Transplantation of CT_{100}-producing neuronal cells into mouse brain caused cortical atrophy and the appearance of the Alz-50 antigen; the deposition of Aβ in the vicinity of the transplantation of human neurons that secrete Aβ into rodent brain did not cause detectable lesions.

Overexpression of normal βAPP *in vitro* can lead to the accumulation of CT_{100}-like C-terminal fragments and to the death of the neuronal cells in which these fragments build up [16].

While massive expression of Aβ in the brains of transgenic mice failed to elicit neuronal loss and an AD phenotype, the CT_{100} -transgenic mice recreated many aspects of the synaptic damage, neuronal death, disruption of the neuronal cytoskeleton and lysosomal abnormalities that are seen in the brains of AD patients. The phenotype of these transgenic animals suggests that the CT fragment is a critical component of the molecular mechanisms of AD neurodegeneration. In PC12 cells and cultured rat cortical neurons, CT_{105} peptide was more potent than Aβ fragments in inducing lactate dehydrogenase release from both types of cells [18].

Ionic effects of CT fragments

Fraser et al. [7] have demonstrated that either extracellular application or intracellular injection of amyloidogenic CT_{105} elicited strong inward currents or toxic effects in *Xenopus* oocytes. The channel-inducing and cytotoxic effects of CT_{105} were much more potent than those of any Aβ fragments.

In addition, CT_{105} induced a large transient inward current in the rat Purkinje neuron; this current was accompanied by a large influx of Ca^{2+} and induced a depression of parallel fiber-Purkinje cell synaptic transmission both at application and at distant sites [9]. This was consistently followed by cell death within 30–40 min of CT_{105} application. In contract, Aβ induced a much smaller transient inward current in the rat Purkinje neuron than the CT fragment, but the current was not accompanied by a large Ca^{2+} influx. From these results in *Xenopus* oocytes and in rat Purkinje neurons and cultured mammalian cells, the toxic effects of CT_{105} may be related to a nonselective channel-inducing or pore-making effect [7, 9, 19].

We also demonstrated that CT_{105} formed cation-selective channels in a planar lipid bilayer composed of palmitoyloleoylphosphatidylethanolamine and palmitoyloleoylphosphatidylcholine [80:20, 25 mg/ml in decane [8]].

The endoplasmic reticulum Mg^{2+}-Ca^{2+} ATPase is the major enzyme involved in sequestering calcium in the endoplasmic reticulum. We investigated the effects of CT_{105} on microsomal calcium uptake to study the effects of CT_{105} on the calcium homeostasis of the neuronal cell.

In this study, we demonstrated that 10 μM CT_{105} inhibits calcium uptake via Mg^{2+}-Ca^{2+} ATPase in rat brain microsomes by $52.0 \pm 7.8\%$, while $Aβ_{25-35}$ and $Aβ_{1-42}$ had no effect on this ATPase at 50 μM. Also, CT_{105} showed the inhibitory effect in a concentration-dependent manner [11]. This result shows that CT_{105} has quite different mechanisms from Aβ for calcium homeostasis.

It is generally agreed that Na^+-Ca^{2+} exchange plays an important role in neuronal calcium homeostasis. Experimental evidence indicates that the Na^+-Ca^{2+} exchanger is a critical component of a cell's repertoire for lowering intracellular calcium. CT_{105} inhibited Na^+-Ca^{2+} exchanger activity in SK-N-SH cells by about 50%, whereas $Aβ_{1-42}$ had no effect at 50 μM. The inhibitory effect of CT_{105} was concentration-dependent [12].

In conclusion, these results demonstrate that CT_{105} inhibits microsomal Ca^{2+} uptake and the Na^+-Ca^{2+} exchanger, which are considered to have a key role in the Ca^{2+} homeostasis of excitable cells and suggest that this inhibitory effect of CT_{105} may contribute to disruption of intracellular calcium concentration and is possibly involved in inducing the neural toxicity characteristic of AD [11, 12].

In the planar lipid bilayer system, CT_{105}-induced channels showed rather higher cation selectivity ($PK/PCl = 10.2$) [8]. Similar high cation selectivity was also known for the ion channel induced by Aβ ($PK/PCl = 11$). An important property of CT_{105}-induced channels was its high selectivity sequence of PCa > PNa > PK > PRb > PLi > PCs > PMg [8], whereas the Aβ-induced channels showed higher selectivity to Cs^+ and Li^+ with the selectivity sequence of PCs > PLi > PCa > PRb > PLi > PNa.

The direct cytotoxic effect of CT_{105} comparable to that of $A\beta$ was also indicated by the following findings: 1) the similar or larger ionophoric effect of CT_{105} than that of $A\beta$ under our experimental conditions and in *Xenopus* oocyte [7] and 2) the higher selectivity of CT_{105}-induced pores to Ca^{2+} and Na^+, the two primary ions involved in various neurodegenerative processes, than that of $A\beta$-induced pores [8]. This nonselective channel-or pore-forming effect seems to be reminiscent of the properties of pore-forming proteins, e.g., perforin, the membrane-attack complex of complement, and eosinophilic cation protein, known to cause membrane damage and cytotoxicity. CT_{105} is not only a precursor of $A\beta$, but also it is considered cytotoxic by itself [20].

Behavioral and memory effects of CT

While higher doses of intracerebroventricular injection of recombinant CT_{105} caused sudden death and a transient paralysis of hindlimbs, lower doses of CT_{105} induced a marked, dose-dependent reduction of step-through latency in the passive avoidance test [13] and of spatial memory in the 8 arm radial maze tests, but the same dose of $A\beta$ didn't (Fig. 1). These results suggest that the CT peptide itself is a neurotoxic substance *in vivo*, although we cannot rule

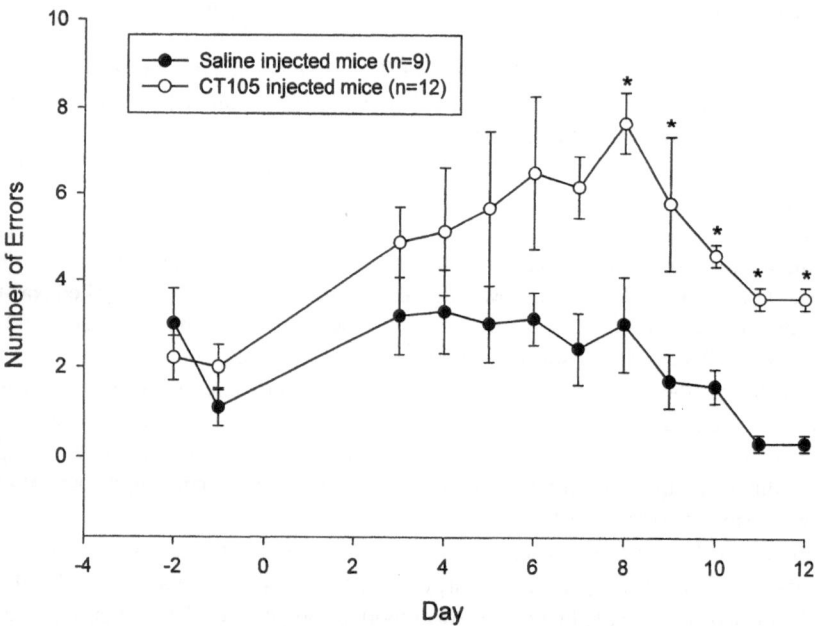

Figure 1. Effects of CT_{105} given by intracerebroventricular (i.c.v.) injection on 8-arm maze test performance. The training trials were carried out before an administration of CT_{105}. The 8 arm maze test was carried out from 3 to 14 after an administration of CT_{105}. Data are mean ± S.E. *; $p < 0.05$, significantly different from the saline injected mice. (ANOVA test)

out the possibility that recombinant CT_{105} may have different structural and functional properties compared with native CT. The lack of effect of boiled CT_{105} on passive avoidance performance indicates the necessity of the specific conformation of CT_{105} in inducing the nerurotoxicity. In line with the results of the present study, CT_{105} greatly shortened the duration of high-frequency stimulation-induced long-term potentiation in the rat hippocampus *in vivo* [10] and in transgenic mice expressing the CT of βAPP [4].

Neuropathologic effects of CT peptides

Extensive neuronal loss with an accompanying increase in reactive astrocytes around senile plaques is one of the well-established pathological findings in AD brains. In the previous study, we used two parameters to assess the effect of CT_{105} on brain histology: glial fibrillary acidic protein (GFAP) immunoreactivity for gliosis and acid fuchsin staining for neuronal degeneration [13]. The neocortex exhibited the most severe neurodegeneration and reactive gliosis. Hippocampal regions showed only modest increases in GFAP immunoreactivity, without a significant increase in the number of acid fuchsin-stained neuronal cells. The thalamus was completely free from CT_{105}-induced changes in reactive gliosis and neuronal degeneration [13]. Either intrinsic differences of various brain regions in the vulnerability to CT_{105}-induced toxicity or regional differences in the distribution of intracerebroventricularly administered CT_{105} may explain these results.

Our study clearly shows that CT_{105} activates astrocytes directly, not through activation of microglial cells, and also our study showed that 100 nM CT-APP induced increased expression of vimentin and GFAP (unpublished observation). Immunoreactivity of transcription factor NF-κB was also increased and its translocation to the nucleus was observed (unpublished observation). Furthermore, CT-APP resulted in activation of mitogen-activated protein kinase (MAPK) pathways including p38, c-Jun N-terminal kinase, and p44/42 extracellular signal responsive kinase (ERK). Pretreatment of PD098059, a specific MAP kinase (MEK) inhibitor and SB203580, an inhibitor of p38 and SEK decreased nitric oxide (NO) production, which may mediate adjacent neurotoxicity and also decreased NF-κB activation. But PDTC, the inhibitor of NF-κB activation, could not affect MAPK's activation whereas NO production and astrocytosis were abolished. Furthermore, conditioned media obtained from CT-APP-treated astrocytes promoted neurotoxicity and pretreatment of NO and peroxynitrite scavengers attenuated its toxicity. These results suggest that CT-APP may participate in Alzheimer's pathogenesis through MAPKs- and NK-κB- dependent astrocytosis and iNOS induction. Taken together, these results suggest another possible mechanism for astrocytosis and roles of CT-APP in Alzheimer's disease pathogenesis. Most probably the combined effects of CT_{105} on both neuronal and glial cells may be responsible for the neuropathological findings of the present study. Consistent with the results of our present study, extensive gliosis and neuronal loss were also reported in transgenic mice that overexpress 104-amino acid CT [4].

Table 1 summarizes the results of experiments for Aβ and CT_{105} on various effects *in vitro* and *in vivo*. In summary, CT_{105} peptide has much more potent toxic effects, channel

Table 1. Summary of effects induced by Aβ and CT fragments of βAPP

	Aβ	CT
Neurotoxicity		
Cultured cells	+	+10
in vivo ICV	+	++++
Channel Effect		
Xenopus *oocyte*	–	+++
Purkinje Cells	–	+++
Lipid Bilayer	+	++
Microsome Ca^{2+}	–	+++
Hippocampus *in vivo*	+	+++
Behavioral and free radical effects of Aβ and CT		
Learning and memory impairment	+	+++
Free radical generation	+++	–
NO generation	+	+++

effects and learning and memory impairment effects than Aβ, whereas Aβ has much greater free radical generation effects than CT_{105}. Several lines of evidence show that Aβ-bearing CT fragments from several different cells may be released into the media or extracellular fluid [14, 18]. It is not yet clear what concentration of CT is present extracellularly *in vivo* for AD brains; however, CT peptides have been found in paired helical filaments, in senile plaques, in microvessels and in choroid plexus from AD brains. Thus, it can be speculated that even a small amount of CT that may be persistently produced in the brains of AD patients for a long interval may potentially alter adjacent neuronal and glial cells, ultimately to play a role in the production of pathologic lesions in AD patients' brains.

We proposed the hypothesis of an etiological role of amyloidogenic CT fragments of βAPP in AD in Figure 2.

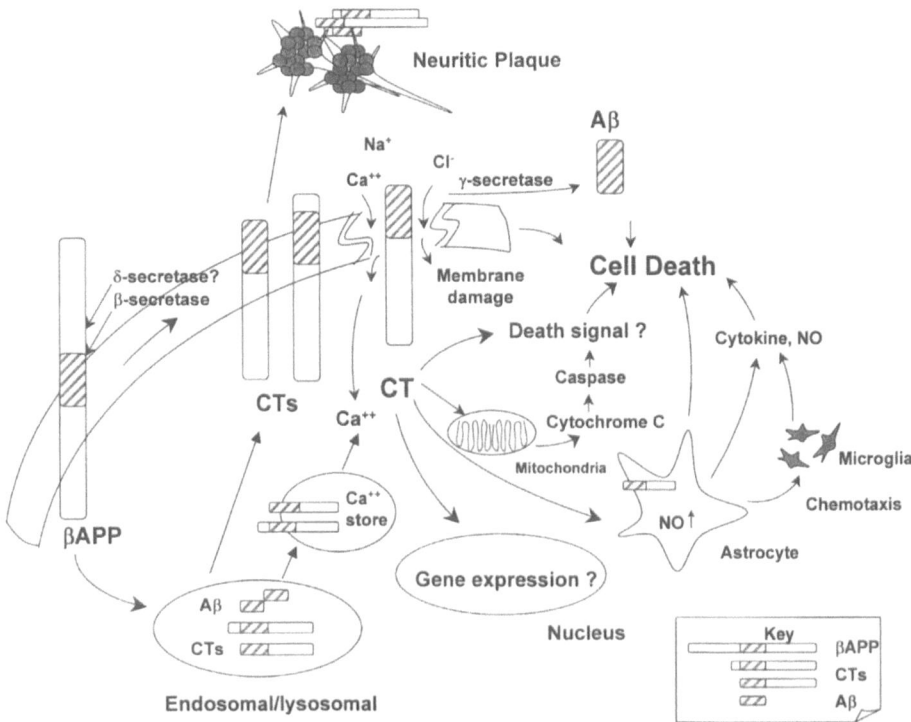

Figure 2. Hypothesis of an etiological role of amyloidogenic CT fragments of βAPP in AD; βAPP is internalized and processed in the endosomes/lysosomes where CT fragments and Aβ are produced. Aβ bearing CT fragments may be rapidly further metabolized to smaller fragments containing Aβ under the normal condition. In AD and related disorders such as Down's syndrome, excessive production of βAPP and/or reduction of some endosomal/lysosomal activities may induce accumulation of amyloidogenic CT fragments of βAPP in the neuron and/or near the membrane. Intracellular CT fragments and Aβ may form ion channels or pores in the cell membrane or puncture holes in Ca^{2+} stores. Both actions could result in a large increase in intracellular Ca^{2+} concentration and cell damage, leading to cell death. CT may attack mitochondria, which lead to the increased release of cytochrome c and the activation of caspase3 and finally to death. In addition, CT may increase the production of NO in astrocytes, which may induce cell death. CT fragments may be released from the cell and/or more easily released from the damaged neurons into the extracellular space. Extracellular CT fragments form de novo ion channels, and this may induce neuronal death from outside the cells. Finally, CT fragments may be deposited as the complete fragment or further broken down into Aβ fragments and deposited as neuritic plaques. Other metabolites of βAPP and other gene products linked with AD, e.g., apolipoprotein E, presenilin 1 and 2, and τ protein, are not included here.

Acknowledgements
This study was supported by grants-in-aid from the Ministry of Health and Welfare (HMP-98-N-6-0002), a grant from Korea Science and Engineering Foundation (1999-2004), Basic Research Fund of Korea Research Foundation (1999~2000), BK21 Human Life Sciences and SNU Hospital (1999~2000).

References

1 Selkoe DJ (1994) Alzheimer's disease: a central role for amyloid. *J Neuropathol Exp Neurosci* 60: 607–619

2 Kang J, Lemaire HG, Unterbeck A, Salbaum MN, Masters CL, Grzeschik KH, Multhaup C, Beyreuther K, Muller-Hill B (1987) The precursor of Alzheimer's disease amyloid A4 protein resembles a cell surface receptor. *Nature* 325: 733–736

3 Clemens JA, Stephenson DT (1992) Implants containing beta-amyloid protein are not neurotoxic to young and old rat brain. *Neurobiol Aging* 13: 581–586

4 Nalbantoglu J, Tirado-Santiago G, Lahsaini A, Poirier J, Gonocalves O, Verge G, Momoli F, Weiner SA, Massicotte G, Jullien JP et al (1997) Impaired learning and LTP in mice expressing the carboxyl terminus of the Alzheimer amyloid precursor protein. *Nature* 387: 500–505

5 Suh YH, Chong YH, Kim SH, Choi W, Kim KS, Jeong SJ, Fraser SP, Djamgoz MBA (1996) Molecular physiology, biochemistry, and pharmacology of Alzheimer's amyloid precursor protein (APP). *Ann N Y Acad Sci* 786: 169–183

6 Kim SH, Suh YH (1997) Meeting report: 1st international Symposium on Alzheimer's disease. *Amyloid. Int J Exp Clin Res* 4: 61–65

7 Fraser S, Suh YH, Chong YH, Djamgoz MA (1996) Membrane currents induced in Xenopus oocytes by the carboxyl terminal fragment of the β-amyloid precursor protein. *J Neurochem* 66: 2034–2040

8 Kim HJ, Suh YH, Lee MH, Ryu PD (1999) Cation selective channels formed by c-terminal fragment of β-amyloid precursor prtoein. *Neuroreport* 10(7): 1427–31

9 Hartell NA, and Suh YH (2000) Effects of fragments of β-amyloid precursor protein on parallel fiber-purkinje cell synaptic transmission in rat cerebellum. *J Neurochem* 74: 1112–1121

10 Cullen WK, Suh YH, Anwyl R, Rowan MJ (1997) Block of late-phase long-term potentiation in rat hippocampus *in vivo* by β-amyloid precursor protein fragments. *Neuroreport* 8: 3213–3217

11 Kim HS, Park CH, Suh YH (1998) C-terminal fragment of amyloid precursor protein inhibits calcium uptake into rat brain microsomes by Mg^{2+}-Ca^{2+} ATPase. *NeuroReport* 9(17): 3875–3879

12 Kim HS, Lee JH, Suh YH (1999) C-terminal fragment of Alzheimer's amyloid precursor protein inhibits sodium/calcium exchanger activity in SK-N-SH cell. *NeuroReport* 10; 113–116

13 Song DK, Won MH, Jung JS, Lee JC, Kang TC, Suh HW, Huh SO, Paek SH, Kim YH, Kim SH et al (1998) Behavioral and neuropathologic changes induced by central injection of carboxyl-terminal fragment of beta-amyloid precursor protein in mice. *J Neurochem* 71(2): 875–878

14 Yankner BA, Dawes LR, Fisher S, Villa Komaroff L, Oster Granite ML, Neve RL (1989) Neurotoxicity of a fragment of the amyloid precursor associated with Alzheimer's disease. *Science* 245: 417–420

15 Fukuchi K, Sopher B, Martin GM (1993) Neurotoxicity of beta-amyloid. *Nature* 361: 122–123

16 Fukuchi K, Kamino K, Deeb SS, Smith AC, Dang T, Martin GM (1992) Overexpression of amyloid precursor protein alters its normal processing and is associated with neurotoxicity. *Biochem Biophys Res Commun* 182: 165–173

17 Higaki J, Quon D, Zhong A, Cordell B (1995) Inhibition of beta-amyloid formation identifies proteolytic precursors and subcellular site of catabolism. *Neuron* 14: 651–659

18 Kim SH, Suh YH (1996) Neurotoxicity of a carboxy-terminal fragment of the Alzheimer's amyloid precursor protein. *J Neurochem* 67: 1172–1182

19 Fraser SP, Suh YH, Djamgoz MBA (1997) Ionic effects of the Alzheimer's disease β-amyloid precursor protein and its metabolic fragments. *Trends Neurosci* 20: 67–72

20 Suh YH (1997) An etiological role of amyloidogenic carboxyl-terminal fragments of the β-amyloid precursor protein in Alzheimer's disease. *J Neurochem* 68: 1781–1791

Neuroscientific Basis of Dementia
C. Tanaka, P.L. McGeer, Y. Ihara (eds)
© 2001 Birkhäuser Verlag Basel/Switzerland

Amyloid β-protein granules in glial cells in Alzheimer's disease brain

Haruhiko Akiyama, Hiromi Kondo, Eiko Tanno and Kenji Ikeda

Tokyo Institute of Psychiatry, Tokyo, Japan

Introduction

Amyloid β-protein (Aβ) is produced continuously in the brain under both physiological and pathological conditions. Overproduction of Aβ may be a cause of some types of familial Alzheimer's disease (FAD) and Alzheimer's disease (AD) associated with Down's syndrome. However, there is little evidence for upregulation of Aβ production in the vast majority of AD cases. In the normal brain, Aβ is removed successfully before it accumulates as amyloid fibrils. Impairment of Aβ removal could therefore be a mechanism of Aβ accumulation in AD. Examination of *post mortem* brain tissues suggests that elimination of Aβ from the brain takes place in multiple pathways.

Phagocytic removal of Aβ by microglia/macrophages

Phagocytic removal of previously deposited Aβ is seen most typically in AD brains which have been subjected to ischemia [1, 2]. In the AD cerebral cortex complicated with recent ischemia, activated microglia and monocyte-derived macrophages remove the necrotic tissue debris. Aβ accumulates in these phagocytic cells as intracellular granules. Aβ in such intracellular granules shows a peculiar immunohistochemical profile. It is negative for antibodies recognizing the amino-terminal portion of Aβ [2].

Aβ elimination has been demonstrated in some animal experiments, in which Aβ injected into rat brain is quickly taken up by reactive microglia [3, 5]. Aβ accumulates in these cells as intracellular granules and then gradually disappears. An *in vitro* study has also shown that Aβ, once deposited on the culture dishes, is partially eliminated by glial cells [4]. Aβ granules in rat microglia and macrophages are 6E10(–)/4G8(+), indicating that processing of the amino-terminal portion occurs similarly to that of the human brain [5].

In the brain of AD without complication, intracellular Aβ granules are found in some microglia that agglomerate in senile plaques. Aβ in these microglia is also 6E10(–)/4G8(+). This suggests that Aβ is taken up by reactive microglia and removed from senile plaques. Heavy accumulation of Aβ in *post mortem* brain tissues indicates that the removal is at best partial, presumably being overwhelmed by addition of newly produced Aβ in many senile plaques. In other words, senile plaques can be resolved if Aβ is not provided continuously. Such a notion seems to explain previous observations that the number of senile plaques at the time of autopsy does not correlate with the duration of illness or the severity of neuronal degeneration [6]. The number of senile plaques in the brain at a given time may reflect the balance between the deposition and elimination of Aβ.

Aβ-containing glial cells

Granular accumulation of Aβ is also found in microglia and astrocytes outside the senile plaques [7]. Aβ in these glial granules is again negative for antibodies to the amino-terminal portion of Aβ. In the brain areas where Aβ-containing glial cells are found in association with extracellular Aβ deposits, the carboxyl termini of Aβ in glial granules coincide with those in extracellular deposits, suggesting a common source for these two forms of Aβ accumulation [7, 8].

Aβ-containing glial cells sometimes appear without apparent association with the extracellular Aβ deposits. Such Aβ-containing glial cells are frequent in the hippocampal area. Another condition in which many Aβ-containing glial cells occur in clusters is a particular type of diffuse Aβ deposit which is amorphous and equally positive for Aβ40 and Aβ42. This contrasts with the ordinary, Aβ42 dominant diffuse deposits. The Aβ40 positive amorphous deposits are found in the neocortex of a number of AD patients [8] as well as some aged control cases [9, 10]. This type of diffuse Aβ deposit stains only weakly for complement proteins, C4d and C3d. The lack of heavy accumulation of complement activation products suggests that the Aβ40 positive amorphous deposits represent a very early stage of Aβ deposition [8].

In vitro studies have shown that Aβ is deposited preferentially onto already existing amyloid plaques [11]. If overproduction of Aβ occurs in areas without preformed Aβ aggregates or fibrils, Aβ may remain soluble for a longer period, raising the local concentration of Aβ in the extracellular milieu. If the extracellular concentration of Aβ is beyond the ability of intracellular degradation, but is within the ability of uptake by glial cells, granular accumulation of Aβ occurs without immunohistochemically detectable deposition of Aβ in the neuropil. If the Aβ concentration becomes even higher and exceeds the capability of uptake, amorphous Aβ deposits may appear in association with Aβ-containing glial cells. Once fibrillar Aβ deposits are formed extracellularly, newly produced Aβ precipitates quickly onto the Aβ deposits. This process, in turn, reduces the local concentration of soluble Aβ. At this point, the uptake of Aβ may still surpass the ability of degradation and Aβ-containing glial cells may persist around such deposits. If the extracellular Aβ concentration further declines and becomes even lower than the ability of Aβ degradation, Aβ deposits are no longer associated with Aβ-containing glial cells.

Multiple pathways of Aβ removal from the brain

Aβ-containing glial cells in the Aβ40 positive amorphous deposits and in areas without extracellular Aβ deposits are quiescent with respect to their expression of a variety of cell surface and cytoplasmic antigens, including HLA-D, Fcγ-receptor-1, LFA-1, CR3, CR4 and

CD68 in the case of microglia, and CD44 and ICAM-1 in the case of astrocytes [8]. This appears to contradict previous *in vitro* studies in which Aβ has been shown to activate cultured microglia. It has to be noted, however, that in such studies Aβ is used at relatively high concentrations and often in combination with other stimuli such as γ-interferon and LPS.

Occurrence of Aβ-containing, resting glial cells contrasts with activated microglia that contain Aβ granules in senile plaques and ischemic lesions. In the periphery, phagocytosis of opsonized particles is known to induce activation of the phagocytic cells. Aβ deposits in senile plaques comprise insoluble, large aggregates of amyloid fibrils. The uptake of such insoluble aggregates is likely to be mediated by phagocytosis, which is consistent with the fact that microglia in senile plaques are highly activated. Aβ accumulation in quiescent microglia in the Aβ40 positive amorphous deposits and in areas without extracellular Aβ deposits suggests that the uptake process under these conditions is mediated by another mechanism which neither requires nor induces microglial activation.

Figure 1 summarizes a hypothetical scheme of Aβ metabolism by microglia and astrocytes. As described above, the brain appears to have at least two pathways for Aβ removal.

Aβ Production, Clearance and Deposition in Alzheimer Brain

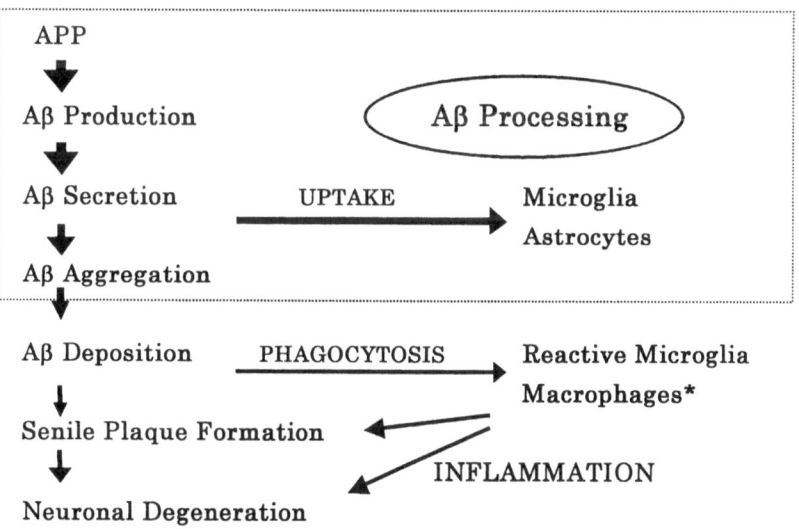

Figure 1. The presumed pathways of Aβ processing in the brain. The boxed area indicates the physiological process in which Aβ is removed by microglia and astrocytes before deposition as insoluble aggregates. This process neither induces nor requires activation of glial cells. Once deposited as insoluble aggregates, Aβ has to be removed by phagocytosis by reactive microglia. The latter process causes activation of microglia and inflammatory responses. Removal of insoluble, aggregated Aβ may be a very slow process in the brain without such complications as ischemia where both reactive microglia and infiltrated macrophages* vigorously phagocytose the Aβ deposits.

The upper half of the scheme (boxed area) illustrates a physiological process in which Aβ produced from its precursor, APP, is taken up and is cleared from brain before it deposits as insoluble aggregates. This non-phagocytic process involves both microglia and astrocytes and is not associated with activation of these cells. The lower half of the figure shows a pathological process that occurs in AD. The latter process is initiated by phagocytosis and is associated with activation of such phagocytic cells as microglia. In addition to the Aβ-immunization to promote phagocytic Aβ removal [12], a quest for a treatment that enhances non-phagocytic removal of Aβ by microglia and astrocytes would be another direction for the development of anti-AD treatments.

References

1 Wisniewski HM, Barcikowska M, Kida E (1991) Phagocytosis of β/A4 amyloid fibrils of the neuritic neo-cortical plaques. *Acta Neuropathol* 81: 588–590
2 Akiyama H, Kondo H, Mori H, Kametani F, Nishimura T, Ikeda K, Kato M, McGeer PL (1996) The amino-terminally truncated forms of amyloid β-protein in brain macrophages in the ischemic lesions of Alzheimer's disease patients. *Neurosci Lett* 219: 115–118
3 Frautschy SA, Cole GM, Baird A (1992) Phagocytosis and deposition of vascular β-amyloid in rat brains injected with Alzheimer β-amyloid. *Amer J Pathol* 140: 1389–1399
4 Shaffer LM, Dority MD, Gupta-Bansal R, Frederickson RCA (1995) Amyloid β protein (Aβ) removal by neuroglial cells in culture. *Neurobiol Aging* 16: 737–745
5 Akiyama H, Kondo H, Tannno E, Ikeda K, McGeer PL (1998) Uptake and intracellular accumulation of amyloid β-protein (Aβ) by microglia. *Soc Neurosci Abstr* 24 (1): 724
6 Hyman BT, Marzloff K, Arriagada PV (1993) The lack of accumulation of senile plaques or amyloid burden in Alzheimer's disease suggests a dynamic balance between amyloid deposition and resolution. *J Neuropathol Exp Neurol* 52: 594–600
7 Akiyama H, Schwab C, Kondo H, Mori H, Kametani F, Ikeda K, McGeer PL (1996) Granules in glial cells of patients with Alzheimer's disease are immunopositive for C-terminal sequences of beta-amyloid protein. *Neurosci Lett* 206: 169–172
8 Akiyama H, Mori M, Saido T, Kondo H, Iked K, McGeer PL (1999) Occurrence of the diffuse amyloid β-protein (Aβ) deposits with numerous Aβ-containing glial cells in the cerebral cortex of patients with Alzheimer's disease. *Glia* 25: 324–331
9 Yamaguchi H, Sugihara S, Ogawa A, Saido TC, Ihara Y (1998) Diffuse plaques associated with astroglial amyloid β protein, possibly showing a disappearing of senile plaques. *Acta Neuropathol* 95: 217–222
10 Funato H, Yoshimura M, Yamazaki T, Saido TC (1998) Astrocytes containing amyloid β-protein (Aβ)-positive granules are associated with Aβ40-positive diffuse plaques in the aged human brain. *Amer J Pathol* 152: 983–992
11 Maggio JE, Stimson ER, Ghilardi JR, Allen CJ, Dahl CE, Whitcomb DC, Vigna SR, Vinters HV, Labenski ME, Mantyg PW (1992) Reversible *in vitro* growth of Alzheimer disease β-amyloid plaques by deposition of labeled amyloid peptide. *Proc Natl Acad Sci USA* 89: 5462–5466
12 Schenk D, Barbour R, Dunn W, Gordon G (1999) Immunization with amyloid-β attenuates Alzheimer-disease-like pathology in the PDAPP mouse. *Nature* 400: 173–177

Neuroscientific Basis of Dementia
C. Tanaka, P.L. McGeer, Y. Ihara (eds)
© 2001 Birkhäuser Verlag Basel/Switzerland

Amyloid β induces phosphorylation and translocation of MARCKS through tyrosine kinase-activated PKC-δ signaling pathway in microglia

Masamichi Nakai[1], Satoshi Tanimukai[1], Keiko Yagi[2, 3], Naoaki Saito[2], Hiroshi Hasegawa[1], Akira Terashima[1] and Chikako Tanaka[1]

[1] Hyogo Institute for Aging Brain and Cognitive Disorders, Himeji, Japan
[2] Biosignal Research Center, Kobe University, Kobe, Japan
[3] Department of Clinical Pharmacy, Kobe Pharmaceutical University, Kobe, Japan

The accumulation of activated microglia containing amyloid β protein (Aβ) around senile plaques is a common pathological feature in patients with Alzheimer disease (AD). Aβ activates microglia and enhances its chemotaxis, phagocytosis, cytokine release and respiratory burst. We reported that protein kinase C (PKC) is involved in Aβ-induced microglial chemotaxis [1]. We also found that Aβ(25-35) and Aβ(1-42) phosphorylate myristoylated alanine-rich C kinase substrate (MARCKS), a specific PKC substrate, which is localized in cultured rat microglia by immunocytochemistry and Western blots using specific antibodies [2].

MARCKS is phosphorylated during neurosecretion, growth factor-dependent mitogenesis and phagocyte activation [3]. MARCKS binds to calcium/calmodulin, crosslinks actin filaments and is activated by phosphorylation followed by translocation from the membrane fraction to the cytosolic fraction [4–6]. In macrophages, MARCKS is distributed in pseudopodia and filopodia, and is phosphorylated when macrophages are activated [7]. These findings led to the suggestion that Aβ phosphorylates MARCKS through a PKC signaling pathway and induces microglial activation.

To identify the intracellular signaling pathway for microglial activation in response to Aβ, we asked how Aβ activates MARCKS in rat microglia. Treatment of microglia with Aβ(25-35) and Aβ(1-42) at 10 nM or 12-O-tetradecanoyl-phorbol-13-acetate (TPA) (1 ng/ml) induced the phosphorylation of MARCKS, an event inhibited by a specific inhibitor for PKC, staurosporine. Phosphorylation of MARCKS by Aβ was inhibited by the tyrosine kinase inhibitors, genistein and herbimycin A, but not by pertussis toxin. Furthermore, PKC-δ was tyrosine-phosphorylated by Aβ(25-35) at 10 nM. As the pretreatment with herbimycin A at 0.1 μM also inhibited the Aβ(25-35)-induced chemotaxis of microglia, we proposed that a tyrosine kinase-activated PKC pathway is involved in the Aβ-induced microglial activation [2].

The localization of PKC isoforms has been studied extensively in the mammalian brain, however, there is no direct evidence of the occurrence of PKC isoforms in microglia. We demonstrated that PKC isoforms α, δ and ε were expressed in cultured rat microglia [2]. Both Aβ(25-35) and Aβ(1-40) significantly increased the kinase activity of only PKC-δ in microglia. PKC is classified into three groups, classical, new and atypical PKC. PKC-δ, which belongs to the new PKC group, is activated and translocated from the cytosol to the membrane fraction upon activation by TPA or diacylglycerol in a Ca^{2+}-independent manner. Moreover, there is another pathway of PKC-δ activation. Some signals are reported to increase the activity of PKC-δ by the phosphorylation of tyrosine residues [8, 9]. These findings suggest that Aβ activated PKC-δ through phosphorylation of tyrosine residue. None of the activators of PKCδ in this manner is, however, found to induce the translocation of PKCδ from the cytosol to the membrane [10].

The affinity of MARCKS for PKC-δ has been shown to be extremely high, as determined using overlay assay [5]. However, there is no evidence for the direct interaction of PKC with MARCKS *in vivo*. We found that MARCKS is co-immunoprecipitated with PKC-δ. These data suggest that PKC-δ is activated through tyrosine phosphorylation and formed a complex with MARCKS in response to Aβ stimulation.

Activation of PKC results in MARCKS phosphorylation and the translocation from the plasma membrane to the perinuclear area where lamp-1 positive lysosome is localized in fibroblasts [6]. As shown in Figure 1, both Aβ(25-35) and Aβ(1-40) translocated MARCKS from the membrane fraction to the cytosolic fraction. Immunocytochemistry with a specific antibody against phospho-MARCKS showed that phosphorylated MARCKS accumulated in

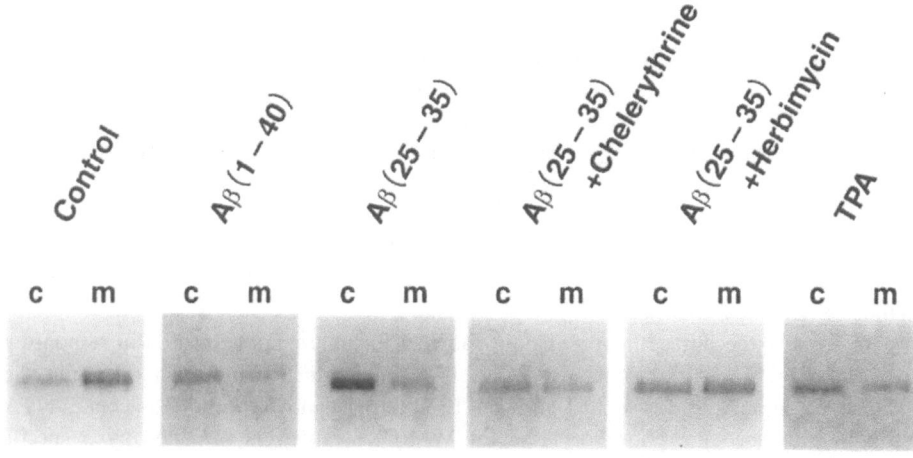

Figure 1. Subcellular fractionation of MARCKS in microglia. Cells were metabolically labeled with [³H]-myristic acid and treated with reagents. Then, the cells were prepared in cytosolic (c) and membranous (m) fractions, fractionated by 10% SDS-PAGE, transferred to PVDF membrane and exposed to X-ray film for one week. Cells were treated with vehicle, Aβ (1–40) at 10 nM, Aβ (25–35) at 10 nM, TPA at 16 nM, Aβ (25–35) at 10 nM for 5 min with prior incubation with chelerythrine at 5 μM for 15 min or herbimycin A at 1 μM for 8 h.

▶ ▶ Figure 2. Confocal microscopic analysis of rat microglia with anti-phospho-MARCKS antibody plus anti-rabbit IgG antibody conjugated with Cy-3. Control microglia showed little phospho-MARCKS immunoreactivity (A). Phase contrast microscopy of control microglia are shown in (B). By stimulation of Aβ (1–40) at 10 nM for 1 min (C) and 5 min (D), Aβ (25–35) at 10 nM for 1 min (E) and 5 min (F), or TPA at 100 nM (G) for 1 min, phospho-MARCKS immunoreactivity appeared around the nuclei of microglia. Scale bar, 10 μm.

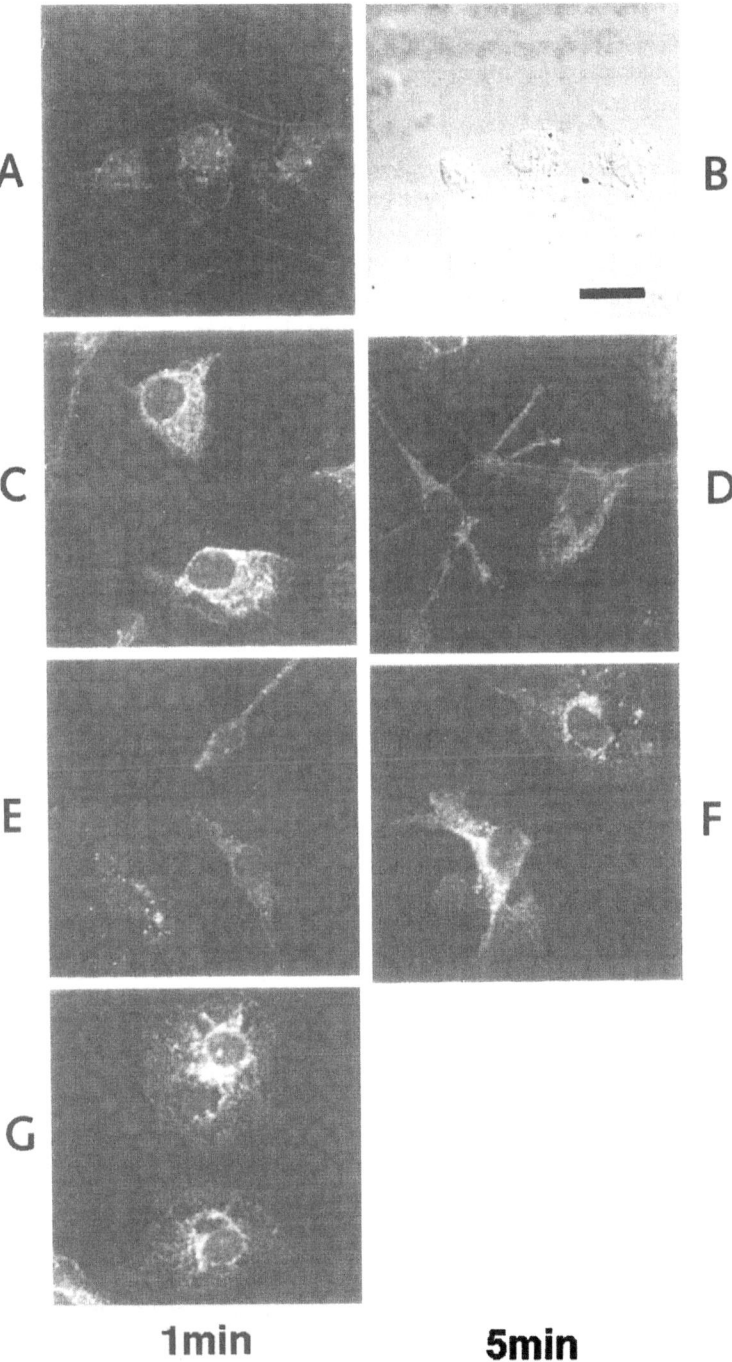

1min   **5min**

the cytoplasm, particularly at the perinuclear region in microglia treated with Aβ (Fig. 2). Considering the activation of PKC-δ by Aβ and the accumulation of phosphorylated MAR-CKS at the perinuclear area in the cytoplasm upon Aβ stimulation, we propose that Aβ induces the translocation of MARCKS through the tyrosine kinase-activated PKC-δ signaling pathway.

Although the downstream path of MARCKS in the Aβ signaling pathway in microglia remains to be clarified, the translocation of phosphorylated MARCKS to the perinuclear area suggests a role in cell motility and phagocytosis. The phosphorylated MARCKS can bind actin filaments [6]. MARCKS is phosphorylated during cell movement, phagocytosis and neurosecretion [11] and is associated with phagosome formation, which is inhibited by PKC inhibitors in macrophages. These findings suggest that Aβ-induced translocation of phosphorylated MARCKS is involved in microglial activation upon Aβ-stimulation.

In summary, stimulation of microglia with Aβ induces the increased kinase activity of PKC-δ and the co-immunoprecipitation of PKC-δ with MARCKS. MARCKS is phosphorylated by Aβ and translocated from the membrane to the perinuclear cytoplasm, which is inhibited by PKC inhibitors and tyrosine kinase inhibitors. From these findings, we propose that Aβ induces phosphorylation and translocation of MARCKS through the tyrosine kinase-activated PKC-δ signaling pathway, resulting in microglia activation.

References

1 Nakai M, Hojo K, Taniguchi T, Terashima A, Kawamata T, Hashimoto T, Maeda K, Tanaka C (1998) PKC and tyrosine kinase involvement in amyloid β (25–35)-induced chemotaxis of microglia. *Neuroreport* 9: 3467–3470

2 Nakai M, Hojo K, Yagi K, Saito N, Taniguchi T, Terashima A, Kawamata T, Hashimoto T, Maeda K, Gschwendt M et al (1999) Amyloid β protein (25–35) phosphorylates MARCKS through tyrosine kinase-activated protein kinase C signaling pathway in microglia. *J Neurochem* 72: 1179–1186

3 Hartwig JH, Thelen M, Rosen A, Janmey PA, Nairn AC, Aderem A (1992) MARCKS is an actin filament crosslinking protein regulated by protein kinase C and calcium-calmodulin. *Nature* 356: 618–622

4 Thelen M, Rosen A, Nairn AC, Aderem A (1991) Regulation by phosphorylation of reverse association of a myristoylated protein kinase C substrate with the plasma membrane. *Nature* 351: 320–322

5 Fujise A, Mizuno K, Ueda Y, Osada S-I, Hirai S-I, Takayanagi A, Shimizu N, Owada MK, Nakajima H, Ohno S (1994) Specificity of the high affinity interaction of protein kinase C with a physiological substrate, myristoylated alanine-rich protein kinase C substrate. *J Biol Chem* 269: 31642–31648

6 Allen LH, Aderem A (1995) Protein kinase C regulates MARCKS cycling between the plasma membrane and lysosomes in fibroblasts. *EMBO J* 6: 1109–1121

7 Rosen A, Keenan KF, Thelen M, Nairn AC, Aderem A (1990) Activation of protein kinase C results in the displacement of its myristoylated, alanine-rich substrate from punctate structures in macrophage filopodia. *J Exp Med* 17: 1211–1215

8 Li W, Yu JC, Michieli P, Beeler JF, Ellmore N, Heidaran MA, Pierce JH (1994) Stimulation of the platelet-derived growth factor β receptor signaling pathway activates protein kinase C-δ. *Mol Cell Biol* 14: 6727–6735

9 Konishi H, Tanaka M, Takemura H, Matsuzaki H, Ono Y, Kikkawa U, Nishizuka Y (1997) Activation of protein kinase C by tyrosine phosphorylation in response to H_2O_2. *Proc Natl Acad Sci USA* 94: 11223–11237

10 Ohmori S, Shirai Y, Sakai N, Fujii M, Konishi H, Kikkawa U, Saito N (1998) Three distinct mechanisms for translocation and activation of the δ subspecies of protein kinase C. *Mol Cell Biol* 18: 5263–5271

11 Wiederkehr A, Staple J, Caroni P (1997) The motility-associated proteins GAP-43, MARCKS, and CAP-23 share unique targeting surface activity-inducing properties. *Exp Cell Res* 236: 103–116

Neuroscientific Basis of Dementia
C. Tanaka, P.L. McGeer, Y. Ihara (eds)
© 2001 Birkhäuser Verlag Basel/Switzerland

Amyloid β-protein accumulation in the human brain during aging

Maho Morishima-Kawashima[1] and Yasuo Ihara[1, 2]

[1] Department of Neuropathology, Faculty of Medicine, University of Tokyo, Tokyo, Japan
[2] Core Research for Evolutional Science and Technology (CREST), Japan Science and Technology Corporation, Saitama, Japan

Introduction

Alzheimer's disease (AD) is the most common dementia among elderly people. About ~90% of AD cases are sporadic, and the remaining ~10% are familial with definite genetic backgrounds. The neuropathological hallmarks of the brain affected by sporadic AD or familial AD are abundant senile plaques and neurofibrillary tangles throughout the cortex and extensive loss of selected populations of the neuron. All the three causative genes associated with familial AD thus far identified are thought to accelerate the formation of senile plaques [1].

Senile plaques are composed mainly of amyloid β-protein (Aβ), a small protein of M_r ~4000. Aβ is derived from β-amyloid precursor protein (APP), a type I transmembrane protein, through sequential cleavages. Two proteases are postulated for generating Aβ from APP: β-secretase which clips the amino-terminus of Aβ [2] and γ-secretase which clips the carboxy-terminus of Aβ. As a result, two major species of Aβ are generated; Aβ40 ending at Val40 and Aβ42 ending at Ala42 with the latter being more amyloidogenic.

To understand the underlying mechanism of Aβ deposition, we have quantified the levels of Aβ in the brains from many autopsied subjects of various ages using sensitive two-site enzyme-linked immunosorbent assay (ELISA) and sought to reconstitute a possible temporal profile of Aβ deposition in a given individual.

Aβ levels correlate with the extent of senile plaque formation

We first quantified Aβ levels in cortices from elderly subjects who exhibited various stages of Aβ deposition and from AD patients [3]. Small pieces of occipital cortex (Brodmann area 17) were extracted with Tris-saline (soluble fraction) and the insoluble residues were extracted with formic acid (insoluble fraction). Aβ levels in both soluble and insoluble fractions were determined by two-site ELISA. Cortical blocks from the adjacent area or from the same locations on the contralateral side were fixed in 10% formalin and subjected to histological and immunocytochemical examinations. The obtained Aβ values were compared with the abundance of senile plaques scored by immunocytochemistry (Fig. 1).

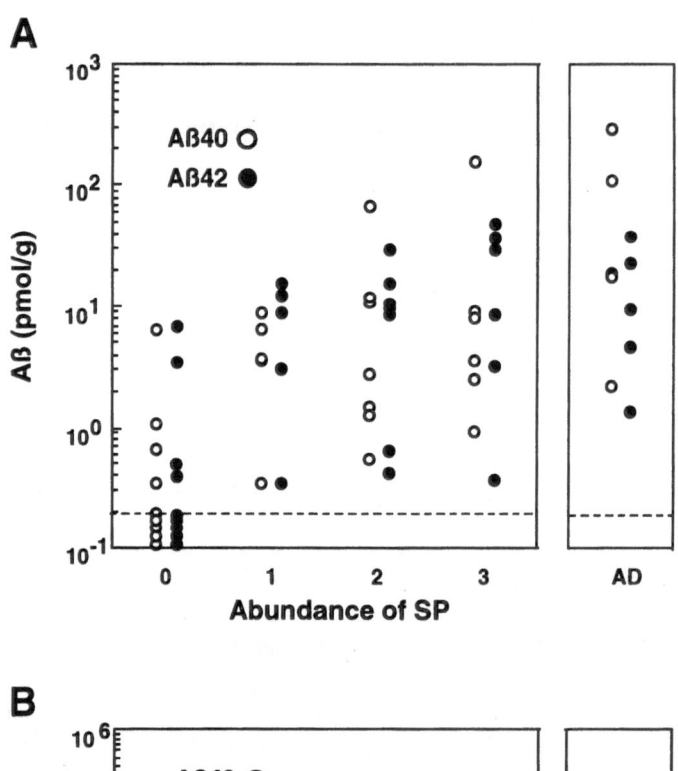

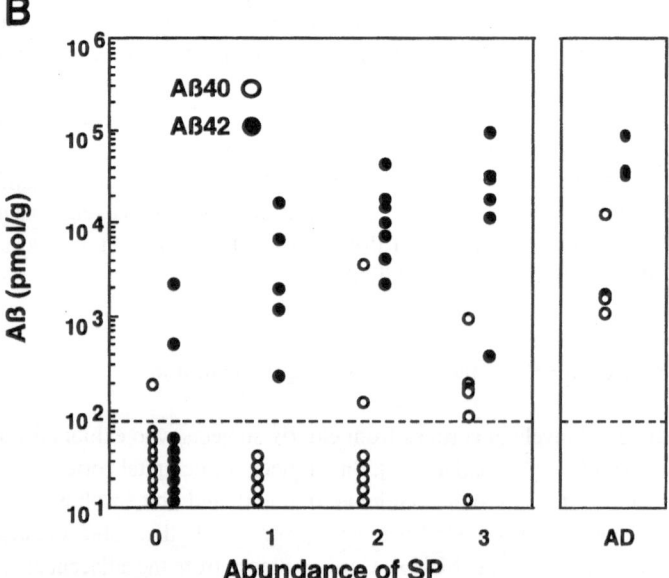

Figure 1. Levels of Aβ in the occipital cortex and the abundance of senile plaques. Aβ40 (○) and Aβ42 levels (●) in the soluble (A) and insoluble fractions (B) of the occipital cortex were quantified by ELISA. The number of senile plaques was counted and their abundance was graded as 0-3. The far-right column on each panel shows the levels of Aβ in AD cortices.

There was a significant correlation between the abundance of senile plaques and the levels of soluble Aβ40 or Aβ42 (Fig. 1A). Soluble Aβ40 and Aβ42 were at very similar levels in the control brains. The ratios of soluble Aβ40/Aβ42 in AD brains were significantly elevated compared with those in control brains.

In contrast to the soluble fraction, the insoluble fraction contained remarkably high levels of Aβ, and there was a marked predominance of Aβ42 over Aβ40 in the insoluble fraction, irrespective of the stage of senile plaque formation (Fig. 1B). The levels of Aβ in the insoluble fraction appeared to be proportional to the abundance of senile plaques. Although the Aβ42 levels in AD cortices were significantly higher than those in control subjects containing low levels of Aβ (stage 0 and 1), there was no significant difference in those levels between control brains containing higher levels of Aβ (stage 3) and AD brains. This may indicate that Aβ42 levels reach the plateau at a later stage of Aβ deposition even in AD subjects.

It should be noted that significant amounts of insoluble Aβ42 were detected by ELISA in some plaque-free brains. This indicates that biochemically detectable Aβ accumulation precedes immunocytochemically detectable Aβ deposition in the occipital cortex.

Age-dependent accumulation of Aβ42

To develop a possible longitudinal profile of Aβ deposition during aging and in AD, and learn of regional differences in Aβ deposition, we next quantified Aβ levels in several brain regions from consecutively autopsied subjects with ages ranging from 24–92 years, and several AD subjects [4, 5].

Pieces of hippocampus CA1 and occipitotemporal cortex T4 at the level of the lateral geniculate body were sampled. The Aβ levels were quantified by two-site ELISA and corresponding areas were immunocytochemically examined [4].

Aβ in CA1 and T4 was characterized by greater insolubility as was also the case in the occipital cortex. There was a strong association of the levels of insoluble Aβ42 in T4 or CA1 with age: a strong tendency toward Aβ42 accumulation was observed between the ages of 50 and 70 years in T4, and a little later in CA1 (Fig. 2 B and D). The levels of soluble Aβ42 in T4 and CA1 showed a similar age-dependent increase, although the levels were much lower than those of insoluble Aβ42. In a given case, the Aβ42 level was consistently higher in T4 than that in CA1. This agrees with the immunocytochemical observation that hippocampus CA1 is the region least affected by senile plaques, while medial occipitotemporal cortex (T4) is among the most vulnerable regions [6]. In this series, the levels of Aβ42 in T4 and CA1 in AD brains were significantly higher than those in corresponding areas in control brains, and the extent of Aβ42 amino-terminal modifications was much greater in AD brains than those in control brains.

Compared with the levels of Aβ42, the levels of insoluble Aβ40 were lower by orders of magnitude and no apparent age dependence of Aβ40 accumulation was observed (Fig. 2 A and C). Increased levels of insoluble (and also soluble) Aβ40 in CA1 and T4 were tightly associated with AD.

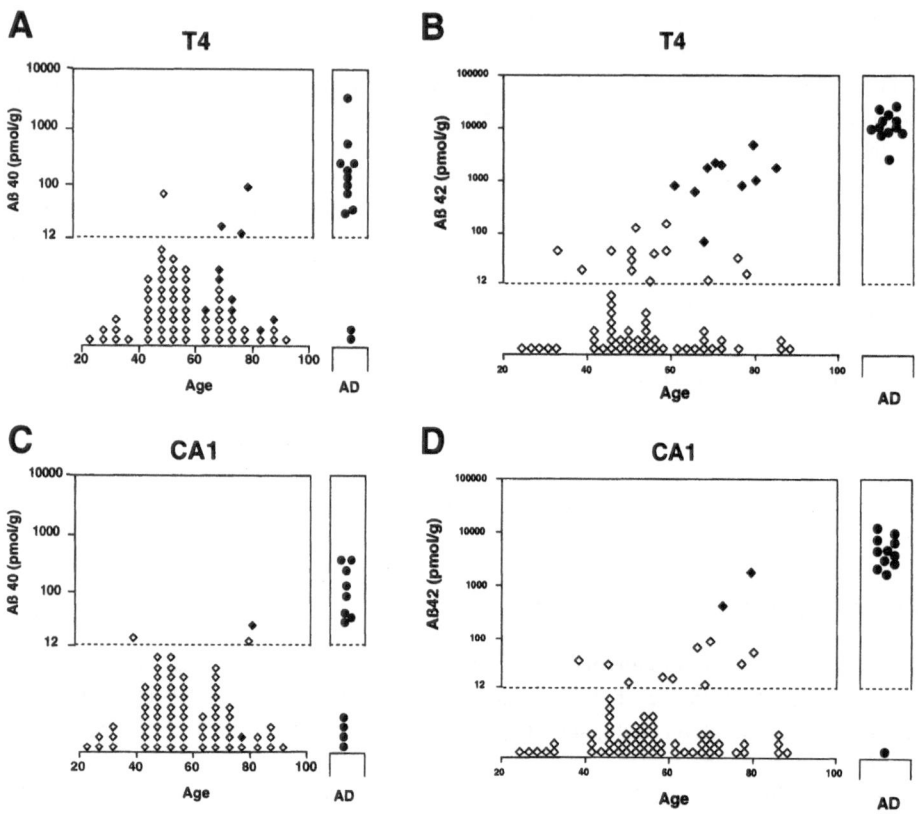

Figure 2. Levels of insoluble Aβ40 (A and C) and Aβ42 (B and D) in T4 (A and B) and CA1 (C and D). Control subjects without senile plaques are indicated by open diamonds and control subjects with senile plaques are by closed diamonds. AD cases are shown by closed circles in the right column of each panel.

The levels of insoluble Aβ42, but not soluble Aβ42, soluble Aβ40, or insoluble Aβ40, were found to be strongly correlated with the amyloid burden, the percentage of amyloid-deposited area to the total area, in T4 and CA1. Even if senile plaques were immunocyto-chemically undetectable, Aβ accumulation was already detectable by ELISA as seen in the occipital cortex (Fig. 2 B and D). Indeed, a regression line showed an apparent critical con-centration of Aβ42 for senile plaque formation; immunocytochemically detectable senile plaques required more than ~400 pmol/g of insoluble Aβ42 in T4 and ~200 pmol/g in CA1. Thus, Aβ accumulation as detected by ELISA precedes the formation of immunocytochem-ically detectable senile plaques, and this histological hallmark represents the consequence of the preceding Aβ accumulation in the cortex.

In contrast to Aβ deposition in the cortex, less attention has been paid to Aβ deposition in deeply located areas in the brain. This is probably because only much smaller numbers of senile plaques are found in the subcortical regions as compared with those in the cortex. We chose the two subcortical regions, the putamen and mammillary body, for the quantification of Aβ levels [5]. Although Aβ accumulated in the two subcortical regions to a much lesser extent than in the cortex of the same subject, the accumulation appeared to occur at the same time as in the cortex. In addition, the Aβ42 levels in the putamen and mammillary body correlated with those in the cortex. This strongly suggests that the levels of Aβ42 in the brain are determined not only by the duration of Aβ accumulation, but also by other as-yet-unidentified regional factors. Thus, it is tempting to speculate that similar, but not the same, mechanisms of increase of Aβ42 levels are activated during the same critical period in all brain regions.

Insoluble Aβ levels increase during aging

In the above studies, we were unable to accurately quantify low levels of Aβ in the brain, and those levels in many subjects fell into "under the detection limit". The reason is that formic acid was used for extraction of insoluble Aβ, which required a large volume of alkaline solution for neutralization, resulting in decreased sensitivity of ELISA. Thus, we have employed a guanidine HCl extraction protocol for the quantification of accurate levels of Aβ in normal brains to learn more about the initial phase of Aβ accumulation. Guanidine HCl was found to be even more effective to extract low levels of Aβ than formic acid and provide reproducible results. Insoluble Aβ was definitely detectable even in the brains from younger subjects using the new protocol. Furthermore, there was a good correlation between Aβ levels determined by formic acid and those by guanidine HCl protocols in case Aβ levels were relatively high.

Insoluble Aβ levels determined by guanidine HCl extraction were plotted against age at death. Both Aβ40 and Aβ42 were detectable in the insoluble fraction of brains from young subjects aged 20–30 years (Fig. 3). Unexpectedly, Aβ40 levels were several-fold higher than Aβ42 levels in those fractions. The Aβ levels appeared to be stable during the next 20 years and started to rise in some brains at the age of late 40 years in an exponential manner. This age-dependent increase was observed in Aβ42 as well as Aβ40, although an increase in the Aβ42 levels was much steeper than that in the Aβ40 levels. When the Aβ40 and Aβ42 levels were compared with each other, both Aβ40 and Aβ42 levels were found to increase in a coordinated manner during aging.

Effects of ApoE allele ε4 on Aβ accumulation

ApoE allele ε4 is known to be a strong genetic risk factor for developing AD and is associated with decreased age of onset [7, 8]. The ApoE genotype is considered to modify the process of Aβ deposition [9, 10] and ApoE ε4 is reported to be associated with an increased extent of Aβ40 deposition in AD brains [11, 12]. Thus, we next investigated the effect of ApoE alleles on Aβ accumulation during aging.

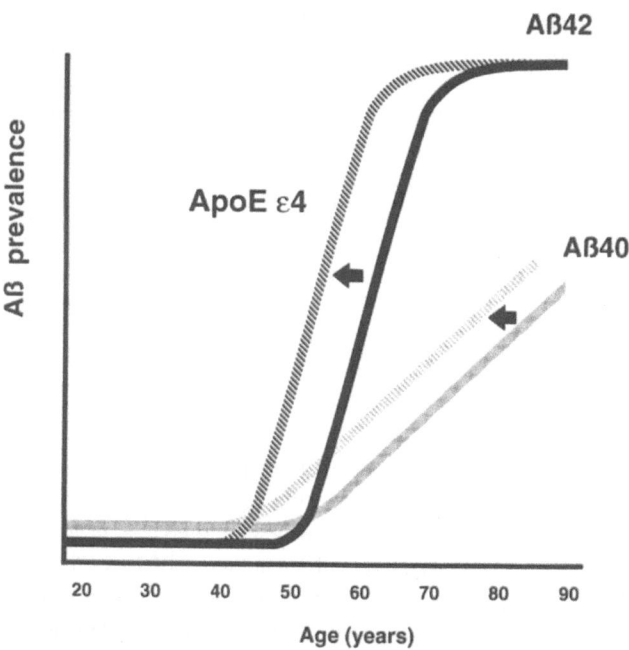

Figure 3. Schematic illustration of Aβ40 and Aβ42 accumulation during aging (fixed lines) and the effects of ε4 allele of ApoE (hatched lines).

When the cut-off line was drawn at 5 pmol/g, both Aβ40 and Aβ42 levels in all the sub-jects (10 cases), irrespective of ApoE genotypes, were below the line under the age of 40 years. Between the ages of 40 and 50 years, Aβ42 levels in 3 out of 8 ε4-carrying subjects (37.5%) were above the line, and those in all of the ε4 noncarrier subjects, except for one subject (13 of 14 cases, 92.9%), were still below the line. Above the age of 50 years, most of the ε4-carrying subjects (15 of 17 cases, 88.2%) significantly accumulated Aβ42 and their Aβ42 levels were above the line. Those from ε4 noncarriers were distributed below and above the line with a greater number (27 of 39 cases, 69.2%) staying below the line between the age of 50 and 70 years. Only above the age of 70 years, Aβ42 levels in a half of ε4 non-carrier (8 of 16 cases, 50%) were above the line. This strongly suggests that ε4 accelerates Aβ42 accumulation, and apparently sets the timing of Aβ42 accumulation earlier during life (Fig. 3).

In contrast to Aβ42, the effect of ε4 on Aβ40 levels was not so obvious. Nevertheless, a similar tendency was observed in Aβ40 accumulation. Between the ages of 40 and 60 years, Aβ40 levels in 5 out of 11 ε4-carrying subjects (45.5%) were above the line, and those in 25 out of 33 ε4 noncarriers (75.8%) were below the line. Above the age of 60 years, Aβ40 lev-els in 11 out of 14 ε4-carrying subjects (78.6%) were above the line. As for the case of ε4 noncarriers, those in 4 of 20 (20.0%) and 8 of 16 (50.0%) were above the line between the age of 60 and 70 years and above the age of 70 years, respectively.

These results indicate that ε4 accelerates the process of Aβ42 accumulation, and presumably that of Aβ40 accumulation, during life (Fig. 3). Thus, ApoE allele ε4, a risk factor for AD, has effects on Aβ deposition similar to the causative genes of AD, APP and presenilins, probably through different underlying mechanisms. Our results are not contradictory to the previous reports describing that the levels of Aβ40 in sporadic AD brains are increased in an ε4 allele dosage-dependent manner [11, 12]. Because Aβ42 levels appear to reach the plateau during aging (see above), there may not be significant differences in the levels of Aβ42 between ε4-carrying AD subjects and ε4-noncarrying ones of older ages. The only difference between them would be the levels of Aβ40: ε4-carrying AD subjects might have started to accumulate Aβ42 earlier and accordingly acquired a greater extent of Aβ40.

The presence of Aβ in the detergent-insoluble, low-density membrane fraction

An important question arises as to the origin of insoluble Aβ in the human brain. We previously showed that in SH-SY5Y cells, significant amounts of Triton X-100 insoluble Aβ are localized to the low-density membrane domain [13]. The domain is rich in glycosphingolipid and cholesterol, and is usually called caveolae because of a characteristic invagination of plasma membrane [14]. But in the central nervous system no similar invagination of plasma membrane is observed, and thus it is referred to here as the detergent-insoluble, low-density membrane domain.

The distinct membrane domains are easily isolated by floating up by sucrose density-gradient centrifugation in the presence of Triton X-100 [15]. According to the protocol, human brain homogenates were fractionated in the presence of Triton X-100. In normal younger cases, significant fractions of Aβ, especially Aβ40, were associated with the low-density membrane domain. During aging, the low-density membrane appeared to accumulate Aβ42 preferentially. This suggests that one pathway to Aβ42 accumulation is mediated through abnormal metabolism of Aβ42 in the low-density membrane domain.

This assumption is supported by our recent observation that the increased levels of Aβ42 are associated with the low-density membrane domain in the brains of transgenic mice harboring mutant *presenilin 2* [16]. Thus, the low-density membrane domain may be involved in the initial phase of Aβ42 accumulation, presumably through altered lipid composition, especially ist cholesterol and sphingolipid contents.

References

1 Scheuner D, Eckman C, Jense M, Song X, Citron M, Suzuki N, Bird TD, Hardy J, Hutton M, Kukull W et al (1996) Secreted amyloid β-protein similar to that in the senile plaques of Alzheimer's disease is increased *in vivo* by the presenilin 1 and 2 and APP mutations linked to familial Alzheimer's disease. *Nat Med* 2: 864–870
2 Vassar R, Bennett BD, Babu-Khan S, Kahn S, Mendiaz EA, Denis P, Teplow DB, Ross S, Amarante P, Loeloff R et al (1999) β-secretase cleavage of Alzheimer's amyloid precursor protein by the transmem-

brane aspartic protease BACE. *Science* 286: 735–741

3 Shinkai Y, Yoshimura M, Morishima-Kawashima M, Ito Y, Shimada H, Yanagisawa K, Ihara Y (1997) Amyloid β-protein deposition in the leptomeninges and cerebral cortex. *Ann Neurol* 42: 899–908

4 Funato H, Yoshimura M, Kusui K, Tamaoka A, Ishikawa K, Ohkoshi N, Namekata K, Okeda R, Ihara Y (1998) Quantification of amyloid β-protein in the cortex during aging and in Alzheimer's disease. *Amer J Pathol* 152: 1633–1640

5 Nakabayashi J, Yoshimura M, Morishima-Kawashima M, Funato H, Miyakawa T, Yamazaki T, Ihara Y (1998) Amyloid β-protein (Aβ) accumulation in the putamen and mammillary body during aging and in Alzheimer disease. *J Neuropathol Exp Neurol* 57: 343–352

6 Braak H, Braak E (1991) Neuropathological staging of Alzheimer-related changes. *Acta Neuropathol* 82: 239–259

7 Strittmatter WJ, Saunders AM, Schmechel D, Pericak-Vance M, Enghild J, Salvesen GS, Roses AD (1993) Apolipoprotein E: high-avidity binding to β-amyloid and increased frequency of type 4 allele in late-onset familial Alzheimer disease. *Proc Natl Acad Sci USA* 90: 1977–1981

8 Corder EH, Saunders AM, Strittmatter WJ, Schmechel DE, Gaskell PC, Small GW, Roses AD, Haines JL, Pericak-Vance MA (1993) Gene dose of apolipoprotein E type 4 allele and the risk of Alzheimer's disease in late-onset families. *Science* 261: 921–923

9 Schmechel DE, Saunders AM, Strittmatter WJ, Crain BJ, Hulette CM, Joo SH, Pericak-Vance MA, Goldgaber D, Roses AD (1993) Increased amyloid β-peptide deposition in cerebral cortex as a consequence of apolipoprotein E genotype in late-onset Alzheimer disease. *Proc Natl Acad Sci USA* 90: 9649–9653

10 Rebeck GW, Reiter JS, Strickland DK, Hyman BT (1993) Apolipoprotein E in sporadic Alzheimer's disease: allelic variation and receptor interactions. *Neuron* 11: 575–580

11 Gearing M, Mori H, Mirra SS (1997) Aβ-peptide length and apolipoprotein E genotype in Alzheimer's disease. *Ann Neurol* 39: 395–399

12 Ishii K, Tamaoka A, Mizusawa H, Shoji S, Ohtake T, Fraser PE, Takahashi H, Tsuji S, Gearing M, Mizutani T et al (1997) Aβ1-40 but not Aβ1-42 levels in cortex correlate with apolipoprotein E ε4 allele dosage in sporadic Alzheimer's disease. *Brain Res* 748: 250–252

13 Morishima-Kawashima M, Ihara Y (1998) The presence of amyloid β-protein in the detergent-insoluble membrane compartment of human neuroblastoma cells. *Biochemistry* 37: 15247–15253

14 Simons K, Ikonen E (1997) Functional rafts in cell membranes. *Nature* 387: 569–572

15 Lisanti MP, Scherer PE, Vidugiriene J, Tang Z, Hermanowski-Vosatka A, Tu YH, Cook RF, Sargiacomo M (1994) Characterization of caveolin-rich membrane domains isolated from an endothelial-rich source: implications for human disease. *J Cell Biol* 126: 111–126

16 Sawamura N, Morishima-Kawashima M, Waki H, Kobayashi K, Kuramochi T, Frosch MP, Ding K, Ito M, Kim T-W, Tanzi RE et al (2000) Mutant-presenilin 2-transgenic mice: A large increase in the levels of Aβ42 is presumably associated with the low-density membrane domain that contains decreased levels of glycerophospholipids and sphingomyelin. *J Biol Chem; in press*

Neuroscientific Basis of Dementia
C. Tanaka, P.L. McGeer, Y. Ihara (eds)
© 2001 Birkhäuser Verlag Basel/Switzerland

Molecular mechanisms underlying initiation of amyloid fibril formation

Katsuhiko Yanagisawa

Department of Dementia Research, National Institute for Longevity Sciences, Obu, Japan

Introduction

Deposition of aggregated amyloid β-protein (Aβ), a proteolytic cleavage product of the amyloid precursor protein (APP), is a fundamental pathological event in the development of Alzheimer's disease (AD) [1]. Although a great deal of effort has been made to clarify the pathogenesis of AD, we are still far from a complete understanding of the molecular mechanism underlying the initiation of amyloid fibril formation. In regard to the mechanism of protein aggregation, the nucleation-dependent polymerization theory is widely accepted [2]. In this model, a long lag-time is required for nucleation. Furthermore, the concentration of a given protein is required to be greater than a critical level for nucleation. In regard to the polymerization of Aβ *in vitro*, Aβ at concentrations below 20 μM is unlikely to aggregate spontaneously. Considering that the physiological concentration of Aβ in biological fluids, including the cerebrospinal fluid, sera and culture media, is as low as 10 nM, even in the case of the expression of some genes responsible for familial AD which could induce increased generation of Aβ [3–6], one would assume that Aβ probably aggregates in brains in the presence of pathological chaperones such as apolipoprotein E4, α1-antichymotrypsin and proteoglycans. Alternatively, a pathological Aβ species with the ability to act as a seed or a template for fibril formation could be generated via a conformational alteration of Aβ, as has been postulated for the development of prion diseases [7]. In this chapter, I describe two novel Aβ species with seeding ability, which were identified in the brains with AD or in culture media in our studies [8–10].

GM1 ganglioside-bound amyloid β-protein in the brain of cases with Alzheimer's disease

To address the molecular mechanism underlying fibril formation of Aβ in the brains of cases with AD, we fractionated cerebral cortices from AD and Down's syndrome brains, showing early pathological changes of AD, by sucrose-density gradient fractionation. We identified a novel Aβ species which was characterized by its tight binding to GM1 ganglioside (GM1 ganglioside-bound Aβ, GM1/Aβ) (Fig. 1). The GM1/Aβ showed unique molecular characteristics, including an extremely high aggregation potential and altered immunoreactivity. Based on these molecular characteristics, we hypothesized that GM1/Aβ adopts an altered

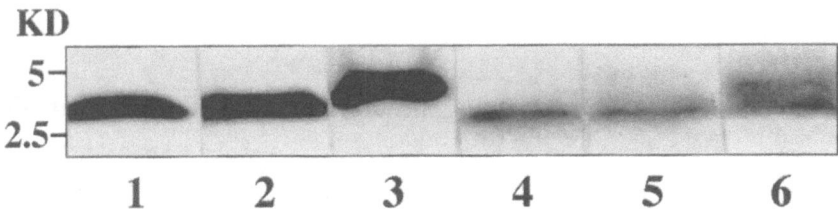

Figure 1. GM1 ganglioside-bound amyloid β-protein in brains. Immunoreactivity of amyloid β-protein (Aβ) (lanes 1–3) and its cholera toxin binding (lanes 4–6) in the membane fractions prepared from the brains of cases showing early pathological changes of Alzheimer's disease (lanes 3 and 6) and in the fractions containig cored senile plaques (lanes 2 and 5) were investigated. Synthetic Aβ1-42 and authentic GM1 ganglioside were loaded in lanes 1 and 4, respectively. Aβ in lane 3 showed retarded mobility. In lane 6, there were two bands that reacted with the cholera toxin. The upper band at ~5KD corresponded exactly to the slowly migrating Aβ in lane 3. The blot was immunoprecipitaed with BC05, a monoclonal antibody specific for Aβ42.

conformation and acts as a "template" that initiates amyloid fibril formation. Interestingly, GM1/Aβ was not detected in normal control brains even though GM1 ganglioside is expressed in abundance on the cell surface and Aβ is secreted into the extracellular space even under physiological conditions. Thus, we assumed that this unique Aβ is generated as a result of abnormalities in the intracellular generation or transport of these molecules.

Subsequent to our initial report, several investigators have reported on the generation and characterization of GM1/Aβ *in vitro*. It has been reported that Aβ adopts an altered secondary structure via GM1 binding [11–13], and that the rate of fibril formation of soluble Aβ is accelerated in the presence of GM1 ganglioside [14]. These results strongly support our hypothesis.

Previous studies have reported that APP is predominantly transported to the basolateral surface of polarized epithelial cells such as Madin-Darby canine kidney (MDCK) cells [15]. Although the intracellular site(s) of generation of Aβ remain(s) to be determined, it was reported that Aβ is predominantly secreted from the basolateral surface of MDCK cells [15, 16]. In contrast, GM1 ganglioside has been reported to be transported to the apical surface of the MDCK cells on the lipid microdomains rich in cholesterol and sphingoglycolipids [17]. Thus, GM1/Aβ may be generated as a result of alterations in the polarized trafficking of intracellular Aβ. Alternatively, evidence is now accumulating that lipid microdomains with high concentrations of cholesterol and sphingoglycolipids contain substantial amounts of Aβ even under physiological conditions [18, 19]; therefore, it is possible that Aβ binds to the GM1 ganglioside in the lipid microdomains as an abnormal event. In this regard, it is noteworthy that the Aβ-GM1 binding depends on the lipid composition of the membranes; the binding is more likely to occur in cholesterol-rich membranes [13]. The other thing to note is that cholesterol asymmetry in synaptic plasma membranes is modulated with age and under knockout of apolipoprotein E [20, 21].

A seeding Aβ generated from apically missorted APP in MDCK cell

To investigate the molecular mechanism of the generation of GM1/Aβ from the viewpoint of intracellular polarized trafficking of proteins and lipids, a culture system of MDCK cells transfected with the full-length and carboxyl-terminal truncated (ΔC) APP cDNA was employed. It has previously been reported that ΔC APP is significantly missorted to the apical surface due to the lack of a sorting signal for the apical direction [15]. The Aβ species secreted in the media of the apical and basolateral compartments of the culture were immunoprecipitated and characterized by Western blotting. An Aβ species with unique molecular characteristics, including its appearance as a smear on immunoblotting and retarded mobility on gel electrophoresis, was identified in the media of the apical compartment of an MDCK cell culture transfected with ΔC APP cDNA (referred to as apical Aβ). The apical Aβ was immunoprecipitated with an antibody specific for the midportion of Aβ; however, it was not with an antibody specific for the amino terminus of Aβ. Interestingly, however, following treatment with formic acid, the apical Aβ was immunoprecipitated with the antibody specific for the amino terminus of Aβ. This result suggests that the apical Aβ may be a conformationally altered isoform of Aβ. To determine why this unique Aβ was secreted only in the media of the apical compartment of the MDCK cell culture, we paid attention to the differences in lipid composition between the apical and basolateral plasma membranes. Since it has previously been reported that the concentration of cholesterol in the apical plasma membranes and apically transported vesicles is substantially higher than those in any other cellular membranes [22], we first examined the possibility that the apical Aβ acquired its unique molecular characteristics in the presence of cholesterol. To examine this possibility, the cells were treated with compactin, an inhibitor of cholesterol synthesis, or filipin, a cholesterol-binding reagent. With these treatments, the unique molecular characteristics of the apical Aβ were lost in a dose-dependent manner through these reagents.

To further characterize the molecular characteristics of the apical Aβ from the viewpoint of seeding ability, synthetic Aβ (Aβ1-40) was incubated with the apical Aβ at a molar ratio of 2000:1, and the resultant amyloid fibril formation was monitored by thioflavin T assay. The fluorescence activity was increased only in the incubation mixture containing apical Aβ obtained from the media of the MDCK cells transfected with ΔC APP cDNA. To confirm that the acquisition of the ability of the apical Aβ to act as a seed was a result of its conformational alteration, the fluoresence activity after treatment of the apical Aβ with formic acid was determined. The enhancement of fibril formation by the apical Aβ was substantially suppressed following the formic acid treatment. Furthermore, the cholesterol-dependence of the acquisition of the seeding ability by the apical Aβ was confirmed by treatment of the cells with compactin and filipin. As shown in Figure 2, the seeding ability was significantly reduced. The acceleration of amyloid fibril formation by apical Aβ was also confirmed by electron microscopy. A helical fibrillar structure with a diameter of 10 nm was detected only in the incubation mixture containing apical Aβ obtained from the media of MDCK cells transfected with ΔC APP cDNA (Fig. 3).

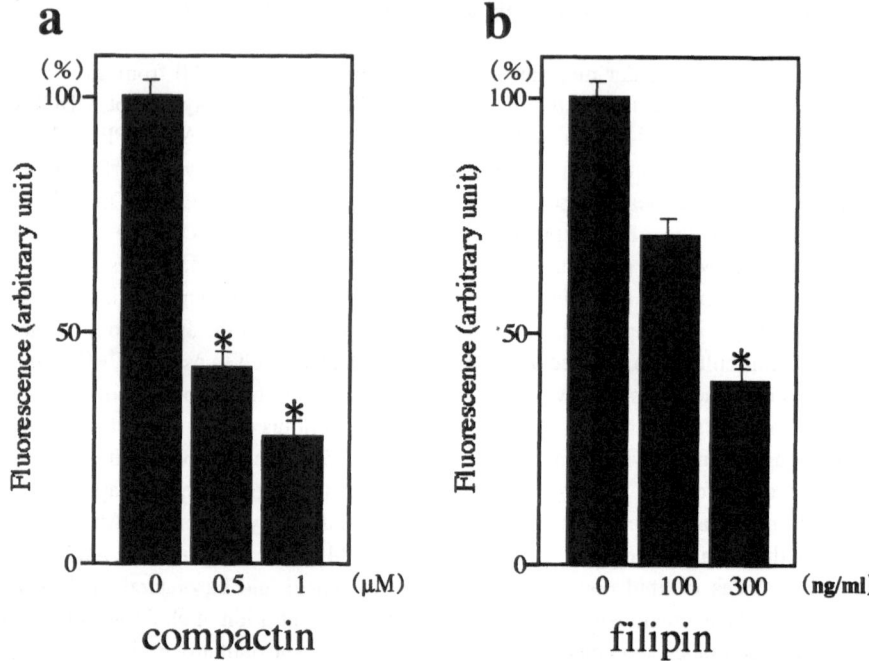

Figure 2 Inhibition of apical Aβ-induced acceleration of fibril formation of synthetic Aβ1-40 following treatment of the cells with compactin or filipin.

Hypothetical model for the generation of novel Aβs with seeding ability: implications of GM1/Aβ in the development of Alzheimer's disease

From the experimental evidence obtained so far, GM1/Aβ and apical Aβ are distinct molecules. However, these two share several molecular characteristics, including the altered immunoreactivity and dependence of their generation on the presence of lipids, GM1 ganglioside or cholesterol. Significantly, both of these novel Aβ species potentially enhance fibril formation of soluble Aβ (Tab. 1). We propose the following mechanism for amyloid fibril formation in AD brains involving these two Aβ species (Fig. 4). In this model, Aβ binds to GM1 ganglioside in detergent-insoluble glycosphingolipid-rich domain (DIG). The mechanism by which Aβ binds to GM1 ganglioside still remains to be determined; however, it is likely to be due to alterations in the lipid composition of the membrane as described above. Following its binding to GM1 ganglioside, Aβ adopts an altered conformation as reported previously [12, 13], and may act as a "template" that generates the "seeding" Aβ. This scenario is based on experimental evidence obtained in our own studies, but it also appears somewhat similar to the postulated mechanism of generation of the amyloidogenic prion protein. It has been suggested that, in the conversion of the prion protein from its normal to

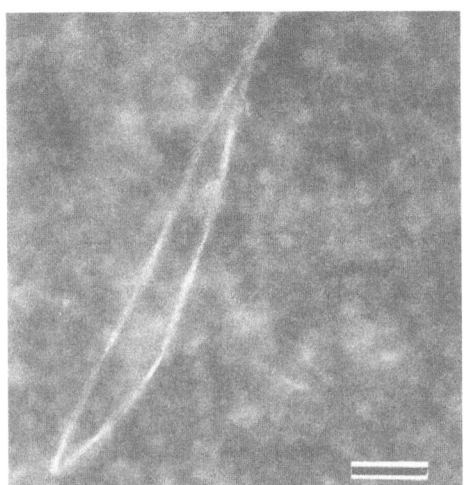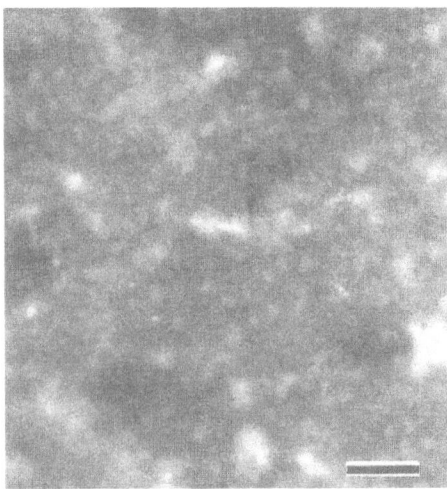

Figure 3 Electron micrographs of mixtures containing 50 μM synthetic Aβ plus the Aβ immunoprecipitated from the media of the apical (left panel) or basolateral (right panel) compartments of the MDCK cells transfected with ΔC APP cDNA incubated for 3 h at 37 °C. Note that typical amyloid fibrils were formed from synthetic Aβ plus apical Aβ, whereas no fibrillar structures were formed from synthetic Aβ plus basolateral Aβ. The bar indicates a length of 100 nm. (Reprinted, with permission, from Mizuno et al., 1999 [copyright: American Society for Biochemistry and Molecular Biology, Inc.])

abnormal form, it interacts with a putative resident molecule in the DIGs, called protein X, and that this protein X-bound prion protein acts as a "template" for the generation of a β-sheet-rich, abnormal form of the protein [7]. It is worthy of note that the conversion of prion protein to its abnormal form is also cholesterol-dependent [23, 24].

Pathogenesis of Alzheimer's disease from the viewpoint of its being a conformational disease

Recently, much attention has been focused on the conformational alterations of constitutive proteins in the brain in various neurodegenerative diseases, referred to as *conformational diseases*. In the process of development of these diseases, a constitutive protein in the brain undergoes minor perturbation of its structure, resulting in an increase in its β-sheet content. Conformational conversion of prion protein with resultant aggregation of its abnormal form is a well-known example of this category of diseases [7]. AD could also be included in the conformational disease group; however, to date, no study has ever shown the generation of conformationally altered Aβ with seeding ability. Our study is the first to present evidence for the generation of a seeding Aβ.

It has been suggested, in some conformational diseases, that a molecule other than a given protein, called *molecular chaperone*, is involved in the the structural conversion of a

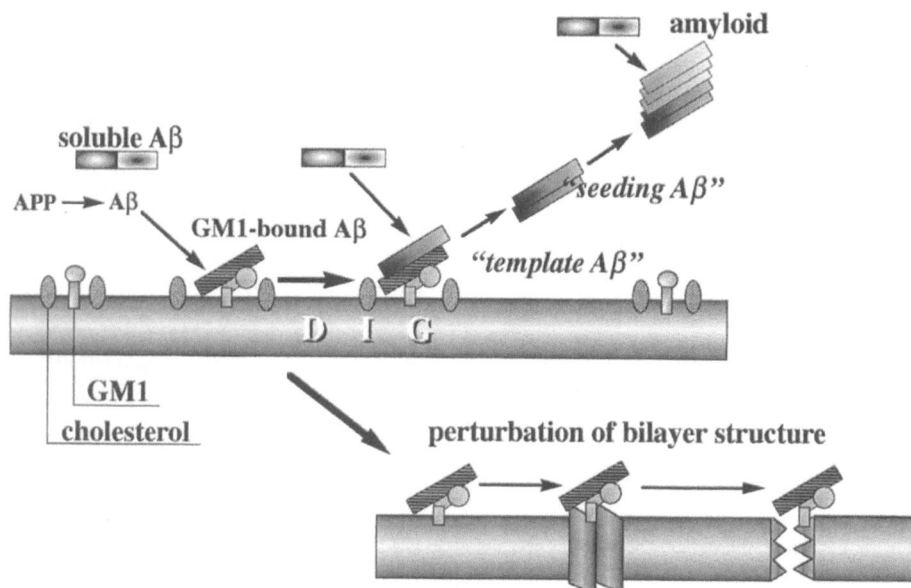

Figure 4 A hypothetical model of the mechanism of generation of "template" and "seeding" Aβ: implications of the generation of GM1/Aβ in the development of Alzheimer's disease.

constitutive protein. As described above, protein X is a molecular chaperone for the conversion of prion protein to its abnormal form. From evidence accumulated from our recent and other studies, it is speculated that the GM1 ganglioside may be a molecular chaperone for the conversion of Aβ.

Acknowledgements
T. Mizuno and H. Naiki contributed to this study on the cholesterol-dependent generation of seeding Aβ. The assistance of Y. Hanai in the preparation of this manuscript is acknowledged.

References

1 Selkoe DJ (1994) Cell biology of the amyloid β-protein precursor and the mechanism of Alzheimer's disease. *Annu Rev Cell Biol* 10: 373–403
2 Jarrett JT, Lansbury PT Jr (1993) Seeding "one-dimensional crystallization" of amyloid: a pathogenic mechanism in Alzheimer's disease and scrapie? *Cell* 73: 1055–1058
3 Citron M, Oltersdorf T, Haass C, McConlogue L, Hung AY, Seubert P, Vigo-Pelfrey C, Lieberburg I, Selkoe DJ (1992) Mutation of the β-amyloid precursor protein in familial Alzheimer's disease increases β-protein production. *Nature* 360: 672–674
4 Cai XD, Golde TE, Younkin SG (1993) Release of excess amyloid β protein from a mutant amyloid β protein precursor. *Science* 259: 514–516
5 Suzuki N, Cheung TT, Cai XD, Odaka A, Otvos L Jr, Eckman C, Golde TE, Younkin SG (1994) An

increased percentage of long amyloid β protein secreted by familial amyloid β protein precursor (βAPP717) mutants. *Science* 264: 1336–1340

6 Scheuner D, Eckman C, Jensen M, Song X, Citron M, Suzuki N, Bird TD, Hardy J, Hutton M, Kukull W et al (1996) Secreted amyloid β-protein similar to that in the senile plaques of Alzheimer's disease is increased *in vivo* by the presenilin 1 and 2 and APP mutations linked to familial Alzheimer's disease. *Nat Med* 2: 864–870

7 Prusiner SB (1997) Prion diseases and the BSE crisis. *Science* 278: 245–251

8 Yanagisawa K, Odaka A, Suzuki N, Ihara Y (1995) GM1 ganglioside-bound amyloid β-protein (Aβ): A possible form of preamyloid in Alzheimer's disease. *Nat Med* 1: 1062–1066

9 Mizuno T, Haass C, Michikawa M, Yanagisawa K (1998) Cholesterol-dependent generation of a unique amyloid β-protein from apically missorted amyloid precursor protein in MDCK cells. *Biochim Biophys Acta* 1373: 119–130

10 Mizuno T, Nakata M, Naiki H, Michikwa M, Wang R, Haass C, Yanagisawa K (1999) Cholesterol-dependent generation of a seeding amyloid β-protein in cell culture. *J Biol Chem* 274: 15110–15114

11 McLaurin J, Chakrabartty A (1996) Membrane disruption by Alzheimer β-amyloid peptides mediated through specific binding to either phospholipids or gangliosides. Implications for neurotoxicity. *J Biol Chem* 271: 26482–26489

12 Choo-Smith LP, Surewicz WK (1997) The interaction between Alzheimer amyloid β (1-40) peptide and ganglioside GM1-containing membranes. *FEBS Lett* 402: 95–98

13 Matsuzaki K, Horikiri C (1999) Interactions of amyloid β-peptide (1-40) with ganglioside-containing membranes. *Biochemistry* 38: 4137–4142

14 Choo-Smith LP, Garzon-Rodriguez W, Glabe CG, Surewicz WK (1997) Acceleration of amyloid fibril formation by specific binding of Aβ (1-40) peptide to ganglioside-containing membrane vesicles. *J Biol Chem* 272: 22987–22990

15 Haass C, Koo EH, Capell A, Teplow DB, Selkoe DJ (1995) Polarized sorting of β-amyloid precursor protein and its proteolytic products in MDCK cells is regulated by two independent signals. *J Cell Biol* 128: 537–547

16 Haass C, Koo EH, Teplow DB, Selkoe DJ (1994) Polarized secretion of β-amyloid precursor protein and amyloid β-peptide in MDCK cells. *Proc Natl Acad Sci USA* 91: 1564–1568

17 Simons K, Ikonen E (1997) Functional rafts in cell membranes. *Nature* 387: 569–572

18 Lee SJ, Liyanage U, Bickel PE, Xia W, Lansbury PT Jr, Kosik KS (1998) A detergent-insoluble membrane compartment contains Aβ *in vivo*. *Nat Med* 4: 730–734

19 Morishima-Kawashima M, Ihara Y (1998) The presence of amyloid β-protein in the detergent-insoluble membrane compartment of human neuroblastoma cells. *Biochemistry* 37: 15247–15253

20 Igbavboa U, Avdulov NA, Schroeder F, Wood WG (1996) Increasing age alters transbilayer fluidity and cholesterol asymmetry in synaptic plasma membranes of mice. *J Neurochem* 66: 1717–1725

21 Igbavboa U, Avdulov NA, Chochina SV, Wood WG (1997) Transbilayer distribution of cholesterol is modified in brain synaptic plasma membranes of knockout mice deficient in the low-density lipoprotein receptor, apolipoprotein E, or both proteins. *J Neurochem* 69: 1661–1667

22 van Helvoort A, van Meer G (1995) Intracellular lipid heterogeneity caused by topology of synthesis and specificity in transport. Example: sphingolipids. *FEBS Lett* 369: 18–21

23 Taraboulos A, Scott M, Semenov A, Avrahami D, Laszlo L, Prusiner SB, Avraham D (1995) Cholesterol depletion and modification of COOH-terminal targeting sequence of the prion protein inhibit formation of the scrapie isoform [published erratum appears in. *J Cell Biol* 1995, 30(2): 501]. *J Cell Biol* 129: 121–132

24 Naslavsky N, Stein R, Yanai A, Friedlander G, Taraboulos A (1997) Characterization of detergent-insoluble complexes containing the cellular prion protein and its scrapie isoform. *J Biol Chem* 272: 6324–6331

Neuroscientific Basis of Dementia
C. Tanaka, P.L. McGeer, Y. Ihara (eds)
© 2001 Birkhäuser Verlag Basel/Switzerland

Catabolism of amyloid-β peptide in brain parenchyma

Takaomi C. Saido and Nobuhisa Iwata

Laboratory for Proteolytic Neuroscience, RIKEN Brain Science Institute, Saitama, Japan

Introduction

Life and disease are like opposite blades of one sword or, more subjectively, disease is a part of life. In Alzheimer's disease (AD), the deposition of amyloid-β peptide (Aβ) plays a central pathogenic role, triggering the pathological cascade spanning decades as established by a series of studies analyzing the phenotypes of gene mutations causing early-onset familial AD (FAD) [1–3]. On the other hand, Aβ is also a physiological peptide constantly secreted from cells throughout life and known to be trophic under certain conditions, although its definite *in vivo* function remains unclear [4, 5]. Additionally, its precursor, amyloid precursor protein (APP), is a source of soluble APP fragments that function as major serine protease and/or metalloprotease inhibitors in the brain [6, 7]. Therefore, the presence of Aβ is a relevant and probably necessary part of living activities.

Human brains start to accumulate Aβ by as yet unknown mechanism(s) upon aging [8]. It should be noted that aging is still the major risk factor even for those carrying the familial AD-causing gene mutations. This suggests that Aβ is, at least in young-enough and healthy brains, fully catabolized immediately after secretion before being deposited. The power of catabolism can also be demonstrated by the animal models of AD [9]; the transgenic mice that overexpress APP nearly 100-fold as compared to control mice require as long as several months to display apparent Aβ deposition, indicating that even unphysiologically high levels of Aβ can be disposed of in these mice during the first several months.

This mechanism of clearing Aβ in the brain is important for two reasons in AD research. First, a reduction in this process would lead to pathological accumulation of Aβ and thus to AD development; only a 30% reduction in the catabolic rate would exert an equivalent effect on the steady-state amount of Aβ as compared to a 50% increase in the anabolic rate, a typical phenotype of FAD-causing gene mutations [10]. Second, we may be able to prevent the disease onset or decelerate its progression by manipulating this process.

However, our current knowledge about the *in vivo* catabolism of Aβ is quite limited as compared with the mass of information available concerning the anabolism of Aβ. Analyzing extracellular peptide catabolism in brain parenchyma or any other tissue in a relevant manner is difficult because the local topology and environment of the catabolic machinery need to be maintained as they are during experimentation in both cellular and anatomical terms. Therefore, such an oversimplified experimental paradigm as incubating Aβ peptide with a given peptidase in test tubes or with cells in culture is likely to produce an artifact. The extracellular situations are so different between the *in vitro* and *in vivo* systems that it is simply inappropriate to apply the knowledge obtained in *in vitro* experimental paradigms to understanding the catabolism of secreted peptides in the brain. It should also be noted that proteolytic systems are often drastically altered when cells are transferred from an organ to culture.

This article reviews the present status of Aβ catabolism studies as well as the current efforts to establish a reliable *in vivo* experimental paradigm in which the actual catabolic processes in the brain can be captured. The possible medical implications will also be discussed.

Physiological and pathological Aβ catabolism

Clearance of Aβ from brain parenchyma can be divided into two major categories (Fig. 1). One is a physiological catabolism, in which secreted soluble Aβ is fragmented under normal conditions. The other is a rather pathological catabolism presumably involved in clearance of fibrillar Aβ that could otherwise develop into diffuse and then cored plaques. This pathological catabolism is likely to involve proteases capable of degrading fibrillar proteins such as those constituting extracellular matrix. However, little is known about the enzymatic and molecular basis of fibrillar Aβ clearance, although morphological studies suggest that astrocytes may participate in this process [11]. The physiological catabolism is primarily focused on in this chapter.

In addition, extracellular protein catabolism is rendered quite complex by the presence of various possible mechanisms. Clearance of a given peptide from the brain may be achieved by (i) transport to vascular systems, which may not after all depend on peptidases in brain parenchyma (ii) proteolysis conducted by extracellular peptidases, or (iii) receptor-mediated endocytosis followed by endosomal-lysosomal proteolysis. We may also need to consider the possible existence of a physiological cofactor that assists or regulates these

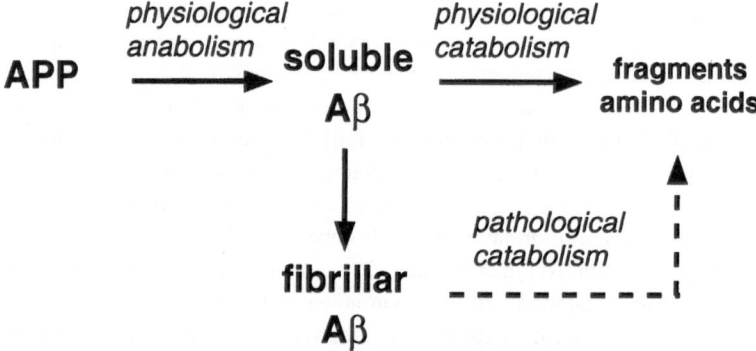

Figure 1. Physiological anabolism, physiological catabolism and pathological catabolism of Aβ. See text.

processes. Again, it is important to employ a relevant *in vivo* experimental paradigm to resolve these issues.

Another important issue is that we may need to distinguish between the two different forms of Aβ, Aβ$_{1-40}$ and Aβ$_{1-42}$, because they may possibly be catabolized in distinct manners. Besides, Aβ$_{1-42}$ is pathogenically more important than Aβ$_{1-40}$ for the following reasons. (i) Although Aβ$_{1-40}$ is physiologically produced in much larger quantities than Aβ$_{1-42}$, the majority of Aβ peptides deposited in human brain is Aβ$_{x-42}$. (ii) Aβ$_{1-42}$ is more hydrophobic and fibrillogenic than Aβ$_{1-40}$, indicating the conformational difference that would alter sensitivity to proteolysis. (iii) FAD-causing gene mutations consistently elevate the Aβ$_{1-42}$. Therefore, the primary target for the catabolism study should be Aβ$_{1-42}$ rather than Aβ$_{1-40}$.

Candidate peptidases

Test tube experiments, in which synthetic Aβ peptides are subjected to proteolysis in the presence of a given peptidase, have shown that various peptidases can proteolyze Aβ. Besides, the primary structure of Aβ suggests that Aβ is a potential substrate for some of those peptidases with specific substrate sequence specificity. For instance, the presence of three basic amino acid residues, one arginine and two lysines, indicates that trypsin and trypsin-like peptidases are likely to be capable of proteolyzing Aβ. Figure 2 summarizes the cleavage sites in Aβ that may be attacked by major peptidases and demonstrates that there are indeed a number of possible candidates. Obviously, the most important would be the one that catalyzes the rate-limiting process.

Several recent studies have emphasized the possible involvement of insulin-degrading enzyme [12]. An unidentified serine protease influenced by metalloendopeptidase 24.15 has also been reported [13]. We should also bear in mind that cells can take up Aβ and digest it intracellularly [14]. All these possibilities arose from essentially *in vitro* experiments

Figure 2. Possible cleavage sites in Aβ$_{1-42}$ by various peptidases. (a) aminopeptidase; (a') dipeptidyl peptidase; (b) chymotrypsin-like endopeptidase; (c) trypsin-like endopeptidase; (d) gelatinase A (MMP-2), α-secretase; (e) endopeptidase 24.11 (NEP, neprilysin); (f) cathepsin D/E, gelatinase B (MMP-9); (g) collagenase; (h) carboxypeptidase; (h') peptidyl dipeptidase; (i) insulin-degrading enzyme. Most of the cleavages sites were determined using Aβ$_{1-40}$ or Aβ$_{1-42}$ as a substrate and the others were assessed based on the specificities of peptidases. See for instance Howell et al. [28], McDermott & Gibson [29], Barrett and McDonald [30] and McDonald and Barrett [31].

employing cultured cells or culture supernatants. Therefore, it is difficult to determine which one is the most relevant candidate as discussed above.

In vivo analysis of Aβ catabolism

In order to capture the actual *in vivo* Aβ catabolism in the brain, we developed a new experimental design in which we injected multiply radio-labeled Aβ peptides into rat hippocampus and analyzed their metabolism by tracing the radioactivities [15]. The peptides were internally radiolabeled with ^{3}H and ^{14}C without altering their chemical structures in order to minimize the risk of producing artifacts. This new approach led to novel findings and medical implications that would otherwise have remained veiled.

The ^{3}H/^{14}C-Aβ$_{1-42}$ peptide underwent proteolytic degradation in the hippocampus as analyzed by high-performance liquid chromatography (HPLC) with a half life of 15–20 min. The possibility that transport to vascular systems may be the primary mechanism of Aβ$_{1-42}$ clearance can be discarded because the injected radioactivity remained inside the hippocampus throughout the catabolic processes. Therefore, brain parenchyma possesses independent and potent proteolytic systems capable of determining the life spans of extracellular peptides, including Aβ. We also succeeded in capturing the major catabolic intermediate, the structure of which was identified as Aβ$_{10-37}$ by chemical sequencing and mass spectrometry. Furthermore, we discovered that metalloendopeptidase inhibitors, phosphoramidon and thiorphan, suppressed the appearance of this catabolic intermediate as well as the entire degradation of the ^{3}H/^{14}C-Aβ$_{1-42}$ peptide. These observations suggest that neutral endopeptidase(s) similar or identical to neprilysin is (are) involved in the initial cleavage of Aβ$_{1-42}$ to produce the catabolic intermediate and that this step is a rate-limiting process (Fig. 3).

We also observed that infusion of neutral endopeptidase inhibitor, thiorphan, into rat hippocampus caused biochemical and pathological deposition of endogenous Aβ. This observation suggests that a thiorphan-sensitive peptidase(s) is indeed involved in physiological Aβ$_{42}$ catabolism and that alternative proteolytic pathway(s) or possible secondary catabolic processes, such as lysosomal proteolysis, that may be mobilized upon Aβ accumulation, are not efficient enough to fully compensate for the suppression of the thiorphan-sensitive degradation. Interestingly, the amount of Aβ$_{40}$ was much less affected by thiorphan administration than that of Aβ$_{42}$. Clearance of Aβ$_{40}$ therefore seems not to depend so much on neutral endopeptidase-catalyzed proteolysis as that of Aβ$_{42}$ does, or it could even be transport-dependent rather than catabolism-dependent.

Medical implications and future perspectives

The *in vivo* amount of any peptide is determined by the balance between anabolism and catabolism. Physiological but potentially amyloidogenic peptides such as Aβ could be turned into pathogenic agents by subtle alterations in this balance over a long period of time. In the majority of early-onset familial AD cases, Aβ accumulation is caused by only an approxi-

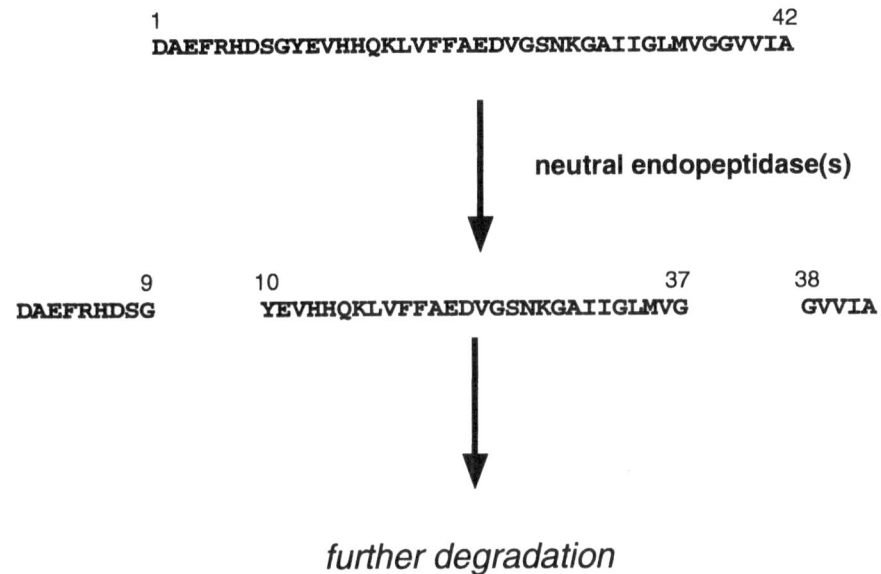

1   42
DAEFRHDSGYEVHHQKLVFFAEDVGSNKGAIIGLMVGGVVIA

neutral endopeptidase(s)

9   10   37   38
DAEFRHDSG   YEVHHQKLVFFAEDVGSNKGAIIGLMVG   GVVIA

further degradation

Figure 3. The major catabolic pathway of $A\beta_{1-42}$. See text.

mately 50% increase in $A\beta_{42}$ anabolism [16–18]. In contrast, the sporadic AD cases do not seem to be caused by such simple overproduction of $A\beta_{42}$ [16]. Therefore, aging-associated reduction of catabolism is a strong candidate mechanism that could account for the accumulation in the senile brain; in a very rough estimation, failure to clear a few % of secreted $A\beta_{42}$ would lead to massive pathological $A\beta$ deposition within a decade [10]. Such a catabolic reduction can be achieved either by down-regulation of the responsible peptidase(s) or by up-regulation of the corresponding inhibitor(s). It is noteworthy that almost every peptidase is under strict regulation by endogenous inhibitor(s) *in vivo* since any uncontrolled excessive proteolytic action would be destructive to cells and tissues. Inhibitors could also be taken up from outside through diet and therefore may possibly pose an environmental risk factor.

Therefore, identification of the major $A\beta$-degrading enzyme(s) would have the following medical implications. Catabolic reduction may possibly be used as a diagnostic marker for certain cases of sporadic AD. We may also be able to up-regulate the responsible peptidase(s) or down-regulate the corresponding endogenous inhibitor(s) and thus to facilitate $A\beta$ clearance through, for instance, gene therapy. Another possible strategy would be to seek for peptidase inhibitors capable of penetrating the blood-brain barrier as environmental risk factors in diets. Avoiding such risk factors could preclude down-regulation of $A\beta$ catabolism in the brain. The advantage of this strategy is that dietary approaches would be much less costly to the society than pharmaceutical approaches.

As the conventional nomenclature of neprilysin and enkephalinase implies, opioids and other neuropeptides are known to be physiological substrates for this peptidase [19]. If neprilysin and similar peptidases are the major Aβ-degrading enzymes, both Aβ and some neuropeptides share, at least in part, a common catabolic pathway; the quantity and localization of Aβ and of opioids may correlate under physiological conditions. Furthermore, if we focus on opiod metabolism, catabolic reduction could function as a kind of pain-control measure through elevating opioid concentrations. Also, a decreased ability to synthesize opioids that could be caused by aging may be compensated for by reduced catabolism. Although these are all speculations that need to be verified, aging-dependent accumulation of Aβ may possibly be a consequence of sacrificing Aβ clearance for maintaining appropriate neuropeptide levels. Aβ accumulation, in turn, may cause disturbed neuropeptide metabolism. It is also possible that, in more pathological circumstances, both the Aβ and neuropeptide levels may be seriously perturbed together, accelerating the disease progression. A possible clinical implication based on these assumptions would be that up-regulation of neuropeptide synthesis in any way, or even exogenous administration, might lessen the need for the catabolic system to compensate for the assumed decline in synthesis. A more speculative presupposition is that such epidemiologically identified anti-risk factors as the chronic administration of non-steroid anti-inflammatory drugs [20] may contribute to up-regulation of Aβ catabolism if it lessens the need for opioids through attenuating pain.

Incidentally, the rate of peptide catabolism may oscillate on a daily basis to regulate the peptide levels according to the circadian rhythm. A number of peptidases, including cathepsin A, C and D [21], dipeptidyl peptidase IV [22], vasopressin-converting aminopeptidase [23], pyroglutamyl peptidase I [24], aspartyl aminopeptidase [25] and endopeptidase 22.19 [26] have been shown to undergo diurnal up- and down-regulation *in vivo*. Such alteration could be achieved through down- and up-regulation, respectively, of a corresponding endogenous inhibitor, which in turn might influence other peptidases in the brain. Sleep abnormalities, a typical aging-associated neurological disorder probably corresponding to distorted circadian rhythm [27], may be related to altered diurnal regulation of peptide metabolism in the brain. If so, the malfunctioning of peptide metabolism might present a common mechanism through which brain aging in general produces various neurological phenotypes.

Verification of these speculations will not be easy because of the temporal distance of decades between the cause and effect in human cases as well as the presence of many risk parameters. Epidemiological studies, in particular longitudinal ones, are necessary to examine all the hypotheses associated with sporadic AD. Establishment and use of "perfect" animal models that represent all the major human pathologies and symptoms should also be useful.

References

1 Hardy J (1997) Amyloid, the presenilins and Alzheimer's disease. *Trends Neurosci* 20: 154–159
2 Selkoe DJ (1998) The cell biology of β-amyloid precursor protein and presenilin in Alzheimer's disease. *Trends Cell Biol* 8: 447–453
3 Price DL, Tanzi RE, Borchelt DR, Sisodia SS (1998) Alzheimer's disease: genetic studies and transgenic

models. *Annu Rev Genet* 32: 461–493
4   Selkoe DJ (1993) Physiological production of the β-amyloid protein and the mechanism of Alzheimer's disease. *Trends Neurosci* 16: 403–409
5   Sisodia SS, Price DL (1995) Role of the β-amyloid protein in Alzheimer's disease. *FASEB J* 9: 366–370
6   Oltersdorf T, Fritz LC, Schenk DB, Lieberburg I, Johnson-Wood KL, Beattie EC, Ward PJ, Blacher RW, Dovey HF, Sinha S (1989) The secreted form of the Alzheimer's amyloid precursor protein with the Kunitz domain is protease nexin II. *Nature* 341: 144–147
7   Miyazaki K, Hasegawa M, Funahashi K, Umeda M (1993) A metalloproteinase inhibitor domain in Alzheimer amyloid protein precursor. *Nature* 362: 839–841
8   Funato H, Yoshimura M, Kusui K, Tamaoka A, Ishikawa K, Ohkoshi N, Namekata K, Okeda R, Ihara Y (1998) Quantification of amyloid β-protein (Aβ) in the cortex during aging and in Alzheimer's disease. *Am J Pathol* 152: 1633–1640
9   Johnson-Wood K, Lee M, Motter R, Hu K, Gordon G, Barbour R, Khan K, Gordon M, Tan H, Games D et al (1997) Amyloid precursor protein processing and Aβ42 deposition in a transgenic mouse model of Alzheimer disease. *Proc Natl Acad Sci USA* 94: 1550–1555
10 Saido TC (1998) Alzheimer's disease as proteolytic disorders: Anabolism and catabolism of β-amyloid. *Neurobiol Aging 19*, S69–S75
11 Yamaguchi H, Sugihara S, Ogawa A, Saido TC, Ihara Y (1998) Diffuse plaques associated with astroglial amyloid β protein, possibly showing a disappearing stage of senile plaques. *Acta Neuropathol* 95: 217–222
12 Qiu WQ, Walsh DM, Ye Z, Vekrellis K, Zhang J, Podlisny MB, Rosner MR, Safavi A, Hersh LB, Selkoe DJ (1998) Insulin-degrading enzyme regulates extracellular levels of amyloid β-protein by degradation. *J Biol Chem* 273: 32730–32738
13 Yamin R, Malgeri EG, Sloane JA, McGraw WT, Abraham CR (1999) Metalloendopeptidase EC 3.4.24.15 is necessary for Alzheimer's amyloid-β peptide degradation. *J Biol Chem* 274: 18777–18784
14 Ida N, Masters CL, Beyreuther K (1996) Rapid cellular uptake of Alzheimer amyloid βA4 peptide by cultured neuroblastoma cells. *FEBS Lett* 394: 174–178
15 Iwata N, Tsubuki S, Takaki Y, Watanabe K, Sekiguchi M, Hosoki E, Kawashima-Morishima M, Lee H-J, Hama E, Sekine-Aizawa Y et al (2000) Identification of the major Aβ1-42-degrading catabolic pathway in brain parenchyma: Suppression leads to biochemical and pathological deposition. *Nat Med* 6: 143–150
16 Scheuner D, Eckman C, Jensen M, Song X, Citron M, Suzuki N, Bird TD, Hardy J, Hutton M, Kukull W et al (1996) Secreted amyloid β-protein similar to that in the senile plaques of Alzheimer's disease is increased *in vivo* by the presenilin 1 and 2 and APP mutations linked to familial Alzheimer's disease. *Nat Med* 2: 864–870
17 Lemere CA, Lopera F, Kosik KS, Lendon CL, Ossa J, Saido TC, Yamaguchi H, Ruiz A, Martinez A, Madrigal L et al (1996) The E280A Presenilin 1 mutation leads to a distinct Alzheimer's disease phenotype: Increased Aβ42 deposition and severe cerebellar pathology. *Nat Med* 2: 1146–1150
18 Duff K, Eckman C, Zehr C, Yu X, Prada CM, Perez-tur J, Hutton M, Buee L, Harigaya Y, Yager D et al (1996) Increased amyloid-β42(43) in brains of mice expressing mutant presenilin 1. *Nature* 383: 710–713
19 O'Cuinn G, O'Connor B, Gilmartin L, Smyth M (1995) Neuropeptide inactivation by peptidases. *In:* O'Cuinn (ed.): *Metabolism of Brain Peptides.* CRC Press, Boca Raton, FL, 99–157
20 Breitner JC (1996) The role of anti-inflammatory drugs in the prevention of Alzheimer's disease. *Annu Rev Med* 47: 401–411
21 Obled C, Arnal M, Valiln C (1980) Variation through the day of hepatic and muscular cathepsin A, C and D activities and free amino acids of blood in rats: influence of feeding schedule. *Brit J Nutr* 44: 61–69
22 Balschun D, Schuh J (1983) Dipeptidyl-peptidase IV (DP IV): rhythmic changes of serum activity in mice. *Pharmazie 38,* 424
23 Liu B, Burbach JP (1988) Circadian variations of vasopressin level and vasopressin-converting aminopeptidase activity in the rat pineal gland. *Peptides* 9: 973–978
24 Ramirez M, Sanchez B, Arechaga G, Garcia S, Lardelli P, Yenzon D, de Gandarias JM (1991) Diurnal variation and left-right distribution of pyroglutamyl peptidase I activity in the rat brain and retina. *Acta Endocrinol* 125: 570–573
25 Ramirez M, Sanchez B, Arechaga G, Lardelli P, Garcia S, Ozaita A, de Gandarias JM (1992) Daily rhythm of aspartate aminopeptidase activity in the retina, pineal gland and occipital cortex of the rat. *Neuroendocrinology* 56: 926–929
26 Ferro ES, Sucupira M, Marques N, Camargo AC, Menna-Barreto L (1992) Circadian rhythm of the endopeptidase 22.19 (EC 3.4.22.19) in the rat brain. *Chronobiol Int* 9: 243–249
27 Ancoli-Israel S, Kripke DF (1991) Prevalent sleep problems in the aged. *Biofeedback Self-Regul* 16:

349–359
28 Howell S, Nalbantoglu J, Crine P (1994) Neutral endopeptidase can hydrolyze β-amyloid(1-40) but shows no effect on β-amyloid precursor protein metabolism. *Peptides* 16: 647–652
29 McDermott JR, Gibson AM (1996) Degradation of Alzheimer's β-amyloid protein by human cathepsin D. *Neuroreport* 7: 2163–2166
30 Barrett AJ, McDonald JK (1980) *Mammalian Proteases. A Glossary and Bibliography. Vol. 1 Endopeptidases.* Academic Press, London
31 McDonald JK, Barrett AJ (1986) *Mammalian Proteases. A Glossary and Bibliography. Vol. 2 Exopeptidases.* Academic Press, London

Diagnosis and therapeutics of dementia

Neuroscientific Basis of Dementia
C. Tanaka, P.L. McGeer, Y. Ihara (eds)
© 2001 Birkhäuser Verlag Basel/Switzerland

Lessons in familial Alzheimer's disease

Martin N. Rossor[1], John C. Janssen[1], Nicholas C. Fox[1], Richard J. Harvey[1], John Stevens[2]
and Elizabeth K. Warrington[1]

[1] *Dementia Research Group, Institute of Neurology and* [2] *Radiology Department, The National Hospital for*
Neurology and Neurosurgery, London, UK

Introduction

Autosomal dominant inheritance has been a clinical hallmark of a variety of neurodegenerative dementias such as Huntington's disease; by contrast, others have traditionally been considered to be sporadic. However, within the group of dementia disorders, familial cases are increasingly recognised with a clinical phenotype very similar to, if not identical to, that of the commoner sporadic disorder. Thus many cases of frontotemporal degeneration, corticobasal degeneration, prion disease, Parkinson's disease and dementia with Lewy bodies have all been reported on a familial basis. Alzheimer's disease (AD), classically considered a sporadic disorder, is now recognised on a clear autosomal dominant basis.

Alzheimer's original case was a 51-year old woman reported in 1907 without any family history. The prevalent view for many years was that it occurred only as a sporadic disorder despite the occasional report of apparent autosomal dominant inherited pedigrees (for review [1]). However, in the late 1970s and early 1980s large pedigrees began to be reported with adequate documentation of clinical features and histopathological confirmation of AD [2]. Such families showed clear autosomal dominant inheritance affecting both males and females with no cases of non-penetrance and moreover a relatively constant age at onset within a given family. Such pedigrees were ideal for genetic linkage studies which initially focused on chromosome 21 due to the known link between trisomy 21 Down syndrome and the development of an AD histopathology. Some of these earlier studies confirmed linkage to markers on chromosome 21 although the data were inconsistent due, as is now known, to genetic heterogeneity within familial AD.

Early linkage studies coincided with dramatic advances in understanding the molecular basis of senile plaques and amyloid angiopathy. Following the isolation and sequencing of the deposited β-amyloid, the amyloid precursor protein (APP) gene was cloned [3]. The first APP mutation associated with familial AD was identified in 1991 in two patients with an APP V717I mutation [4]. This was followed by reports of APP V717G [5] and V717F [6]. Another mutation identified in a single large Swedish pedigree was a double mutation at 670/671 [7], i.e., the mutations all framed the β-amyloid domain of the APP molecule. More recently additional mutations have been described at codons 665, 713, 715, 716 and 723

[8–12]. Mutations at 692 and 693 adjacent to the β secretase site are associated predominantly with amyloid angiopathy and a somewhat different phenotype [13, 14].

Although the first to be described, APP mutation pedigrees are extremely rare and it soon became apparent that other genetic loci must exist. Linkage to chromosome 14 was reported in 1992 [15] and subsequent gene cloning revealed a hitherto unknown gene called presenilin 1 [16]. A similar homologous gene, presenilin 2, was found to account for AD in the Volga German pedigrees [17].

The identification of APP and presenilin mutation familial AD has made a major contribution to understanding the molecular pathology of familial AD and by implication, AD. Both APP 717 and presenilin1 & 2 mutations appear to increase the proportion of $A\beta_{1-42}$, which can be observed in plasma, cerebrospinal fluid (CSF), *in vitro* and in transgenic models. By contrast to $A\beta_{1-40}$, $A\beta_{1-42}$ has a greater propensity to form fibrillary β-amyloid deposits [18]. Indeed, there is increasing evidence that deposition of $A\beta_{1-42}$ is the first event and indeed may precipitate subsequent $A\beta_{1-40}$ deposition [19]. There is, however, some controversy as to whether the formation of senile plaques and fibrillary amyloid deposits are an epiphenomenon rather than directly pathogenic. However, the existence of APP mutations which frame the amyloid domain and which result in the full panoply of Alzheimer histopathology is persuasive evidence that abnormalities in amyloid metabolism can be a primary event in the pathophysiology of AD.

It is probable that other genetic loci exist as not all autosomal dominant familial AD pedigrees identified so far have established APP or PS mutations. Presenilin 1 mutations account for the majority of early-onset familial AD, APP and presenilin 2 being exceedingly rare. In our own series of 57 pedigrees with an onset below the age of 65, 3 were due to APP and 18 due to presenilin mutations.

In addition to advancing our understanding of the pathophysiology of Alzheimer's disease, the study of familial AD also allows us to answer important questions about phenotype/genotype correlations and about onset and progression of the disease.

Phenotype/genotype correlations

A striking observation with respect to clinical and histopathological features of familial AD is that apart from the family history and age at onset, the disease is remarkably similar to sporadic AD. Memory impairment is the early salient feature followed by more widespread cognitive deficits. A pedigree with M139V PS1 mutation presented with an early speech production deficit [20], but these appear to be exceptions rather than the rule. Group comparisons of familial and sporadic AD have revealed subtle differences [21]. In our own studies, we found relative preservation of naming in the familial AD group [22]. Non-cognitive features emerge later, although myoclonus can be early and tends to be more common than in sporadic AD and in some families can be sufficiently florid to resemble Creutzfeldt-Jakob disease [23].

Age at disease onset is relatively constant within families although this may in part be due to selection bias and only prospective studies will establish this. Certainly outliers in some families occur and there have been examples of apparent non-penetrance [24]. ApoE genotype has an epistatic effect and reduces age at onset in APP mutation cases, but less so in presenilin. Other genetic factors are likely to be important; for example, two separate pedigrees with an identical M139V mutation had a consistent within-family age at onset with a statistically significantly different between-family age at onset [25]. An early analysis of presenilin and APP families suggested that the former had an earlier age at onset. Whilst some very early-onset PS1 families have been reported, a number of later-onset cases have been identified as further mutations have been discovered, suggesting that the earlier observation may be partly biased due to the selection of early-onset cases for genetic studies [26].

Prospective longitudinal studies

A prospective longitudinal study of at-risk individuals from FAD pedigrees provides an ideal opportunity to identify the earliest and even presymptomatic clinical features. Most studies of the cognitive and neuroimaging profiles of Alzheimer's disease have been undertaken on patients with a clear diagnosis, i.e., the disease is already well-established. To try and identify earlier features, those with mild memory impairment or minimal cognitive impairment have been selected but again, by implication, the disease is already established. The other approach is to take people who are at high risk by virtue of age, but in view of the relatively low incident rate of AD in the general population, large numbers have to be studied and thus only limited investigations can be performed. By contrast, offspring of affected parents from FAD pedigrees who carry a 50% risk can be studied prospectively from before the average age at onset at a time when they are entirely well. We established a prospective longitudinal study of at-risk individuals from 19 pedigrees in 1992 and have followed individuals with annual neuropsychometric assessment and neuroimaging to identify the earliest changes and subsequent progression.

Earliest cognitive deficits

Impairment of memory is the earliest symptom reported by patients and carers. This has been confirmed using recognition memory tests. Thus Newman et al. [27] reported a single individual who when first entering the study was in full-time employment and asymptomatic with an MMSE of 30/30 [28]. At the first assessment their performance on a recognition memory test for words [29] was already below the 5th percentile and later assessments revealed a deficit in visual memory before more widespread cognitive impairment intervened. A similar pattern has been identified in other individuals [30].

The early deficit in verbal memory has also been shown in a group study by Fox et al. [31]. Over a 6-year period 63 subjects underwent serial assessment, during which time 10 subjects developed symptoms of episodic memory loss and subsequently fulfilled criteria for diagnosis of AD. The mean (SD) time between first assessment to appearance of symptoms was 2.6 (1–4) years. At first assessment, those who subsequently became affected already had significantly lower verbal memory recognition test and WAIS performance IQ scores.

Neuroimaging

Both functional position emission tomography (PET) and single photon emission computerised tomography (SPECT) and structural magnetic resonance imaging (MRI) have been performed on individuals at risk of familial AD. Functional neuroimaging theoretically may be able to demonstrate changes in metabolism prior to changes on structural imaging which are presumed to reflect neuronal loss. Indeed, group studies of at-risk patients do show a lower global cerebral metabolic rate for glucose using FDG [32]. The distribution of the metabolic defect is similar to that of affected individuals, i.e., a bitemporal, biparietal pattern. Similar results have been obtained in individuals at risk due to the apoE4 genotype [33, 34]. Although group studies show an overall deficit, variability in the technique makes it less robust at an individual level and there is a limitation to the number of serial scans that can be performed.

We have used MRI as being the most robust methodology which can be used safely on a repeated basis. The acquisition of volume images allows precise regional volumetry and early loss of hippocampal volume can be demonstrated in at-risk individuals at a very early symptomatic or even presymptomatic stage [35]. These changes are similar to those reported in minimal cognitive impairment and patients with very mild memory impairment [36].

Regional volumetry, however, requires a prior decision of which areas to study. Thus, while there are good reasons to examine the hippocampus in view of the early memory impairment, other areas may be involved and there is an inevitable bias towards choosing areas that can be easily outlined. Moreover, there is quite marked variation in segmentation, making such techniques less reliable for longitudinal studies. In view of this, an alternative approach was developed by Fox et al. [37, 38], which utilises the surface complexity of the brain to positionally match and register serially acquired volume images. Following registration, areas of difference between the two scans can be identified and using the brain boundary shift integral, changes in volume can be quantified [39]. It is assumed that loss of brain tissue volume reflects neuronal loss and indeed there is a close correlation between overall functional decline as measured by the MMSE and the rate of total brain tissue loss [40].

Using this technique it is possible to demonstrate a rate of tissue loss which falls outside the control range in individuals at risk of familial AD. Follow up of such individuals shows subsequent development of FAD. Rates of change are more valuable than individual scans which frequently fall within the normal range. Using this robust method it is also possible to track individual patients and to show that tissue loss is discernible 3 or 4 years prior to the development of symptoms (see Fig. 1 [41]).

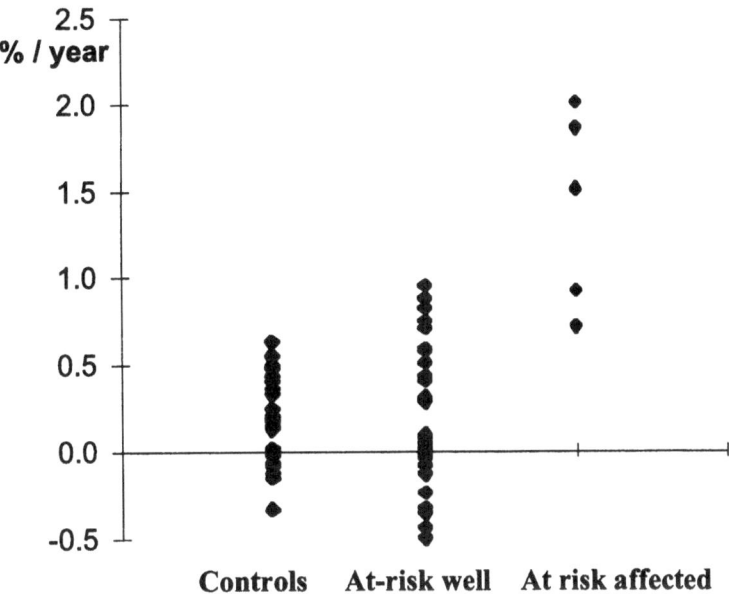

Figure 1. Rates of change in global cerebral volume on MRI for healthy controls, individuals at risk of familial Alzheimer's disease who remained well and at-risk individuals who subsequently became clinically affected during follow-up. T1-weighted volumetric imaging was performed on a 1.5 T scanner yielding 124 1.5 mm
slices thorough the head with a 128×256 image matrix. Baseline and first follow-up scans were registered
using an automated technique that utilises the complex structure of the brain to produce precise positional
alignment. Change in cerebral volume was computed directly from the registered scans by calculation of the
brain boundary shift integral [39]. Reproduced from Fox et al. [41].

Conclusion

In conclusion, the similar, although not identical, clinical and neuropathological features of
familial AD and sporadic AD suggest that lessons learnt from the former can be applied to
the more general and common sporadic disorder. Not only has this generated insights into
the underlying molecular pathology but it has allowed us to discern the earliest changes in
the disorder and to identify potential influences on subsequent progression. A presymptomatic period of tissue loss can be identified, which clearly presents a window of opportunity for therapeutic intervention.

References

1 Rossor MN (1992) Familial Alzheimer's Disease. *In*: MN Rossor (ed.): *Unusual Dementias*. 3rd ed.
   Bailliere Tindall, London, 517–534

2 Nee LE, Polinsky RJ, Eldridge R, Weingarter H, Smallberg S, Ebert M (1983) A family with histologically confirmed Alzheimer's disease. *Arch Neurol* 40: 203–208

3 Kang J, Lemaire H-G, Unterbeck A, Salbaum JM, Master CL, Grzeschik K-H, Multhaup G, Beyreuther K, Muller-Hill B (1987) The precursor of Alzheimer's disease amyloid A4 protein resembles a cell-surface receptor. *Nature* 325: 733–737

4 Goate A, Chartier Harlin M-C, Mullan M, Brown J, Crawford F, Fidani L, Guiffra L, Haynes A, Irving N, James L et al (1991) Segregation of a missence mutation in the amyloid precursor protein gene with familial Alzheimer's disease. *Nature* 349: 704–706

5 Chartier-Harlin MC, Crawford F, Houlden H, Warren A, Hughes D, Fidani L, Goate A, Rossor M, Roques P, Hardy J et al (1991) Early-onset Alzheimer's disease caused by mutations at codon 717 of the beta-amyloid precursor protein gene. *Nature* 353: 844–846

6 Murrell J, Farlow M, Ghetti B, Benson M (1991) A mutation in the amyloid precursor protein associated with hereditary Alzheimer's disease. *Science* 253: 97–98

7 Mullan M, Crawford F, Axelman K, Houlden H, Lilius L, Winblad B, Lannfelt L (1992) A pathogenic mutation for probable Alzheimer's disease in the APP gene at the N-terminus of B-amyloid. *Nat Genet* 1: 345–347

8 Peacock ML, Murman DL, Sima AA, Warren JT Jr, Roses AD, Fink JK (1994) Novel amyloid precursor protein gene mutation (codon 665Asp) in a patient with late-onset Alzheimer's disease. *Ann Neurol* 35: 432–438

9 Carter DA, Desmarais E, Bellis M, Campion D, Clerget DF, Brice A, Agid Y, Jaillard SA, Mallet J (1992) More missense in amyloid gene [letter]. *Nat Genet* 2: 255–256

10 Ancolio K, Dumanchin C, Barelli H, Warter JM, Brice A, Campion D, Frebourg T, Checler F (1999) Unusual phenotypic alteration of beta amyloid precursor protein (betaAPP) maturation by a new Val-715 → Met betaAPP-770 mutation responsible for probable early-onset Alzheimer's disease. *Proc Natl Acad Sci USA* 96: 4119–4124

11 Eckman CB, Mehta ND, Crook R, Perez TJ, Prihar G, Pfeiffer E, Graff RN, Hinder P, Yager D, Zenk B et al (1997) A new pathogenic mutation in the APP gene (I716V) increases the relative proportion of A beta 42(43). *Hum Mol Genet* 6: 2087–2089

12 Kwok JB, Li QX, Hallupp M, Milward L, Whyte S, Schofield PR (1998) Novel familial early-onset Alzheimer's disease mutation (Leu723Pro) in amyloid precursor protein (APP) gene increases production of 42(43) amino-acid isoform of amyloid beta peptide. *Neurobiol Aging* 19: S91

13 Hendriks L, Van Duijn C, Cras P, Van Hul P, Harskamp F, Warren A, McInnis M, Antonarakis S, Martin J-J, Hofman A et al (1992) Presenile dementia and cerebral haemorrhage linked to a mutation at codon 692 0f the beta amyloid precursor protein gene. *Nat Genet* 1: 218–221

14 Kamino K, Orr HT, Payami H, Wijsman EM, Alonso ME, Pulst SM, Anderson L, O'dahl S, Nemens E, White JA et al (1992) Linkage and mutational analysis of familial Alzheimer disease kindreds for the APP gene region. *Amer J Hum Genet* 51: 998–1014

15 Schellenberg GD, Bird TD, Wijsman EM, Orr HT, Anderson L, Nemens E, White JA, Bonnycastle L, Weber JL, Alonso ME et al (1992) Genetic linkage evidence for a familial Alzheimer's disease locus on chromosome 14. *Science* 258: 668–671

16 Sherrington R, Rogaev EI, Liang Y, Rogaeva EA, Levesque G, Ikeda M, Chi H, Lin C, Li G, Holman K et al (1995) Cloning of a gene bearing mis-sense mutations in early-onset familial Alzheimer's disease. *Nature* 375: 754–760

17 Levy Lahad E, Wijsman EM, Nemens E, Anderson L, Goddard KA, Weber JL, Bird TD, Schellenberg GD (1995) A familial Alzheimer's disease locus on chromosome 1 [see comments]. *Science* 269: 970–973

18 Scheuner D, Eckman C, Jensen M, Song X, Citron M, Suzuki N, Bird TD, Hardy J, Hutton M, Kukull W et al (1996) Secreted amyloid beta-protein similar to that in the senile plaques of Alzheimer's disease is increased *in vivo* by the presenilin 1 and 2 and APP mutations linked to familial Alzheimer's disease [see comments]. *Nat Med* 2: 864–870

19 Iwatsubo T, Mann DM, Odaka A, Suzuki N, Ihara Y (1995) Amyloid β protein (Aβ) deposition: Aβ42(43) precedes Aβ40 in Down syndrome. *Ann Neurol* 37: 294–299

20 Kennedy AM, Newman SK, Frackowiak RSJ, Cunningham VJ, Roques P, Stevens J, Neary D, Bruton CJ, Warrington EK, Rossor MN (1995) Chromosome 14 linked familial Alzheimer's disease: A clinico-pathological study of a single pedigree. *Brain* 118: 185–205

21 Swearer JM, O'Donnell BF, Drachman DA, Woodward BM (1992) Neuropsychological features of familial Alzheimer's disease. *Ann Neurol* 32: 687–694

22 Warrington EK, Agnew SK, Kennedy AM, Rossor MN (1999) Neuropsychological profiles of familial

Alzheimer's disease associated wtih mutation in the presenilin 1 gene. *J Neurol; in press*

23 Haltia M, Viitanen M, Sulkava R, Ala-Hurula V, Poyhonen M, Goldfarb L, Brown P, Levy E, Houlden H, Crook R et al (1994) Chromosome 14-encoded Alzheimer's disease: genetic and clinicopathological description. *Ann Neurol* 36: 362–367

24 Rossor MN, Fox NC, Beck J, Campbell TC, Collinge J (1996) Incomplete penetrance of familial Alzheimer's disease in a pedigree with a novel presenilin-1 gene mutation [letter]. *Lancet* 347: 1560

25 Fox NC, Kennedy AM, Harvey RJ, Lantos PL, Roques PK, Collinge J, Hardy J, Hutton M, Stevens JM, Warrington EK et al (1997) Clinicopathological features of familial Alzheimer's disease associated with the M139V mutation in the presenilin 1 gene. Pedigree but not mutation specific age at onset provides evidence for a further genetic factor. *Brain* 120: 491–501

26 Cruts M, VanBroeckhoven C (1998) Presenilin mutations in Alzheimer's disease. *Hum Mutat* 11: 183–190

27 Newman SK, Warrington EK, Kennedy AM, Rossor MN (1994) The earliest cognitive change in a person with familial Alzheimers disease – presymptomatic neuropsychological features in a pedigree with familial Alzheimers disease confirmed at necropsy. *J Neurol Neurosurg Psychiat* 57: 967–972

28 Folstein M, Folstein S, McHughs P (1975) The "Mini mental state": a practical method for grading the cognitive state of patients for the clinician. *J Psychiat Res* 12: 189–198

29 Warrington EK (1984) *Manual for the Recognition Memory Test for words and faces*. NFER-Nelson, Windsor

30 McFie J (1975) *Assessment of organic intellectual impairment*. Academic Press, New York

31 Fox NC, Warrington EK, Seiffer AS, Agnew SK, Rossor MN (1998) Presymptomatic cognitive deficits in individuals at risk of familial Alzheimer's disease. A longitudinal prospective study. *Brain* 121: 1631–1639

32 Kennedy AM, Frackowiak RSJ, Newman SK, Bloomfield PM, Seaward J, Roques P, Lewington G, Cunningham VJ, Rossor MN (1995) Deficits in cerebral glucose metabolism demonstrated by positron emission tomography in individuals at risk of familial Alzheimer's disease. *Neurosci Lett* 186: 17–20

33 Reiman EM, Caselli RJ, Yun LS, Chen K, Bandy D, Minoshima S, Thibodeau SN, Osborne D (1996) Preclinical evidence of Alzheimer's disease in persons homozygous for the epsilon 4 allele for apolipoprotein E [see comments]. *N Engl J Med* 334: 752–758

34 Small GW, Mazziotta JC, Collins MT, Baxter LR, Phelps ME, Mandelkern MA, Kaplan A, Larue A, Adamson CF, Chang L et al (1995) Apolipoprotein-e type-4 allele and cerebral glucose-metabolism in relatives at risk for familial Alzheimer-disease. *J Amer Med Assoc* 273: 942–947

35 Fox NC, Warrington EK, Freeborough PA, Hartikainen P, Kennedy AM, Stevens JM, Rossor MN (1996) Presymptomatic hippocampal atrophy in Alzheimer's disease: a longitudinal MRI study. *Brain* 119: 2001–2007

36 Jack CR, Petersen RC, Xu YC, Obrien PC, Smith GE, Ivnik RJ, Boeve BF, Waring SC, Tangalos EG, Kokmen E (1999) Prediction of AD with MRI-based hippocampal volume in mild cognitive impairment. *Neurology* 52: 1397–1403

37 Freeborough PA, Fox NC (1997) The boundary shift integral: an accurate and robust measure of cerebral volume changes from registered repeat MRI. *IEEE Trans Med Imaging* 16: 623–629

38 Fox NC, Freeborough PA, Rossor MN (1996) Visualisation and quantification of atrophy in Alzheimer's disease. *Lancet* 348: 94–97

39 Fox NC, Freeborough PA (1997) Brain atrophy progression measured from registered serial MRI: validation and application to Alzheimer's disease. *JMRI* 7: 1069–1075

40 Fox NC, Scahill RI, Crum WR, Rossor MN (1999) Correlation between rates of brain atrophy and cognitive decline in AD. *Neurology* 52: 1687–1689

41 Fox NC, Warrington EK, Rossor MN (1999) Serial magnetic resonance imaging of cerebral atrophy in preclinical Alzheimer's disease. *Lancet* 353: 2125

Neuroscientific Basis of Dementia
C. Tanaka, P.L. McGeer, Y. Ihara (eds)
© 2001 Birkhäuser Verlag Basel/Switzerland

Biological markers for differential diagnosis of Alzheimer's disease and related disorders

Masatoshi Takeda[1], Takashi Kudo[1], Yu Nakamura[1], Toshihisa Tanaka[1], Takashi Nishikawa[1], Kazuhiro Shinosaki[1] and Tsuyoshi Nishimura[2]

[1] Department of Clinical Neuroscience, Osaka University Graduate School of Medicine, Osaka, Japan
[2] Department of Human and Cultural Sciences, Koshien University, Takarazuka, Japan

Introduction

The concept of "Alzheimer's disease" has been changed several times since the original proposal. A case reported by Alois Alzheimer in 1905 was named "Alzheimer's disease" by Emil Kraepelin in his "Textbook of Psychiatry 5th Edition", which was initially proposed to differentiate presenile-onset dementia due to unknown causes from progressive paralysis and cerebrovascular dementia. Since then, the term "Alzheimer's disease" has been used to describe the typical presenile-onset neurodegenerative dementia like Pick's disease, which is another typical presenile onset neurodegenerative dementia. Alzheimer's disease and Pick's disease had been categorized as idiopathic dementias in presenile age for a long time.

Until the 1980s, the term "Alzheimer's disease" was used to describe dementia in the forties and fifties, which was regarded as different from so-called senile dementia. However, the progress in neuropathological and biochemical study of dementia in the 1980s and 1990s has demonstrated that both Alzheimer's disease and senile dementia are characterized by an abundant presence of neurofibrillary tangles and senile plaques in the neocortex and limbic cortex of the affected brain. These two diseases are difficult to distinguish from each other and they are both termed as dementia of Alzheimer type (DAT). The term DAT was mainly used to differentiate these diseases from cerebrovascular dementia, and probably included all neurodegenerative dementias other than cerebrovascular dementia.

Since the 1990s, the heterogenous nature of Alzheimer's disease has been recognized owing to the development of symptomatology of dementia, brain-imaging and diagnostic tools. Several classifications of Alzheimer's disease have been widely used in clinical settings, such as early-onset *versus* late-onset, familial *versus* non-familial (sporadic), typical *versus* atypical.

It is now more important to notice that several primary neurodegenerative dementias are regarded as independent disorders from Alzheimer's disease, which are generally categorized as "non-Alzheimer neurodegenerative dementias". They include diffuse Lewy body disease, frontotemporal dementia, corticobasal degeneration, and others. The schematic diagram is shown in Figure 1. It is required to give a correct diagnosis of these non-Alzheimer neurodegenerative disorders in clinical settings.

It is easy to give a diagnosis of Alzheimer's disease for typical cases; however, it is also true that there are cases of atypical Alzheimer's disease which are difficult to diagnose correctly. Clinicians are facing more or less the same situation when they are expected to diagnose one of the non-Alzheimer dementias. In this sense the need for biological diagnostic markers for the diagnosis is urgent.

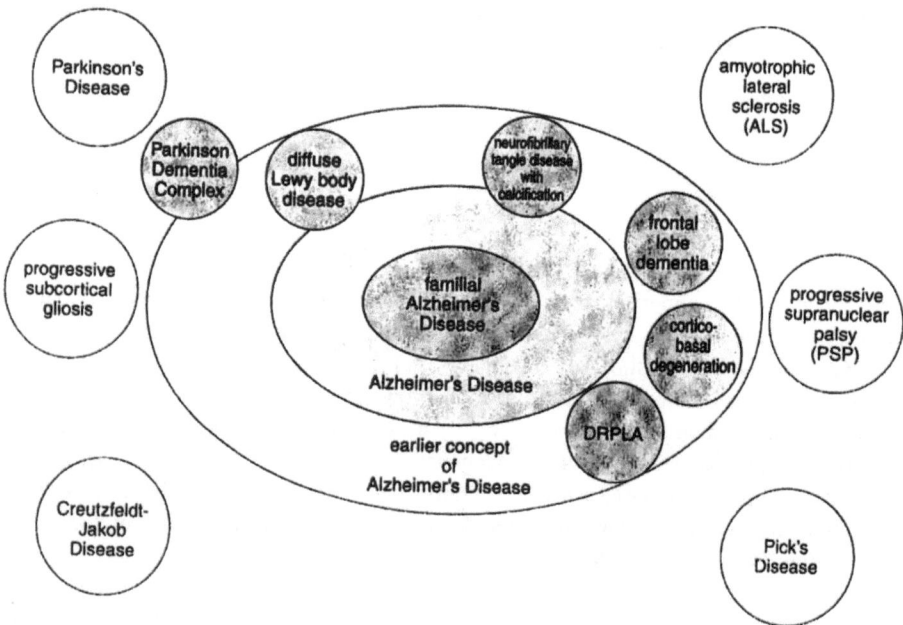

Figure 1. Primary neurodegenerative dementing disorders.

Differential diagnosis of dementia

Neurodegenerative dementia disorders should be differentiated from Alzheimer's disease and some characteristics of each non-Alzheimer dementia are briefly described here.

Diffuse Lewy body disease is the second most prevalent dementia following Alzheimer's disease and is characterized by abundant Lewy bodies in the neocortex. The cognitive impairment is sometimes variable depending upon the patient's condition, showing variable levels of performance in cognitive examination by date and place depending on the degree of concentration. Visual delusion and related hallucinations are characteristics of patients with diffuse Lewy body disease. Patients often show syncope, falling or temporary unconsciousness like transient ischemic attack, and most typically signs and symptoms of Parkinsonism are observed at some time in the course of the duration of the disease.

Frontal lobe dementia or frontotemporal dementia is characterized by decreased spontaneity, decreased motivation and abnormal behavior. The uniqueness in symptoms of frontal lobe dementia as compared with those of Alzheimer's disease is the lack of focal signs and symptoms attributed to dysfunction of the local neocortex. The patients are more obviously afflicted with impaired judgement or loss of motivation rather than memory impairment. About half of the patients show familial occurrence. Pick's disease is characterized by pre-senile dementia with personality changes, abnormal behavior and loss of morality. The patients' social behavior has deteriorated but the memory function is relatively preserved. Pick's disease is included as a subtype of frontotemporal dementia.

Corticobasal degeneration starts by motor clumsiness of one side limb and progressively shows impaired coordination of the upper limb. Impaired fixation of his own body, sign of alien hand are well recognized signs of this disease. Along with impaired motor function of the limbs, dementia occurs insidiously.

Differential diagnosis among non-Alzheimer degenerative dementia is possible to a certain extent in typical cases. However there are many cases in which the correct diagnosis by clinical observation is difficult. Biological diagnostic markers are needed to give the final diagnosis of Alzheimer's disease and of other dementias. It is to be noticed that signs and symptoms of dementia are different for each individual, and are unique to each patient. Considering these situations, biological markers for dementia are essential, in addition to neuroimaging and other examination.

Biological markers for Alzheimer's disease

Many parameters have been studied as candidates for possible biological markers for diagnosis of Alzheimer's disease, some of which are listed in Table 1. Some cases of Alzheimer's disease are due to familial Alzheimer's disease (FAD), in which causative genes have been identified, It is possible to do genetic screening for FAD with the identified gene mutations.

Genetic diagnosis

About 10–30 percent of AD are familial and caused by genetic factors. At present, three causative genes have been identified. Amyloid precursor protein (APP) gene is the first causative gene identified with FAD in 1991. Val in 717 residue of APP is substituted by Gly, Phe, or Ileu (London type mutation), which is located 3–5 amino acids apart from the gamma-secretase cleavage site of the molecule. Another mutation in the APP gene is the 670/671 double mutation, known as the Swedish-type mutation. This mutation is located near the N terminal of beta-peptide which is to be cleaved by beta-secretase. Since amyloid beta-peptide is cleaved from APP, these mutations are believed to cause AD by the mechanism related to increased production of amyloid beta-peptide.

The subsequent studies have, however, revealed that the frequency of mutations in the APP gene is much less than expected, accounting for less than 1% of all Alzheimer's disease. The subsequent search for a new gene for FAD has disclosed presenilin-1 on chromosome 14 and presenilin-2 on chromosome 1 in 1995. At present more than 55 mutations have

Table 1. Biological markers for AD diagnosis

CSF	Total tau protein level
	Phosphorylated tau protein level
	Ubiquitin level
	α1-antichymotrypsin level
	amyloid β-protein level (β40 and β42(43))
	amyloid β42(43) level
	secreted form of APP
	SOD activity
	Cathepsin D activity
Serum	α1-antichymotrypsin
	apolipoprotein E
	APP mutation
	Presenilin–1 mutation
	Presenilin-2 mutation
	Presenilin-1 polymorphism
	Anti-GFAP titer
Cultured fibroblast	Aberrant cytoskeletal arrangement
	Abnormal Ca mobilization
	APP isomer
	SOD activity
	Cell adhesion
Cutaneous tissue	Deposition of amyloid β protein
	Change in elasticiy of cutaneous tissue

been identified with presenilin-1, accounting for about 70–80% of FAD cases. Because these mutations in APP and PSs show almost 100% penetrance, a carrier of one of these mutations is expected to be subject to AD in the onset age determined by the mutation. Theoretically, it is possible to give a genetic diagnosis of the FAD victim before the onset of clinical symptoms; however, we should be very cautious about giving the genetic diagnosis to the possible carrier of the FAD mutation, because the declaration of Alzheimer's disease is quite a disaster to most individuals. It should be considered that declaration of genetic disease is stressful and sometimes cannot be tolerated by the possible victims and their family, especially when the disease is long-lasting progressive dementia. Still, social discussion should be awaited for when the genetic diagnosis of FAD can be given to the possible patients and to their family.

Apolipoprotein E4

Apolipoprotein E is a haloprotein in serum playing an important role in lipid metabolism and transport. A single gene on chromosome 19 codes for it and three isomers are known: E2, E3 and E4. The wild-type is E3 whose 112th residue is Cys and 158th residue is Arg.

Apolipoprotein E4 is substituted by Arg in the 112th residue, and apolipoprotein E2 is substituted by Cys in the 158th residue of the molecule of apolipoprotein E3. The frequency of these three isomers is slightly different by race showing e2: e3: e4 = 0.08: 0.77: 0.15 for Caucasians, and e2: e3: e4 = 0.05: 0.86: 0.09 for Japanese population. The frequency of e4 in the Japanese population is about half of the Caucasians. In Alzheimer patients, e4 frequency is 0.52, which is significantly higher than that of the general population, indicating the possibility that individuals with e4 are more prone to Alzheimer's disease. The higher frequency of e4 has been originally demonstrated with patients of early-onset Alzheimer's disease, and then the higher frequency has been also demonstrated with late-onset Alzheimer's disease. The following studies have revealed that higher e4 frequency is observed with patients of many other diseases causing dementia, such as diffuse Lewy body disease, Pick's disease and cerebrovascular dementia, indicating that apolipoprotein E4 can be a risk factor of age-associated dementia in general.

Apolipoprotein E4 is not only a risk factor of dementia, but it may also accelerate the progression of Alzheimer pathogenesis. When the onset age is compared with the number of Apo E4 alleles, it is clearly demonstrated that subjects with two Apo E4 alleles disclose earlier onset of AD than subjects with one Apo E4 allele [1]. Subjects with one ApoE4 allele show earlier onset than those with no Apo E4 allele. It is concluded that Apolipoprotein E4 is a risk factor of Alzheimer's disease but of course there are patients with AD but without ApoE4 allele and subjects with the two ApoE4 alleles can live their lives without AD occurrence. However it can be used to evaluate the risk for being affected by AD in a population [2].

Biological markers in cerebrospinal fluid

Tau protein

Tau is one of the microtubule-associated proteins, composing the core cytoskeletal system in neurons and in other cells. Tau protein is playing an important role in neuronal functions, stimulating microtubule polymerization. Tau protein is hyperphosphorylated in AD brain and phosphorylation of tau results in decreased tubulin-binding capacity, causing diminished stabilization of polymerized microtubules.

Neurofibrillary tangles are composed of paired helical filaments (PHF) and they are known to be composed of phosphorylated tau. Tau protein level in AD brain is increased, and the level of phosphorylation is kept at a much higher level than the control brain.

Based on the knowledge described above, tau protein level in cerebrospinal fluid (CSF) has been studied for a possible biological marker for diagnosis of AD. It is shown that tau protein level is significantly increased in AD groups over other groups. As shown in Figure 2, the AD group shows higher tau levels than non-Alzheimer dementia, cerebrovascular disorder, or neurological control [3]. However, some AD patients show lower tau protein levels and some non-AD patients show higher levels, indicating the insufficient specificity of tau protein levels as the diagnostic marker of AD. There is a certain overlap between the AD group and other dementia groups. Another important finding is that some non-Alzheimer

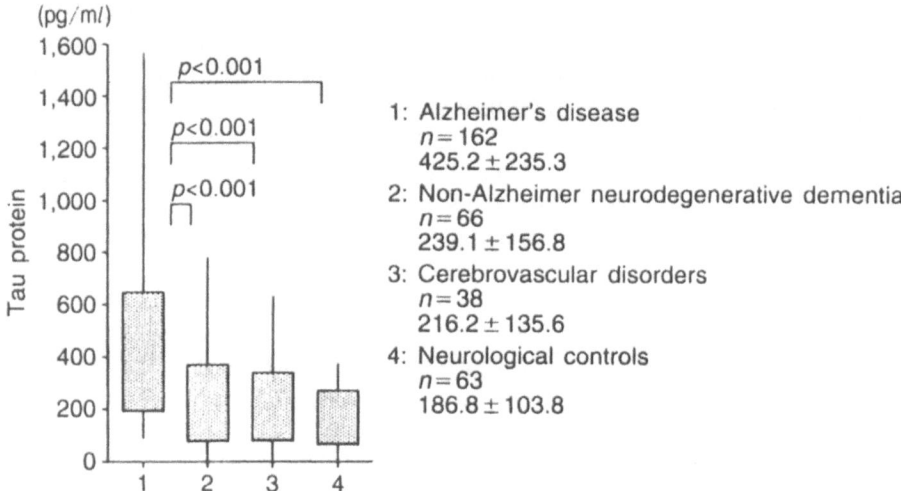

Figure 2. Tau protein level in CSF of patients with (1) Alzheimer's disease, (2) Non-Alzheimer neurodegenerative dementia, (3) Cerebrovascular disorders and (4) Neurological controls.

dementia patients also show increased tau levels in CSF. Patients with frontotemporal lobe dementia, Pick's disease and corticobasal degeneration show increased tau levels in CSF.

We studied the correlation between tau level and ApoE4 alleles, showing no significant correlations between tau level in CSF and the number of ApoE4 alleles [4]. The tau level in CSF is already increased in the initial stage of AD and it shows a relatively constant level along with the progression of the disease, demonstrating no significant increase in tau level with progression of AD staging or duration of AD [5].

Amyloid beta-protein

Amyloid beta-protein is deposited in the brain parenchyma and vascular wall of AD patients. Amyloid fibril is composed of amyloid beta-protein which is cleaved from APP. Amyloid beta-protein is composed of 39–43 amino acid residues. Longer amyloid beta-protein with 42 [43] residues are more prone to be aggregated and also more cytotoxic than amyloid beta-protein with 40 residues.

The level of amyloid beta-protein was measured in CSF from AD patients, There was no difference between the AD and the healthy control group, when amyloid beta-protein with 42 residues (Aβ42) and amyloid beta-protein with 40 residues (Aβ40) are measured together. Now utilization of more specific antibodies, specific for only Aβ42(43), has made it possible to measure the level of Aβ42(43) in CSF by the ELISA system. The Aβ42(43) level is significantly decreased with AD patients. The combination of the lower level of Aβ42(43) and the higher level of tau in CSF can be used as the diagnostic marker for AD [6].

Ubiquitin

Ubiquitin is a protein functioning in the degradation of unnecessary/abnormal proteins by the ATP-dependent proteosome system. The targeted protein is marked by ubiquitination and the ubiquitinated protein is recognized by the protease. In this way ubiquitination is important in the scavenging system of the aged protein. Ubiquitin is known to be localized in neurofibrillary tangles of AD brain. It is shown that ubiquitin level is increased in AD brain tissue. It is also shown that ubiquitin level in CSF of AD patients is significantly increased, which can be used as another biological marker for AD diagnosis [7].

Significance of biological markers

As shown above, tau protein levels, Aβ42(43) levels, and ubiquitin levels in CSF can be used as biological diagnostic markers for AD. The interesting point is that these three parameters show rather independent change by patients: some AD patients show more pronounced high tau levels, others show more decreased Aβ(42) levels, and others show more increased ubiuitin levels. These parameters can be utilized as a parameter to further classify AD patients. The characteristics of these parameters, corresponding to clinical signs and symptoms of each AD patients, remain to be studied

References

1 Corder EH, Saunders AM, Strittmatter WJ, Schmechel DE, Gaskell PC, Small GW, Roses AD, Haines JL, Pericak-Vance MA (1993) Gene dose of apolipoprotein E type 4 allele and the risk of Alzheimer's disease in late-onset families. *Science* 261: 921–923

2 Nishimura T, Takeda M, Shinosaki K, Nishikawa T, Nakamura Y, Yoshida Y, Sasaki H, Arai H, Hirai S, Shouji M et al (1998) Basic and clinical studies on ApoE gene typing by line probe assay (LiPA) as a biological marker for Alzheimer's disease and related disorders: Multicenter study in Japan. *Meth Find Exp Clin Pharmacol* 20: 793–799

3 Nishimura T, Takeda M, Nakamura Y, Yosbida Y, Arai H, Sasaki H, Shouji M, Hirai S, Khise K, Tanaka K et al (1998) Basic and clinical studies on the measurement of tau protein in cerebrospinal fluid as a biological marker for Alzheimer's disease and related disorders: multicenter study in Japan. *Meth Find Exp Clin Pharmacol* 20: 227–235

4 Takeda M, Shinosaki K, Nishikawa T, Nakamura Y, Nishimura T, Yoshida Y, Arai H, Sasaki H, Shouji M, Hirai S et al (1998) Tau protein levels in cerebrospinal fluid and apolipoprotein E genotyping in differential diagnosis of dementia. *Ann Psychiat* 7: 135–146

5 Morihara T, Kudo T, Ikura Y, Kashiwagi Y, Miyamae Y, Nakamura Y, Tanaka T, Shinozaki K, Nishikawa T, Takeda M (1998) Increased tau protein level in postmortem cerebrospinal fluid. *Psychiat Clin Neurosci* 52: 107–110

6 Motter R, Vigo-Pelfrey C, Kholodenko D, Barbour R, Johnson-Wood K, Galasko D, Chang L, Miller B, Clark C, Green R (1995) Reduction of beta-amyloid peptide 42 in the cerebrospinal fluid of patients with Alzheimer's disease. *Ann Neurol* 38: 643–648

7 Kudo T, Iqbal K, Ravid R, Swaab DF, Grundke-Iqbal I (1994) Alzheimer disease; Correlation of cerebrospinal fluid and brain ubiquitin levels. *Brain Res* 639: 1–7

Dietary factors and the risk of Alzheimer's disease: a low fish consumption and a relative deficiency of ω-3 polyunsaturated fatty acids

Akira Ueki[1], Mieko Otsuka[1], Satosi Sasaki[2], Yoshio Nanba[3], Yasuyoshi Ouchi[3] and Kazuhiko Ikeda[4]

[1] *Department of Neurology, Omiya Medical Center, Jichi Medical School, Saitama, Japan*
[2] *Department of Epidemiology, National Cancer Research Center, Chiba, Japan*
[3] *Department of Geriatric Medicine, Graduate School of Medicine, Tokyo University, Tokyo, Japan*
[4] *Department of Ultrastructure, Tokyo Metropolitan Institute of Psychiatry, Tokyo, Japan*

Introduction

Multiple risk factors, both genetic and environmental, are involved in the development of late-onset Alzheimer's disease (AD). Though diet is a potentially modifiable factor, few studies have reported on the association between AD and diet. Recently, Kalmijn and colleagues [1] have found in a large prospective study in Rotterdam that fish consumption is inversely related to the incidence of AD.

In this study, we assessed dietary habits of Japanese patients with AD and compared them with those of age-matched healthy Japanese elderly. Special interest was focused on the intake balance of polyunsaturated fatty acids (PUFA). There are two series of PUFAs called n-3 (ω3) PUFA and n-6 (ω6) PUFA, which are not synthesized in mammals and are also called essential fatty acids. Eicosapentaenoic acids (EPA) and docosahexaenoic acid (DHA) are n-3 PUFA enriched in fish oil, and linoleic acid (LA) and arachidonic acid (AA) are n-6 PUFA abundant in meat. Many chronic diseases in the elderly are associated with increased ratios of n-6/n-3 PUFA; i.e., excess n-6 and relative n-3 deficiency, mainly caused by preferential intake of more meat than fish [2]. We also assessed nutritional factors in the function of apolipoprotein E (apoE) genotype to observe gene-environmental interactions.

Methods

Participants were 37 Japanese AD patients (72.8 ± 8.6 years) and 70 healthy elderly (76.1 ± 10.6 years). AD was diagnosed according to the criteria of DSM-TV [3] with cranial computed tomography (CT) scan. Controls were recruited from cognitively healthy spouses of AD patients or cognitively healthy patients without neurological diseases, hyperlipidemia or diabetes mellitus. Informed consent was obtained from all subjects and/or their family members for the study.

Dietary habits were assessed using a self-administered diet history questionnaire (DHQ) developed for dietary surveys in Japanese subjects [4]. The DHQ was designed to assess one-month dietary habits which were recorded by spouses or daughters of both AD patients and controls. Nutrients were calculated using an *ad hoc* computer program developed for DHQ and were expressed by 1,000 kcal energy intake.

ApoE was genotyped using the conventional PCR-based method.

Results

Dietary habits in patients with AD

Dietary habits of AD patients markedly deviated from those of controls. Most controls took traditional Japanese meals consisting of fish, vegetables and rice. However, as many as 57.1% of AD patients disliked fish and 60.5% of AD patients exclusively took meat as a protein source. AD patients also disliked green-yellow vegetables. These dietary habits began in their 20 s or 30 s, and they became more exaggerated after the onset of dementia.

Food groups

Out of 18 food groups investigated, three food groups, fish, meat and green-yellow vegetables were significantly different between AD patients and controls. In AD patients, fish consumption was significantly lower (40.2 ± 20.0 vs 61.9 ± 28.1 g: $p < 0.0001$) and meat consumption was significantly higher (27.3 ± 16.4 vs 20.5 ± 15.6 g: $p = 0.037$) than controls (43.2 ± 28.1 vs 74.6 ± 62.8 g: $p = 0.005$).

Nutrients

Energy-adjusted intake of nutrients showed that AD patients took significantly lower amounts of calcium ($p = 0.0016$), vitamin C ($p = 0.0046$) and carotene ($p = 0.0104$).

Lipid profiles

Lipid profiles showed no difference in the intake of cholesterol, saturated fatty acid (SFA), and monounsaturated fatty acids (MUFAs) between AD patients and controls (Tab. 1). AD patients took significantly lower amounts of n-3 PUFAs and higher amounts of n-6 PUFAs than controls, though the magnitude of significance is bigger in n-3 PUFAs. The n-6/n-3 ratio was significantly higher in AD patients than controls.

Table 1. Lipid profile

		AD (n = 37)	Controls (n = 70)	P value
cholesterol	(mg)	155.7 ± 91.1	161.6 ± 67.8	0.7094
	ε4+	185.8 ± 117.0	161.0 ± 73.4	NS
	ε4−	135.5 ± 68.5	154.1 ± 62.5	NS
SFA	(en%)	7.3 ± 2.0	7.4 ± 2.9	NS
MUFA	(en%)	8.9 ± 2.6	8.2 ± 2.7	NS
PUFA	(en%)	6.7 ± 1.8	6.5 ± 2.0	NS
n-3	(en%)	1.3 ± 0.4	1.6 ± 0.6	0.0042
n-6	(en%)	5.4 ± 1.6	5.0 ± 1.5	0.1597
n-6/n-3		4.4 ± 1.1	3.3 ± 1.0	<0.0001

SFA: Saturated fatty acid
MUFA: monounsaturated fatty acid
PUFA: polyunsaturated fatty acid

Nutrients and apolipoprotein E genotype

The n-6/n-3 ratio was the lowest in non-ε4 carriers of controls (2.94 ± 0.68) and the ratio increased in the following order: non-ε4 carriers of controls (3.59 ± 1.06), ε4 carriers of AD (3.98 ± 0.69) and non-ε4 carriers of AD (4.74 ± 2.26). Calcium, vitamin C and carotene were significantly lower only in AD patients without ε4 (Tab. 2).

Table 2. Nutrients and Apo E genotype

		n6/n3	Ca (mg)	Vit.C (mg)	Carotene (mg)
‹ε4+›					
AD	(n = 14)	4.1 ± 0.9	361.6 ± 152.2	67.2 ± 30.0	1187.5 ± 807.2
Controls	(n = 13)	2.8 ± 0.7	432.9 ± 200.0	82.7 ± 37.7	1346.8 ± 789.7
		p = 0.0003	p = 0.3046	p = 0.2452	p = 0.6092
‹ε4−›					
AD	(n = 22)	4.6 ± 2.0	279.6 ± 103.2	65.1 ± 39.2	874.2 ± 585.6
Controls	(n = 49)	3.5 ± 1.0	441.0 ± 158.8	91.4 ± 38.6	1774.0 ± 1272.6
		p = 0.0012	p < 0.0001	p = 0.0101	p = 0.0023

Discussion

Dietary habits of AD patients who took more meat and less fish and vegetables contrasted clearly with those of healthy Japanese elderly who took traditional Japanese foods. Though our study design is cross-sectional, present results agree with the previous findings by Kalmijin et al. [1] that fish consumption is inversely related with the incidence of AD. One question still exists, i.e., whether these dietary habits are the cause or the consequence of dementia. Generally, dietary habits are very conservative and unchangeable, and most patients in the present study had never changed their dietary habits since their youth, far before the onset of dementia. Once dementia starts, their dietary habits become prominent and then a vicious cycle might begin. So, the dietary factor plays at least an exacerbating role.

The n-6/n-3 ratio of PUFA was significantly increased in AD patients compared with controls. It is known that an increased ratio of n-6/n-3 in dietary PUFA is associated with many chronic diseases in the elderly, such as western-type cancers (colorectal, esophageal or breast cancer), cardiovascular and cerebrovascular diseases and also for allergic hyper-reactivity [2]. Eicosanoids derived from n-6 PUFA, such as AA, tend to promote thrombotic, atherosclerotic and inflammatory processes. In contrast, eicosanoids derived from n-3 PUFA, such as EPA and DHA, have opposite effects to these processes [3]. So, excess intake of n-6 PUFA and/or less intake of n-3 PUFA link with thrombotic, atherosclerotic and inflammatory processes, which are assumed to be involved in the pathological process of AD [5, 6].

The n-6/n-3 ratio was the lowest in apoE-ε4 carriers who escaped from dementia. This result implies that genetic risk of AD might be overcome by strict restriction of animal fat intake. The nutritional condition of AD patients without apoE-ε4 is more complex. In addition to the increased n-6/n-3 ratio, vitamin C, carotene and antioxidant levels were significantly lower in this subgroup. Antioxidant level is also important since it is associated with cognitive function in very old men [7]. If our results are confirmed, nutritional intervention according to apoE genotype will become useful for prevention and therapy of AD.

References

1 Kalmijn S, Launer LJ, Ott A, Witteman JCM, Hofman A, Breteler MMB (1997) Dietary fat intake and the risk of incident dementia in the Rotterdam study. *Ann Neurol* 42: 776–782
2 Okuyama H, Kobayashi T, Watanabe S (1997) Dietary fatty acids-The n-6/n-3 balance and chronic elderly diseases. Excess linoleic acid and relative n-3 deficiency syndrome seen in Japan. *Prog Lipid Res* 35: 409–457
3 American Psychiatric Association (1994) Diagnostic and statistical manual of mental disorders, 4th edn, DSM-IV. American Psychiatric Association, Washington D.C.
4 Sasaki S, Yanagibori R, Amano K (1998) Self-administered diet history questionnaire developed for health education: A relative validation of the test-version by comparison with 3-day diet record in women. *J Epidemiol* 8: 203–215
5 Hofman A, Ott A, Breteker MMB, Bots ML, Slooter AJC, van Harskamp F, van Duijn CM, Van Broekhoven C, Grobbee DE (1997) Atherosclerosis, apolipoprotein E and the prevalence of dementia and Alzheimer's disease in the Rotterdam Study. *Lancet* 349: 151–154
6 McGeer EG, McGeer PL (1995) The inflammatory system of brain: implications for therapy of Alzheimer and other neurodegenerative disorders. *Brain Res* 21: 195–218
7 Kalmijn S, Feskens EJM, Launer LJ, Kromhout D (1997) Polyunsaturated fatty acids, antioxidant, and cognitive function in very old men. *Amer J Epidemiol* 145: 33–41

Neuroscientific Basis of Dementia
C. Tanaka, P.L. McGeer, Y. Ihara (eds)
© 2001 Birkhäuser Verlag Basel/Switzerland

Risk factors for dementia

Shigenobu Nakamura[1], Hirofumi Maruyama[1], Hiromasa Toji[1], Hideshi Kawakami[1], Michiko Yamada[2] and Yasuyo Mimori[1]

[1] *Third Department of Internal Medicine, Hiroshima University School of Medicine,* [2] *Radiation Effects Research Foundation, Hiroshima, Japan*

Introduction

Because the frequency of Alzheimer's disease (AD) and other dementias increases according to advancing age, the importance of dementia is recognized more in the society where the aged occupy a larger part of the population. Although the cause or risk factor for dementia has not been fully elucidated, the elucidation of risk factors may provide a useful tool for the prevention of dementia. Risk factors are divided into 3 groups, including age, genetic factors and environmental factors. Age is an uncontrollable factor which greatly influences the incidence of dementias. However, the research on genetic or environmental risk is essential to resolve important clinical issues and to suggest sites for therapeutic intervention. In this article, I will describe some aspects of genetic and environmental risk factors of dementia, especially the radiation effect after the atomic bomb.

Genetic risk factors

Apolipoprotein E (ApoE) promoter gene

A significant portion of patients with AD clearly shows genetic susceptibility. Among various genetic risk factors, the ApoE4 genotype appears to be the most closely related genetic factor for late-onset AD [1]. Other genetic factors within the ApoE gene region may be associated with the development of AD. Bullido reported that homozygosity of a common variant (-491A) in the ApoE promoter region was associated with increased risk of AD in Spanish and North American subjects and that the association was independent of ApoE4 status [2]. Furthermore, the -491A allele was exhibited with higher constitutive levels of ApoE promoter activity than the -491T allele. Accordingly, the -491A allele was suggested as contributing directly to risk of AD through the promoter's efficacy,which prompted us to study the biallelic polymorphism (-491 A/T) frequency of the gene in a Japanese population [3].

The subjects consisted of 118 Japanese sporadic AD patients and 118 healthy controls. Probable AD was diagnosed from clinical findings, according to the National Institute of

Neurological and Communicative Disorders and Strokes-Alzheimer's Disease and Related Disorders Association (NINCDS-ADRDA) criteria.

Genomic DNA was prepared from peripheral white blood cells and ApoE genotyping was performed as previously reported [4]. ApoE promoter genotyping was performed as previously described [3].

The AD group as a whole showed the expected higher frequency of the ApoE ε4 allele with a frequency of 35% in the AD group compared to that of 8% in the control population ($\chi^2 = 48.8$, p < 0.0001). The ε3 allele frequency was low in the AD group ($\chi^2 = 37.8$, p < 0.0001), although the ε2 allele frequency unchanged ($\chi^2 = 1.98$, p = 0.17).

The distribution of ApoE promoter genotypes and alleles in AD cases followed the Hardy-Weinberg equilibrium and did not significantly differ from that of controls. The AA genotype was not overexpressed in AD group compared with controls (87.3% vs 90.7%). The percentage of Japanese controls with AA genotype (90.7%) was significantly higher than Spanish (55.0%, $\chi^2 = 40.00$, p < 0.0001) or North American controls (71.0%, $\chi^2 = 16.60$, p < 0.0001). In individuals without ApoE ε4 alleles, the AA genotype was not overexpressed in the AD group ($\chi^2 = 0.014$, p = 0.91). Furthermore, this tendency was also observed in late-onset AD patients.

A recent study suggested a common polymorphism in the regulatory region of ApoE by modifying the risk of AD, independent from the ApoE ε4 allele in Spanish and North American groups [2]. Therefore, we examined ApoE promoter polymorphism in Japanese AD patients. The distribution of ApoE promoter genotypes and alleles in AD was not significantly different from that of controls. The ApoE promoter genotype frequencies differed markedly between the Japanese and Spanish or North American controls, suggesting that genotype frequencies vary, depending on the ethnic background. Thus, the association of ApoE promoter polymorphism, observed in Spanish and North American AD patients, may be masked due to the ethnic background. In short, we could not confirm the association of ApoE promoter polymorphism with AD in this Japanese population.

Estrogen receptor gene

Estrogen has been reported to improve the cholinergic hypofunction and cerebral blood flow, increase the metabolism of amyloid precursor protein, facilitate neuronal repair, reduce neuronal injury and stimulate glucose transport and metabolism. All these features might encourage an estrogen therapy in AD.

Associations have been shown between estrogen receptor α (ERα) gene polymorphisms and breast cancer, spontaneous abortion, hypertension in women or osteoporosis. The non-transcribed portion of this gene and its mutations are important, since it may be involved in the regulatory domain. Restriction fragment-length polymorphisms (RFLPs) of the ERα gene have been associated with a low bone-mineral density in postmenopausal women. These polymorphisms were a combination of two single nucleotide polymorphisms, located in intron 1 and associated with AD in a Japanese population [5]. As the polymorphisms exist in intron 1, RFLPs would change the enhancer activity of this gene. To determine the contribution of this polymorphism to AD, we examined the enhancer activity of ERα gene among various RFLPs and the relationship between ERα gene RFLPs and AD [6].

ERα polymorphism was analyzed by the method of Maruyama et al. [6]. Studies were performed after obtaining informed consent from patients or their families and approval from local institutional human subject boards. We collected blood from 183 sporadic Japanese AD patients clinically diagnosed by NINCDS-ADRDA criteria, and control blood from 133 healthy Japanese volunteers. Caucasian AD patients (n = 156) were confirmed by examination at autopsy and Caucasian controls (n = 120) were autopsy-examined non-AD cases, both from Cambridge.

There was no significant difference in ERα gene polymorphism distribution between AD and controls (Japanese genotype, $\chi^2 = 6.4$, p = 0.27; Japanese PX allele, $\chi^2 = 1.4$, p = 0.49; Japanese P allele, $\chi^2 = 1.2$, p = 0.28; Japanese X allele, $\chi^2 = 1.0$, p = 0.31; Caucasian genotype, $\chi^2 = 5.1$, p = 0.74; Caucasian PX allele, $\chi^2 = 0.6$, p = 0.91; Caucasian P allele, $\chi^2 = 0.3$, p = 0.62; Caucasian X allele, $\chi^2 = 0.3$, p = 0.59).

The X allele occurred more frequently in the Caucasians than the Japanese ($\chi^2 = 13.1$, p = 0.003). The ApoE gene did not influence ERα gene polymorphism distribution (Caucasians, $\chi^2 = 20.9$, p = 0.64; Japanese, $\chi^2 = 14.5$, p = 0.49).

Women are at greater risk of developing AD than men and estrogen-replacement therapy seems to reduce the risk. We have detected a weak, but significant enhancer activity of the 1.3- kbp fragment in intron 1 and exon 2 of ERα gene, and demonstrated different enhancer activity among haplotypes with higher enhancer activity in x than X allele [6]. A possible regulation of ERα expression through its polymorphisms has led us to examine the frequency of ERα gene RFLPs in AD patients.

An association was reported between ERα gene polymorphisms and Japanese AD patients (49.4% P allele in AD vs 36.3% P allele in control group and 29.1% X allele in AD vs 16.7% X allele in controls) [5]. But we failed to demonstrate this relationship also in Caucasian subjects with pathologically confirmed AD. The reason is not clear as to the discrepant results in Japanese patients. Gender distribution might not contribute to different findings. The P allele frequency in control subjects was lower in Isoe's study (36.3%) [5] than ours (40.6%) [6] or another (43.6%) [7], which would have contributed partly to the discrepancy. In contrast, the X allele distributions in controls did not differ among 3 studies.

Our study used 2 different ethnic groups and pathologically confirmed AD patients. A racial difference existed in the ERα gene; the X allele was more frequent in the Caucasians than the Japanese. Racial differences have been reported in other genes; A0 allele frequency of the tau gene is higher in Japanese than Caucasians, A allele of the -491A/T polymorphism in ApoE promoter gene higher in Japanese than Spanish [3]. ApoE4 genotype is a weaker risk factor for AD among African-Americans or Hispanics than Caucasians or Japanese. Consequently, studies with different ethnic groups are necessary to establish a relationship between polymorphisms and diseases [3, 6].

ERα gene polymorphism distribution does not seem to differ between AD patients and controls, either in Japanese or Caucasians. Further analysis of RFLPs in this region in relation to AD is not warranted.

Environmental risk factors

The prevalence of dementia in the Japanese population was almost same as in the Caucasian, but the ratio of AD to vascular dementia (VD) differed between 2 populations [8]. During the 1980 s, many studies in Japan revealed a predominance of VD, but an increased prevalence of AD has recently been reported [9]. Our study with participants in the Adult Health Study (AHS) of the Radiation Effects Research Foundation (RERF) were performed to determine the prevalence of dementia and its subtypes in the population [10]. Because radiation exposure seems to introduce effects on the brain, such as small head size and mental retardation in the RERF population of prenatally exposed atomic-bombs survivors, the relationship was examined between radiation dose and prevalence of dementia.

Between 1992 and 1996, 2934 men and women aged over 60 years in Hiroshima were examined in the AHS. A screening test for cognitive impairment was conducted by trained nurses during the period of the present study. After excluding those who refused the test, 2222 AHS members (637 men, 1585 women), 75.7% of the eligible AHS population, were screened for dementia.

Cognitive Ability Screening Instrument (CASI) [11] was used to assess cognitive impairment. CASI includes items in Hasegawa's Dementia Scale (HDS) or Mini-Mental State Examination (MMSE) and also an item on judgment, ranging 0 to 100. CASI was conducted completely for 2052 members as previously described [10].

Women were 71.3% of subjects and the female majority reflects the longer average life span, but the ratio for women was twice that for men at the inception of the AHS because men in military service outside Japan at the time of the atomic bombing were not selected as AHS cohort members. Five men and 43 women lived in institutions and 30% could not visit RERF. Those subjects, 4.7% of men and 8.8% of women, underwent examinations at home or other institutions.

We diagnosed 156 subjects with low CASI scores (40 men and 116 women) and no subjects selected randomly with high scores as having dementia, based on DSMIIIR. Four subjects with high CASI scores and suspect of cognitive impairment were diagnosed as having mild dementia. The observed sex-age-specific prevalence rates by subtypes revealed the equal number of AD and VD for men. For women, the number of AD was larger than VD. Mixed-type dementia and equivocal cases were classified as other dementia or unknown type (16.3%).

The overall crude prevalence of dementia was 7.2% {95% CI (6.1–8.3)} in total, 6.6% (4.7–8.5) in men, and 7.4% (6.1–8.7) in women. Dementia prevalence was related significantly to age ($p < 0.01$), and suggested a relationship to sex ($p = 0.065$) and sex-age interaction ($p = 0.065$) [10]. The slope of prevalence with age in the logistic model was steeper in women than in men, indicating a higher probability of dementia for women than for men after 80 years. The crude prevalence of AD was 2.0% (1.0–3.1) in men and 3.9% (2.9–4.7) in women, and that of VD, 2.0% (1.0–3.1) in men and 1.8% (1.2–2.5) in women. The estimated sex-age-specific prevalence of AD depends on age ($p < 0.01$), especially after age 80 and appears to be higher in women than men without an statistical sex difference ($p = 0.34$).

The prevalence of dementia was 4.0%, 35%, 63% or 88% for those examined at the foundation, home, hospital, or institution, respectively. Five men and 43 women (2.2%) lived in hospitals or institutions such as nursing homes. The prevalence was 5.8% and 68.8% for community-dwelling and institutional populations. A considerable proportion (20.6%) of demented cases lived in hospitals or institutions.

The relationship of sex, age, educational history, radiation dose and history of 10 diseases to the prevalence of AD or VD was sought by multivariate logistic linear regression analysis. AD increased more remarkably with age than VD. The prevalence of AD was higher in subjects with a history of head trauma and lower with that of cancer. VD was more prevalent among those with history of stroke and hypertension. Reported histories of other diseases were not related to AD or VD. Neither type of dementia showed any significant effect of sex or radiation exposure. Educational history showed a significant effect in the prevalence of AD, which decreased among those with higher attained education. In contrast, there was no relationship between VD and educational history (Tab. 1).

Table 1. Odds ratios (OR)[#] of Alzheimer's disease and vascular dementia

Risk factor		Alzheimer's disease		Vascular dementia	
		OR	95% CI	OR	95% CI
Sex	(female/male)	1.4	(0.7–3.1)	1.2	(0.5–2.7)
Age	10-year increments	6.3[*]	(4.3–9.6)	2.0[*]	(1.2–3.2)
Education	3-year increments	0.6[*]	(0.4–0.9)	0.9	(0.5–1.6)
Radiation dose	1gray	1.1	(0.7–1.6)	1.0	(0.5–1.6)
Cancer	(+/–)			0.3[*]	(0.05–0.98)
Head trauma	(+/–)			7.4[*]	(1.4–3.03)
Stroke	(+/–)			35.7[*]	(16.6–82.5)
Hypertension	(+/–)			4.0[*]	(1.8–9.8)

[#] The null hypothesis of OR = 1 was tested using multivariate logistic regression analysis.
95% CI: 95% confidence interval. [*]$p < 0.05$ [10].

Our study without population-based sampling showed that CASI was comparable to MMSE and HDS in sensitivity and specificity for dementia [10]. Since health examinations are carried out biennially in the AHS, longitudinal follow-up of cognitive impairment can be attained through physical examination, history recording and other inquiries. A longitudinal study is presumably more sensitive to mild deterioration of cognitive function. We diagnosed 4 subjects with mental deterioration as having mild dementia when compared with the previous examination [10].

The prevalence of dementia was found to increase exponentially with age. The age-specific prevalence rate presented here [10] is the highest among the rates reported in Japan which were obtained only from the aged living in the community. In contrast, 2.2% of the participants in our study were inpatients of hospitals or institutions, and 20.6% of dementia cases lived in these institutions. Our results approximate the age-specific prevalence detected in western Japan [9], which included hospitalized or institutionalized persons. The prevalence of dementia as a whole was predominant in women after their mid-eighties, similar to those reported for Japan and for Rochester, but not in Seattle [12].

VD has been reported more frequently than AD in Japan, and AD is more frequent in European and North American countries [8]. Recent epidemiological studies in Japan have revealed greater rates of AD and lower rates of VD [9]. In the RERF study cohort, the increase in rates of AD was marked in participants aged 80 years and older [10]. Multivariate logistic analysis confirmed a notable increase in the prevalence of AD with advancing age, implying the increasing prevalence of AD where the number of the aged older than 80 years increases as in Japan and the US.

Similarly designed prevalence studies on the Japanese-American population in Seattle and Honolulu followed the same methodology developed by the Ni-Hon-Sea Dementia Project [10, 12, 13]. The prevalence of AD was slightly lower at every age in Seattle than in Hiroshima. The Honolulu study on Japanese-American men reported a higher prevalence of AD than our study.

Overall prevalence rates for AD was almost equal to VD in men, but predominant in women. The predominance of AD in women is consistent in numerous studies in various Japanese areas, probably due to a longer average life span for Japanese women than men.

The Stroke Data Bank cohort study revealed the probability of new-onset dementia/year as 5.4% for stroke patients aged 60 years and 10.4% for stroke patients aged 90 years. The stroke incidence is 3 times higher in men, as reported from RERF cohort in 1992, while multi-variate logistic analysis showed no difference in sexes for VD in our study [10].

The predominance of VD in Japan has been attributed to a high incidence of stroke. We discussed interpretation of DSMIIIR passage "evidence of significant VD" in a Ni-Ho-Sea Dementia Project meeting. The prevalence of VD in those under 80 years of age was higher in our study [10] and its prevalence over ages 80 was higher in the Honolulu study [13]. The incidence of stroke in Japan has been remarkably higher than that in the Japanese-Americans in Hawaii. The difference in survival rate from stroke or severity of stroke at these 2 sites should be investigated in future.

Effects of education were observed only in AD even when adjustment was made for age [10], but this tendency became weak and inconsistent when rates were age-stratified [12]. An association was seen between a prior history of head trauma and AD. The history of disease based on examinations over many years in AHS may offer more accurate information on head trauma.

The radiation effect on the brain in adulthood is unrelated to prevalence of dementia. The negative relationship of cancer to AD needs to be discussed as to whether it is due to a protective effect or to differential survival bias.

Our study involves several limitations; the AHS cohort does not represent the overall Japanese population, because subjects in our study survived the atomic bombing and we excluded a generation of men in military duty at the time. Although subjects unable to visit RERF were examined at their residence, we cannot rule out a bias by exclusion from the

AHS examination. Instead of these obstacles, ongoing biennial medical examinations in the AHS can effectively detect mild demented cases, and this study provides a unique opportunity to address frequently noted differences in rates of AD and VD in Japan and the US and to look for risk factors of dementia.

This is the first study of Japanese dementia rates carried out with a protocol similar to a US study to allow meaningful comparisons. The present study showed more similar rates between Japan and US than in many previous reports, revealing associations between age, attained education, history of head trauma or absence of cancer and AD. Age and history of stroke or hypertension are risks of VD.

Conclusion

Risk factors of AD are age, head trauma, ApoE4 and absence of cancer. Those of VD are age and hypertension in addition to a past history of diabetes mellitus, fibrinogen, factor VIII and von Willebrand factor [14]. The management of those factors is essential to prevent AD or VD, except for age.

Acknowledgements
I deeply express my gratitude to Dr. Yamada (RERF), Drs. Yasuda and Tanaka (Hyogo Institute for Aging Brain and Cognitive Disorders), Dr. Harrington (Aberdeen), Dr. Emson (Cambridge), Dr. Sasaki (Kinoko Espoir Hospital), Drs. Ohmura and Arai (Juntendo University), Drs. Sudo, Ikeda and Izumi in our Department.

References

1 Corder EH, Saunders AM, Strittmatter WJ, Schmechel DE, Gaskell PC, Small GW, Roses AD, Haines JL, Pericak-Vance MA (1993) Gene dose of apolipoprotein E type 4 allele and the risk of Alzheimer's disease in late-onset families. *Science* 261: 921–923
2 Bullido MJ, Artiga MJ, Recuero M, Sastre I, Garcia MA, Aldudo J, Lendon C, Han SW, Moris JC, Frank A et al (1998) A polymorphism in the regulatory region of ApoE associated with risk for Alzheimer's dementia. *Nat Genet* 18: 69–71
3 Toji H, Maruyama H, Sasaki K, Nakamura S, Kawakami H (1999) Apolipoprotein E promoter polymorphism and sporadic Alzheimer's disease in a Japanese population. *Neurosci Lett* 259: 56–58
4 Toji H, Kawakami H, Kawarai H, Nakayama T, Komure O, Kuno S, Nakamura S (1998) No association between apolipoprotein E alleles and olivopontocerebellar atrophy. *J Neurol Sci* 158: 110–112
5 Isoe K, Ji Y, Urakami K, Adachi Y, Nakashima K (1997) Genetic association of estrogen receptor gene polymorphisms with Alzheimer's disease. *Alzheimer Res* 3: 195–197
6 Maruyama H, Toji T, Harrington CR, Sasaki K, Izumi Y, Ohmura T, Arai H, Yasuda M, Tanaka C, Emson PC et al (2000) Lack of an association of estrogen receptor α gene polymorphisms and transcriptional activity with Alzheimer's disease. *Arch Neurol* 57: 236–245
7 Kobayashi S, Inoue S, Hosoi T, Ouchi Y, Shiraki M, Orimo H (1996) Association of bone mineral density with polymorphism of the estrogen receptor gene. *J Bone Miner Res* 11: 306–311
8 Graves A, Larson EB, White L, Teng EL, Homma A (1994) Opportunities and challenges in international collaborative epidemiologic research of dementia and its subtypes: Studies between Japan and the U.S. *Int Psychogeriat* 6: 209–223
9 Ogura C, Nakamoto H, Uemura T, Yamamoto K, Yonemori T, Yoshimura T (1995) Prevalence of senile dementia in Okinawa, Japan. *Int J Epidemiol* 24: 373–379
10 Yamada M, Sasaki H, Mimori Y, Kasagi F, Sudoh S, Ikeda J, Hosoda Y, Nakamura S, Kodama K (1999)

Prevalence and Risks of Dementia in the Japanese Population: RERF's Adult Health Study Hiroshima Subjects. *J Amer Geriat Soc* 47: 189–195

11 Teng E, Hasegawa K, Homma A, Imai Y, Larson E, Graves A, Sugimoto K, Yamaguchi T, Sasaki H et al (1994) The cognitive abilities screening instrument (CASI): A practical test for cross-cultural epidemiological studies of dementia. *Int Psychogeriat* 6: 45–58

12 Graves A, Larson EB, Edland SD, Bowen JD, McCormick WC, McCurry SM, Rice MM, Wenzlow A, Uomoto JM (1996) Prevalence of dementia and its subtypes in the Japanese American population of King Country, Washington State. *Amer J Epidemiol* 144: 760–771

13 White L, Petrovitch H, Ross GW, Masaki KH, Abbott RD, Teng EL, Rodriguez BL, Blanchette PL, Havrik RJ, Wergowske G et al (1996) Prevalence of dementia in older Japanese-American men in Hawaii: The Honolulu-Asia aging study. *JAMA* 276: 955–960

14 Kohriyama T, Yamaguchi S, Tanaka E, Yamamura H, Nakamura S (1996) Coagulation and fibrinolytic parameters as predictors for small-vessel disease revealed by magnetic resonance imaging of the brain. *Clin Neurol* 36: 640–647

Neuroscientific Basis of Dementia
C. Tanaka, P.L. McGeer, Y. Ihara (eds)
© 2001 Birkhäuser Verlag Basel/Switzerland

New therapeutic approaches to Alzheimer's disease

Toshitaka Nabeshima and Kiyofumi Yamada

Department of Neuropsychopharmacology and Hospital Pharmacy, Nagoya University Graduate School of Medicine, Nagoya, Japan

Introduction

Alzheimer's disease (AD) is a neurodegenerative disorder that is neuropathologically characterized by the presence of numerous senile plaques and neurofibrillary tangles accompanied by neuronal loss. The extracellular senile plaques are composed of amyloid β-peptides (Aβ), 40–42 amino acid peptide fragments of the β-amyloid precursor protein (APP), whereas the intracellular neurofibrillary tangles are composed of highly phosphorylated tau proteins [1]. Because of a rapid increase in the elderly population in the world, the development of therapeutic strategies for AD is one of the most important themes in the medical and pharmaceutical sciences. In this article, we review the recent progress in pharmacotherapy for AD, as well as possible future therapeutic strategies (Tab. 1).

Table 1. Possible therapeutic approaches in Alzheimer's disease

Cholinergic therapy
Non-cholinergic neurotransmitter therapy
Anti-inflammatory therapy
Estrogen replacement therapy
Anti-oxidant therapy
Anti-amyloid therapy
Anti-tau phosphorylation therapy
Neurotrophic factor/immunophilin ligand therapy
Vaccine therapy
Gene therapy

Cholinergic therapy

Cholinergic neurons in the nucleus basalis of Meynert are reduced early in the course of the disease, and the dysfunction of cholinergic neurons is believed to be primarily responsible for cognitive deficits in AD. Therapeutic efforts to improve cognitive performance in AD have focused on augmenting cholinergic function either with acetylcholinesterase (AChE) inhibitors or selective muscarinic and nicotinic cholinergic agonists. Tacrine was the first AChE inhibitor to be approved for the treatment of AD. Two more AChE inhibitors (donepezil and rivastigmine) have since been approved. Compared with tacrine, donepezil and rivastigmine produced mild adverse side effects and very low hepatotoxicity.

Direct activation of muscarinic and nicotinic cholinergic receptors has also been considered a therapeutic strategy for AD. Muscarinic M1 receptors are predominantly present in the frontal cortex and hippocampus while M2 and M3 receptors predominate peripherally, where they mediate effects on cardiovascular, respiratory and secretory systems. Accordingly, a specific M1 receptor agonist could potentially ameliorate cognition impairment without the peripheral side-effects associated with stimulation of M2 and M3 receptors. Several M1 receptor agonists such as SB 202026 are currently in clinical trials [2]. Activation of nicotinic receptors with selective agonists such as GTS-21 may also provide some benefits in AD since it shows cognition-enhancing effects as well as neuroprotective effects in animal studies [2].

It is demonstrated that activation of cholinergic receptors potentially decreases amyloido-genic processing of APP by activating protein kinase C activity. Accordingly, cholinergic therapy with AChE inhibitors and selective muscarinic receptor agonists not only improves cognitive deficits by compensating for the cholinergic deficit, but also may slow down the worsening of clinical symptoms in AD by decreasing amyloid deposition in the brain [3].

Non-cholinergic neurotransmitter therapy

In addition to cholinergic deficits, dysfunction of other neurotransmitter systems such as serotonergic and noradrenergic neurons is likely to contribute to behavioral and psychiatric symptoms such as mood, abnormal behavior, sleep disturbances and depression. Glutamate is one of the excitatory amino acids (EAA) in the brain, acting as a neurotransmitter and playing a role in learning and memory. An excess amount of EAA produces neurodegeneration whereas the blockade of N-methyl-D-aspartate (NMDA) receptors protects neurons from ischemic and EAA-induced brain injury. Accordingly, modulation of glutamatergic function has been proposed as a therapeutic strategy for AD [4]. A glutamatergic blockade with NMDA receptor antagonists such as memantine could prevent EAA-induced toxicity, while the enhancement of the neurotransmission could improve intracortical connections and favor learning and memory. Other non-cholinergic neurotransmitter therapeutics, including GABAergic drugs and neuropeptides, have also been investigated in clinical trials [2].

Anti-inflammatory therapy

Inflammatory mechanisms contribute to the neurodestructive process occurring in the senile plaques in the AD brains. The acute phase proteins, cytokines and other proteins associated with inflammation are found in brains of AD patients but not in brains from age-matched controls. The presence of complement components, cytotoxic C5b-9, suggests that complement-mediated neuronal lysis occurs in brains of AD patients. Activated microglial cells, which are associated with senile plaques, may be the source of these mediators. Epidemiological studies demonstrated that previous use of corticosteroid or nonsteroidal anti-inflammatory drugs (NSAID) reduces the risk of AD. McGeer and Rogers (1992) have proposed anti-inflammatory therapy as a therapeutic approach to AD [5]. In a clinical trial, indomethacin has been shown to protect mild to moderately impaired AD patients [6].

Estrogen replacement therapy

Epidemiological studies have indicated that the prevalence of AD after age 65 is two to three times higher in women than men, suggesting gender difference as a risk of AD. Studies also indicated that replacement therapy with estrogens in postmenopausal women delays the onset and decreases the risk of AD. Although the mechanisms by which estrogens affect the pathogenic processes in AD are still unknown, it is demonstrated that estrogens modulate cholinergic neuronal activity, monoamine metabolism and the expression of brain-derived neurotrophic factor (BDNF) mRNA in the brain. Estrogens have also been shown to attenuate excitotoxicity, oxidative injury and $A\beta$ toxicity, to regulate APP metabolism, and to reduce the neuronal generation of $A\beta$ [7].

We recently demonstrated that $A\beta$-induced working memory deficits in a water maze task were significantly potentiated in ovariectomized rats compared with sham-operated rats when mnemonic ability was examined 3 months after ovariectomy. Replacement therapy with β-estradiol partially prevented some aspects of the $A\beta$-induced impairment of performance in ovariectomized rats. Our results support the estrogen replacement therapy in AD [8].

Anti-oxidant therapy

Several lines of evidence suggest that oxidative stress is important in the pathogenesis of AD and that antioxidants may protect neurons from $A\beta$-induced toxicity. We have recently provided the first *in vivo* evidence that oxidative stress is involved in $A\beta$-induced learning and memory impairment. Repeated administration of antioxidants, idebenone and β-tocopherol prevented learning and memory deficits in rats continuously infused with $A\beta$ into the cerebral ventricle [9].

Idebenone showed a clear dose-related antidementia activity in AD [10]. A recent controlled trial showed that treatment with selegiline and/or β-tocopherol in patients with moderately severe impairment from AD slows the progression of the disease [11].

Anti-amyloid therapy

Accumulated evidence supports the idea that the formation and deposition of $A\beta$ in the brain, especially the more hydrophobic/fibrillogenic 1-42 form of $A\beta$ ($A\beta$1-42), are early and key pathological events in the development of AD [12]. Mutations in all three genes known to cause familial AD (APP, presenilin 1 (PS1) and PS2) result in an increase in Aβ1-42 production relative to Aβ1-40 or the total Aβ levels. Transgenic mice with mutant forms of APP develop fibrillar plaques that contain predominantly Aβ1-42 and show many of the pathological characterizations associated with AD. Aβ1-42, not Aβ1-40, is the initially deposited species in brains of AD patients. Diffuse plaques, the earliest forms of senile plaques, consist exclusively of Aβ1-42. These findings support the amyloid cascade hypothesis of AD [12], and the anti-amyloid therapy for AD, which refers to intervention to stop the deposition of amyloid in the brain, for the treatment of AD. Recent progress in understanding the molecular mechanisms of cytotoxic Aβ formation offers a number of potential targets for novel therapeutic strategies for AD (Fig. 1).

Firstly, inhibition of β- and/or γ-secretases could decrease Aβ production although the identity of these enzymes remains unknown. A number of small peptide aldehydes that are known to inhibit cysteine and serine proteinases have been shown to inhibit Aβ formation, probably by inhibiting β-secretase activity [13]. Alternatively, diversion of APP metabolism from an amyloidogenic to a non-amyloidogenic processing pathway may be possible by using muscarinic agonists that increase the percentage of APP molecules cleaved by α-secretase [3]. Thirdly, inhibition of the aggregation of Aβ, which leads to the formation of a cytotoxic form of Aβ, would make for a novel pharmacotherapy [13]. Finally, activation of the enzyme activity of extracellular proteases, which are capable of degrading Aβ and related peptides, may be a potential therapeutic strategy for preventing or slowing the pathogenic mechanism of AD [14].

Anti-tau phosphorylation therapy

Neurofibrillary tangles consist of paired helical filaments (PHF) and related straight filaments. The major constituent of PHF is hyperphosphorylated tau. Aβ fibril induces tau phosphorylation, resulting in the loss of microtubule-binding capacity and somatodendritic accumulation [15]. It is also demonstrated that Aβ can increase the activity of several different protein kinases, such as tau protein kinase I/glycogen synthase kinase 3β (TPKI/GSK-3β), which can potentially phosphorylate tau, and that Aβ-induced neurotoxicity is prevented by TPKI/GSK-3β antisense oligonucleotides [16]. These findings suggest that Aβ fibril formation causes an abnormal phosphorylation of tau through the activation of TPKI/GSK-3β and destabilization of microtubules, resulting in neuronal death. Accordingly, inhibition of TPKI/GSK-3β and possibly other protein kinases that can phosphorylate tau may prevent the progression of AD. Recently it was demonstrated that the prolyl isomerase Pin 1 can restore the ability of phosphorylated tau to bind microtubules and promote microtubule assembly *in vitro*, providing a new insight into the pathogenesis of AD. Furthermore, Pin 1 might be used to derive new therapies for AD and related neurodegenerative diseases [17].

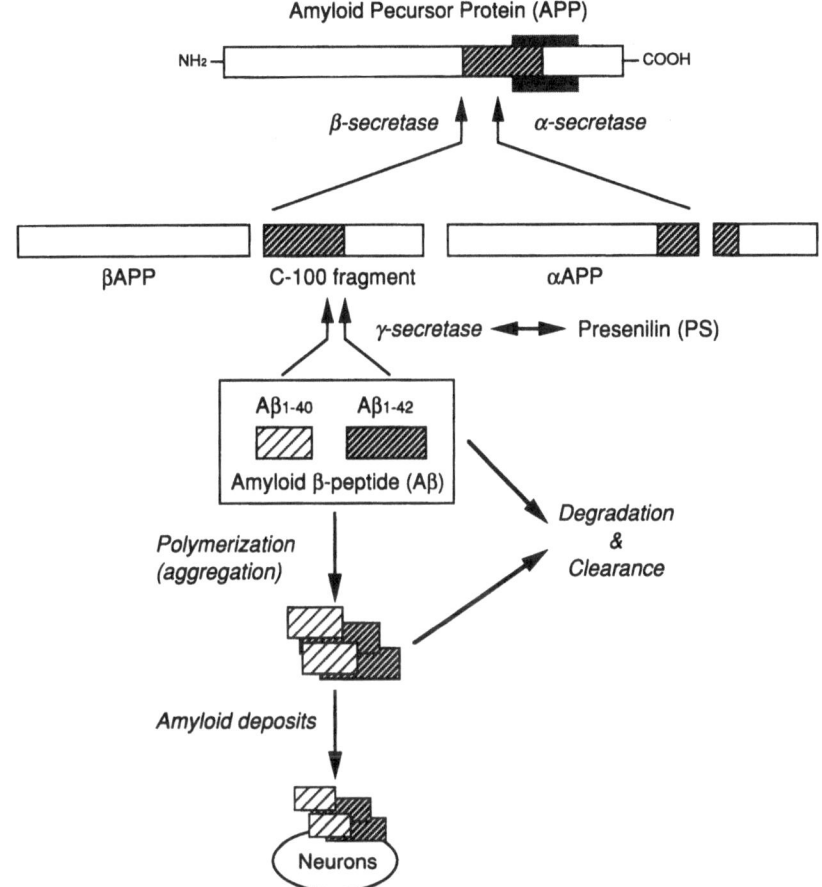

Figure 1. Proteolytic processing of APP and Aβ formation. The cleavage of APP by α-secretase within the Aβ sequence generates non-amyloidogenic soluble βAPP and C-terminal APP fragment. On the other hand, the proteolytic processing of APP by β-secretase, followed by γ-secretase, results in the formation of Aβ. The position where the γ-secretases cleave in this amyloidogenic pathway is critical to determining the levels of the hydrophobic/fibrillogenic 1-42 form of Aβ (Aβ1-42) relative to Aβ1-40. Aβ1-42, not Aβ1-40, is the initially deposited species. The aggregation and fibrillar Aβ formation is required for the neurotoxicity. Aβ is constitutively degraded by a metalloprotease released by microglia.

Neurotrophic factor/immunophilin ligand therapy

Neurotrophic factors promote the maintenance and survival of numerous peripheral and central nervous system (CNS) neurons. Nerve growth factor (NGF) provides a trophic support on cholinergic neurons of the basal forebrain. Although it is not clear whether the memory dysfunction and neuronal loss in AD are caused by a lack of NGF, NGF has been shown to prevent the loss of basal forebrain neurons following axotomy in experimental animals.

Clinically, three AD patients were treated with β-NGF (purified from male mouse submandibular glands) administered continuously into the lateral cerebral ventricle [18]. In this clinical research no clear cognitive amelioration was demonstrated in any patient. An increase in nicotine binding was found in some brain areas in these patients 3 months after the end of NGF treatment. The amount of slow-wave cortical activity was reduced in the first 2 patients but not the third one. Two negative side-effects, a dull constant back pain and a marked weight loss, accompanied NGF treatment. It was concluded that long-term intracerebroventricular (i.c.v.) NGF administration may cause certain potentially beneficial effects, but the i.c.v. route of administration is also associated with negative side-effects that appear to outweigh the positive effects [18].

We have been proposing that orally active NGF synthesis stimulators have potential as therapeutic agents for AD, and may have some advantages compared with NGF itself in terms of quality of life for the patient, since such treatment does not require the insertion of an NGF delivery catheter into the brain [19]. We investigated the effects of idebenone and propentofylline as prototypes of NGF synthesis stimulators. These two compounds partially restored the age-associated decrease of NGF levels in the frontal and parietal cortices, and ameliorated learning and memory impairment induced by bilateral forebrain lesions, a continuous infusion of anti-NGF antibody into the septum and a continuous i.c.v. infusion of Aβ.

Immunophilins are the receptor proteins for the major immunosuppressant drugs cyclosporin A, FK506 and rapamycin. There is evidence that immunophilin ligands are neurotrophic for numerous classes of damaged neurons, both in culture systems and intact animals. Furthermore, the immunosuppressive properties of FK506 and cyclosporin A can be functionally dissociated from their neurotrophic effects. One of these compounds, GPI-1046, has neuroprotective effects in a number of CNS degenerative models including the N-methyl-4-phenyl-1,2,3,6-tetrahydropyridine (MPTP) and 6-hydroxydopamine models of Parkinson's disease [20].

Vaccine therapy

Recently, a novel and exciting result of vaccine therapy for AD was reported by Schenk et al. (1999) [21]. PDAPP transgenic mice, which overexpress mutant human APP, were immunized with Aβ either before the onset of AD-like neuropathologies or at an older age when Aβ deposition and other neuropathological changes were established. Immunization of the young animals, which produced high titres of serum antibody, almost completely prevented the development of Aβ deposition, dystrophic neurites, astrocytosis and microgliosis. Immunization of the older animals also markedly reduced the extent and progression of the AD-like brain lesions.

Gene therapy

Gene therapy can deliver single or several well-defined molecules in a highly specific spatial and temporal fashion. Transplantation of fibroblasts, modified to secrete NGF, into the basal forebrain protected cholinergic neurons and promoted functional recovery. Similarly, when primary fibroblasts, modified to express choline acetyltransferase (ChAT), the synthetic enzyme of ACh, were transplanted into the cerebral cortex, the learning and memory impairment caused by the basal forebrain lesion was ameliorated [22].

Acknowledgments
This study was supported, in part, by Grants-in-Aid for Science Research from the Ministry of Education, Science, Sports and Culture of Japan (No. 07557009, 10897005 and 97450), an SRF Grant for Biomedical Research and by grants from the Suzuken Memorial Foundation and the Research Foundation for Pharmaceutical Sciences, and by a COE Grant.

References

1 Selkoe DJ (1994) Cell biology of amyloid β-protein precursor and the mechanism of Alzheimer's disease. *Annu Rev Cell Biol* 10: 373–403
2 Hasegawa M, Noda Y, Maeda Y, Yamada K, Nabeshima T (1999) The 1998 domestic state of development of cognitive enhancers. *Folia Pharmacol Japonica* 114: 327–336
3 Giacobini E (1997) From molecular structure to Alzheimer therapy. *Jpn J Pharmacol* 74: 225–241
4 Lawlor BA, Davis KL (1992) Does modulation of glutamatergic function represent a viable therapeutic strategy in Alzheimer's disease? *Biol Psychiat* 31: 337–350
5 McGeer PL, Rogers J (1992) Anti-inflammatory agents as a therapeutic approach to Alzheimer's disease. *Neurology* 42: 447–449
6 Rogers J, Kirby LC, Hempelman SR, Berry DL, McGeer PL, Kaszniak AW, Zalinski J, Cofield M, Mansukhani L, Willson P et al (1993) Clinical trial of indomethacin in Alzheimer's disease. *Neurology* 43: 1609–1611
7 Simpkins JW, Green P, Gridley KE, Singh M, de Fiebre NC, Rajakumar G (1997) Role of estrogen replacement therapy in memory enhancement and prevention of neuronal loss associated with Alzheimer's disease. *Amer J Med* 103 (3A): 19S–25S
8 Yamada K, Tanaka T, Zou L-B, Senzaki K, Yano K, Osada T, Ana O, Ren X, Kameyama T, Nabeshima T (1999) Long-term deprivation of oestrogens by ovariectomy potentiates β-amyloid-induced working memory deficits in rats. *Brit J Pharmacol* 128: 419–427
9 Yamada K, Tanaka T, Han D, Senzaki K, Kameyama T, Nabeshima T (1999) Protective effects of idebenone and β-tocopherol on β-amyloid-(1-42) induced learning and memory deficits in rats: implication of oxidative stress in β-amyloid-induced neurotoxicity *in vivo*. *Eur J Neurosci* 11: 83–90
10 Gutmann H, Hadler D, Erzigkeit H (1997) Long-term treatment of Alzheimer's disease with idebenone. *In*: K Iqbal, B Winblad, T Nishimura, M Takeda, HM Wisniewski (eds): *Alzheimer's Disease: Biology, Diagnosis and Therapeutics*. John Wiley and Sons, Chichester, 687–705
11 Sano M, Ernesto C, Thomas RG, Klauber MR, Schafer K, Grundman M, Woodbury P, Growdon J, Cotman CW, Pfeiffer E et al (1997) A controlled trial of selegiline, alpha-tocopherol, or both as treatment for Alzheimer's disease. *N Engl J Med* 336: 1216–1222
12 Hardy H (1997) Amyloid, the presenilins and Alzheimer's disease. *Trends Neurosci* 20: 154–159
13 Allsop D, Howlett D, Christie G, Karran E (1998) Fibrillogenesis of β-amyloid. *Biochem Soc Trans* 26: 459–463
14 Qiu WQ, Ye Z, Kholodenko D, Seubert P, Selkoe DJ (1997) Degradation of amyloid β-protein by a metalloprotease secreted by microglia and other neural and non-neural cells. *J Biol Chem* 272: 6641–6646
15 Busciglio J, Lorenzo A, Yeh J, Yankner BA (1995) β-Amyloid fibrils induce tau phosphorylation and loss of microtubule binding. *Neuron* 14: 879–888
16 Takashima A, Noguchi K, Sato K, Hoshino T, Imahori K (1993) Tau protein kinase I is essential for amy-

loid β-protein-induced neurotoxicity. *Proc Natl Acad Sci USA* 90: 7789–7793
17 Lu P-J, Wulf G, Zhou XZ, Davies P, Lu KP (1999) The prolyl isomerase Pin 1 restores the function of Alzheimer-associated phosphorylated tau protein. *Nature* 399: 784–788
18 Jönhagen ME, Nordberg A, Amberla K, Bäckman L, Ebendal T, Meyerson B, Olson L, Seiger A, Shigeta M, Theodorsson E et al (1998) Intracerebroventricular infusion of nerve growth factor in three patients with Alzheimer's disease. *Dement Geriatr Cogn Disord* 9: 246–257
19 Yamada K, Nitta A, Hasegawa T, Fuji K, Hiramatsu M, Kameyama T, Furukawa Y, Hayashi K, Nabeshima T (1997) Orally active NGF synthsis stimulators: potential therapeutic agents in Alzheimer's disease. *Behav Brain Res* 83: 117–122
20 Snyder SH, Sabatini DM, Lai MM, Steiner JP, Hamilton GS, Suzdak PD (1998) Neural actions of immunophilin ligands. *Trends Pharmacol Sci* 19: 21–26
21 Schenk D, Barbour R, Dunn W, Gordon G, Grajeda H, Guido T, Hu K, Huang J, Jonson-Wood K, Khan K et al (1999) Immunization with amyloid-β attenuates Alzheimer-disease-like pathology in the PDAPP mouse. *Nature* 400: 173–177
22 Winkler J, Thal LJ, Gage FH, Fisher LJ (1998) Cholinergic strategies for Alzheimer's disease. *J Molec Med* 76: 555–567

Subject index